CAN THAT BE RIGHT?

BOSTON STUDIES IN THE PHILOSOPHY OF SCIENCE

Editor

ROBERT S. COHEN, *Boston University*
MARX W. WARTOFSKY † (*Editor 1960–1997*)

Editorial Advisory Board

THOMAS F. GLICK, *Boston University*
ADOLF GRÜNBAUM, *University of Pittsburgh*
SYLVAN S. SCHWEBER, *Brandeis University*
JOHN J. STACHEL, *Boston University*

VOLUME 199

ALLAN FRANKLIN
University of Colorado

CAN THAT BE RIGHT?

Essays on Experiment, Evidence, and Science

KLUWER ACADEMIC PUBLISHERS
DORDRECHT / BOSTON / LONDON

A C.I.P. Catalogue record for this book is available from the Library of Congress.

ISBN 0-7923-5464-8

Published by Kluwer Academic Publishers,
P.O. Box 17, 3300 AA Dordrecht, The Netherlands.

Sold and distributed in North, Central and South America
by Kluwer Academic Publishers,
101 Philip Drive, Norwell, MA 02061, U.S.A.

In all other countries, sold and distributed
by Kluwer Academic Publishers,
P.O. Box 322, 3300 AH Dordrecht, The Netherlands.

Printed on acid-free paper

All Rights Reserved
©1999 Kluwer Academic Publishers
No part of the material protected by this copyright notice may be reproduced or
utilized in any form or by any means, electronic or mechanical,
including photocopying, recording or by any information storage and
retrieval system, without written permission from the copyright owner

Printed in the Netherlands.

TABLE OF CONTENTS

PREFACE vii

ACKNOWLEDGMENTS ix

INTRODUCTION: CONSTRUCTIVISM, POSTMODERNISM, AND SCIENCE 1

CASE STUDIES

CHAPTER 1. HOW TO AVOID THE EXPERIMENTERS' REGRESS 13

CHAPTER 2. THE APPEARANCE AND DISAPPEARANCE OF THE 17- KEV NEUTRINO 39

CHAPTER 3. INSTRUMENTAL LOYALTY AND THE RECYCLING OF EXPERTISE 97

CHAPTER 4. THE RISE OF THE "FIFTH FORCE" 133

THE ROLES OF EXPERIMENT

CHAPTER 5. THERE ARE NO ANTIREALISTS IN THE LABORATORY 149

CHAPTER 6. DISCOVERY, PURSUIT, AND JUSTIFICATION 163

CHAPTER 7. THE RESOLUTION OF DISCORDANT RESULTS 183

CHAPTER 8. CALIBRATION 237

CHAPTER 9. LAWS AND EXPERIMENT 273

POSTSCRIPT 283

REFERENCES 285

INDEX 303

PREFACE

It seems to me that publishing a collection of one's essays is an act of *hubris*. My defense is that someone else first suggested the idea to me. Nevertheless I enthusiastically adopted the idea. A major reason for this is that I wanted to argue against what has become known as the postmodern or constructivist criticism of science and I thought that collecting these essays together adds a weight of evidence not present in the separate essays.

When I first began working in the history and philosophy of science more than twenty years ago things seemed simpler and more collegial. My first work on modern physics, a study of the discovery of parity nonconservation, was intended as an answer to Thomas Kuhn's claim that crucial experiments did not exist. It seemed to me an argument that would be settled solely within the academy. Now, as indicated in the introduction, discussions of the nature of science have become both more important and considerably less collegial. I believe that the views of science presented by postmodern critics to both our students and to society are wrong. They bear little, if any, resemblance to the science I practiced for fifteen years, to that practiced by my colleagues, or to the history of science. These views can, I believe, also cause harm. Many important practical applications in our lives and many social decisions depend on a knowledge of both science and its practice. In order to make reasonable decisions we need to know and understand science. I worry that the postmodern view will both mislead society about the nature of scientific knowledge and also discourage the study of science. If science cannot be trusted, why expend the effort to learn it?

My work collected in this volume is a defense of science as a reasonable enterprise based on valid experimental evidence and critical discussion. Science does provide us with knowledge of the physical world. This volume is also a challenge to the critics of science. Let them show where I have made errors in either fact or in interpretation. Let them offer alternative explanations of the episodes and issues discussed.

One further thought--in preparing these essays for publication I have noticed an unfortunate amount of repetition. The same episodes are used several times to illustrate different issues. For that I apologize. Nevertheless, my work has sometimes been criticized for being too technical. Perhaps repetition can breed understanding.

ACKNOWLEDGMENTS

In an ideal academic world one would have colleagues who always had time to discuss issues, to read your work, and to offer constructive criticism. Fortunately for me, the real world has been a reasonable approximation to the ideal one. In my previous work I have acknowledged the help of those who assisted in the specific work. In this book, in which most of the material has been previously published, I would like to recognize more general debts. These are friends and colleagues who have made working in the history and philosophy of science a joy. I thank Robert Ackermann, Jed Buchwald, Robert Cohen, John Earman, Tim Gould, Colin Howson, Noretta Koertge, John Norton, Joseph Pitt, Alan Shapiro, Howard Smokler, Roger Stuewer, and Noel Swerdlow for being ideal colleagues. David Bartlett, John Cumalat, Bill O'Sullivan, John Price, Mike Ritzwoller, and Chuck Rogers, colleagues in the Physics Department at the University of Colorado have taken time from actually doing science to make important contributions to my work. Finally, my gratitude to Cyndi Betts, whose support and encouragement have been invaluable.

Most of the essays in this book have been published previously. The original publications are: 1994. How to Avoid the Experimenters' Regress. *Studies in the History and Philosophy of Science* 25: 97-121; 1995. The Appearance and Disappearance of the 17-keV Neutrino. *Reviews of Modern Physics* 67: 457-490; 1993. The Rise of the Fifth Force. *Correspondence, Invariance and Heuristics: Essays in Honour of Heinz Post*. S. French and H. Kamminga. Dordrecht, Kluwer Academic Publishers: 83-103; 1996. There Are No Antirealists in the Laboratory. *Realism and Anti-Realism in the Philosophy of Science*. R. S. Cohen, R. Hilpinen and Q. Renzong. Dordrecht, Kluwer Academic Publishers: 131-148; 1993. Discovery, Pursuit, and Justification. *Perspectives on Science* 1: 252-284; 1995. The Resolution of Discordant Results. *Perspectives on Science* 3: 346-420;1997. Calibration. *Perspectives on Science* 5: 31-80; and 1995. Laws and Experiment. *Laws of Nature*. F. Weinert. Berlin, De Gruyter: 191-207.

INTRODUCTION: CONSTRUCTIVISM, POSTMODERNISM, AND SCIENCE

In these postmodern times, when scholars proclaim the impossibility of knowledge, except, of course, their own, I would like to defend a rather old-fashioned notion. This is the idea that science provides us with knowledge about the world which is based on experimental evidence and on reasoned and critical discussion.[1] I further suggest that we can, and do, have reasonable grounds for belief in that evidence. In short, I believe that science is a reasonable enterprise.[2]

Even a generation ago this idea would have seemed obvious and in no need of defense. Things have, however, gotten worse since C.P. Snow bemoaned the lack of communication between the two cultures, the literary and the scientific, and tried to encourage mutual understanding and interaction. Although Snow's humanists were both ignorant of science and also quite happy about that lack of knowledge, they did not claim that science wasn't knowledge. It just wasn't valuable knowledge. Noting the self-impoverishment of scientists, who pay little attention to literature and history, Snow asked, "But what about the other side? They are impoverished too--perhaps more seriously, because they are vainer about it. They still like to pretend that the traditional culture is the whole of 'culture,' as though the natural order didn't exist. As though the explanation of the natural order is of no interest either in its own value or its consequences. As though the scientific edifice of the physical world was not in its intellectual depth, complexity and articulation the most beautiful and wonderful collective work of the mind of man" (Snow 1959, p. 15). Snow went on to discuss the then recent discovery of the nonconservation of parity in the weak interactions, or the violation of left-right symmetry (for details see Franklin 1986, Chapter 3). "It is an experiment of the greatest beauty and originality, but the result is so startling that one forgets how beautiful the experiment is. It makes us think again about some of the fundamentals of the physical world. Intuition, common sense--they are neatly stood on their heads. The result is usually known as the contradiction [nonconservation] of parity. If there were any serious communication between the two cultures, this experiment would have been talked about at every High Table in Cambridge. Was it?" (p. 17) Snow suspected then, as I do now, that it wasn't.

The attitude of many humanists toward science has changed from indifference to distrust, and even hostility. Today science is under attack from many directions, and each of them denies that science provides us with knowledge. Whether it is because of inherent gender bias, Eurocentrism, or the social and career interests of scientists, science is untrustworthy and fatally flawed. "The radical feminist position holds that the epistemologies, metaphysics, ethics, and politics of the dominant forms of science are androcentric and mutually supportive; that despite the deeply ingrained Western cultural belief in science's intrinsic progressiveness, science today serves primarily regressive social tendencies; and that the social structure of science, many of its applications and technologies, its modes of defining research problems and designing experiments, its ways of constructing and conferring meanings are not only sexist but also racist, classist, and culturally coercive" (Harding 1986, p. 9). One might ask how such sexism, racism, or classism determine the content of modern physics. What influence could they have on the decision as to whether or not experiments on atomic-parity violation confirm or refute the Weinberg-Salam unified theory of electroweak interactions? How would they enter into the decision as to whether or not a new elementary particle, the 17-keV neutrino, existed? In my view, as shown in the essays below, these issues were decided on the basis of valid experimental evidence and reasoned discussion.

Recent critics of science do not seem to believe that a knowledge of science is necessary in order to criticize it. Consider a recent work entitled, *Strange Weather: Culture, Science, and Technology in the Age of Limits* (Ross 1991). In a book which claims to deal with science, the author, Andrew Ross, boasts of his ignorance of the subject. "This book is dedicated to all of the science teachers I never had. It could only have been written without them."[3]

Ross also summarizes what he regards as the prevailing humanist view of science.

> It is safe to say that many of the founding certitudes of modern science have been demolished. The positivism of science's experimental methods, its axiomatic self-referentiality, and its claim to demonstrate context-free truths in nature have all suffered from the relativist critique of objectivity. Historically minded critics have described natural science as a social development, occurring in a certain time and place; a view that is at odds with science's self-presentation as a universal calculus of nature's laws. Feminists have also revealed the parochial bias in the masculinist experience and ritual of science's 'universal' procedures and goals. Ecologists have drawn attention to the environmental contexts that fall outside of the mechanistic purview of the scientific world-view. And anthropologists have exposed the ethnocentrism that divides Western science's unselfconscious pursuit of context free *facts* from what it sees as the pseudoscientific *beliefs* of other cultures. The cumulative result of these critiques has been a significant erosion of scientific institutions' authority to proclaim and authenticate truth (Ross 1991, p.11).

Ross' view is quite mild in comparison with Sandra Harding's characterization of Newtonian mechanics. "In that case, why is it not as illuminating and honest to refer to Newton's laws as 'Newton's rape manual' as it is to call them 'Newton's mechanics'" (Harding 1986, p. 113). To anyone who has looked, in an even cursory way, at Newton's *Principia* this statement is neither illuminating nor honest. The level of hostility between the two cultures has increased markedly.

A recent issue of *Social Text,* a leading cultural studies journal, was devoted to "Science Wars." This is the critics' term for the ongoing discussion, between scientists and their critics, particularly responses to the criticism by scientists such as Gross and Levitt (1994) and Wolpert (1992). The critics of science attribute the hostile reaction of scientists to a decline in funding and a need to place blame. Thus, George Levine remarks, "The new counter-aggression of scientists hostile to 'postmodernism' is surely the consequence of an economic pinch hurting them as well as humanists and social scientists" (Levine 1996a). Steven Weinberg notes, however, that "In years of lobbying for federal support of scientific programs, I have never heard anything remotely postmodern or constructivist" (Weinberg 1996). It does not seem to occur to the critics that scientists might actually care about their discipline and how it is portrayed.

That issue of *Social Text* also illustrates some of the problems associated with cultural studies of science. The volume contained an article by Alan Sokal, a physics professor at New York University, entitled "Transgressing the Boundaries: Toward a Transformative Hermeneutics of Quantum Gravity" (Sokal 1996a). The article claimed to show that certain developments in modern physics could be used to further a progressive political agenda. The only problem was that the article was a hoax. (This was revealed in an article in another journal *Lingua Franca* (Sokal 1996b)). It contained precious little quantum gravity and included errors in science that should have revealed its nature to anyone with even a rudimentary knowledge of physics. As illustrated above, Andrew Ross, who is also one of the editors of *Social Text*, doesn't seem to think that knowledge of science is necessary for those who wish to comment on it. Sokal had tried an experiment. He asked, "Would a leading North American journal of cultural studies... publish an article consisting of utter nonsense if (a) it sounded good and (b) it flattered the editors' ideological preconceptions. The answer, unfortunately, is yes" (Sokal 1996b, p. 62). Sokal included comments such as "It has thus become increasingly apparent that physical 'reality'no less than social 'reality' is at bottom a social and linguistic construct" (1996a, p. 217). As Sokal himself remarked he was not discussing theories of physical reality, which might be called social and linguistic constructs, but reality itself. He invited those who believed that the law of gravity was merely a social construct to step out of his twenty first floor window. As Richard Dawkins remarked, "Show me a cultural relativist at thirty thousand feet and I'll show you a hypocrite. Airplanes built according to scientific principles work" (Dawkins 1995, p. 32).

Sokal also claimed that "The π of Euclid and the G of Newton, formerly thought to be constant and universal, are now perceived in their ineluctable historicity" (1996a, p. 222). Whereas one might legitimately argue that G is, in a sense, historical, because it depends on the mathematical form of Newton's Law of Universal Gravitation, and its value depends on our choice of standard units of distance and mass, that can hardly be applied to π, which is determined by the properties of Euclidean space. The article contains many more nonsensical statements.[4]

What then is the import of this hoax? Certainly Sokal has shown a certain intellectual sloppiness on the part of the editors of *Social Text*. Their response that they trusted that Sokal was giving his honest views seems less than convincing. Was it not their responsibility to make sure that anything published in their journal met at least minimal standards of evidence, coherence, and logic? Perhaps, more importantly, it demonstrates the intellectual arrogance of some scholars in cultural studies of science, that they need not know any science to comment on it. Recall Ross' previous gratitude to his nonexistent science teachers. Despite Ross' lack of knowledge of quantum gravity, he did not think it necessary to find someone who had such knowledge to referee the paper. Had he done so, he would have saved himself, and his co-editors, considerable embarrassment.

Sokal's attribution of a political agenda to the critics of science is not misplaced. "Concepts such as objectivity, rationality, good method, and science need to be reappropriated, reoccupied, and reinvigorated for democracy-advancing projects" (Harding 1996, p. 18). It seems clear that if anyone is trying to inject politics and other social views into science it is the critics. The dangers to both society and to science of such tampering should be obvious from the episodes during the 1930s of Lysenkoism in the Soviet Union (Joravsky 1970) and Aryan science in Germany (Beyerchen 1977). This is not to say that science, or its applications in society, should be immune from criticism. It is far too important in our lives for that. I do believe, however, that such criticism should be informed.[5]

There is, I believe, a more serious, if currently somewhat less fashionable, challenge to my view of science. This is the challenge provided by the sociologists of scientific knowledge or social constructivists.[6] Social constructivists imply, however much they may disclaim it, that science does not provide us with knowledge. In a recent work Collins and Pinch (1993) have described science as a golem, a creature that is clumsy at best, evil at worst.[7] In constructivist case studies of science, experimental evidence never seems to play any significant role. In their view the acceptance of scientific hypotheses, the resolution of discordant results, as well as the acceptance of experimental results in general, is based on "negotiation" within the scientific community, which does not include evidence or epistemological or methodological criteria. Such negotiations do include considerations such as career interests, professional commitments, prestige of the scientists' institutions, and the perceived utility for future research.[8] For example, Pickering states, "Quite simply, particle physicists accepted the existence of the neutral current because they could see how to ply their trade more profitably in a world in which the neutral current was real" (Pickering 1984b, p. 87).[9] The emphasis on career interests and future utility is clear.

Part of the problem is that the constructivists conflate pursuit, the further investigation of a theory or of an experimental result, with justification, the process by which that theory or result becomes accepted as scientific knowledge.[10] No one would deny that the considerations suggested by the constructivists enter into pursuit, along with other reasons such as the recycling of expertise, instrumental loyalty, and scientific interest. I suggest that these considerations do not enter into justification.

This anti-evidence view of science is illustrated in Gerald Geison's recent work on Pasteur:

> Historians, philosophers, and sociologists of science have become increasingly reluctant to explain such discoveries as Pasteur was about to make by simply pointing to the empirical evidence at hand--to the "nature of things" or to "the real world out there." Indeed this chapter and much of the rest of the book are meant to suggest just how pliable this supposedly hard evidence of the natural world can be. And yet, for all of that, we do sometimes bump up against situations that ask us to give credence to our historical actors' perception of the empirical world. For no obvious reason to be found in his a priori theoretical commitments or other interests, Pasteur became convinced that he could detect left- and right-handed crystals in the sodium-ammonium paratartrate.
>
> In conceding this point, however, we should not ignore the extent to which Pasteur constructed the empirical world in which he made his first major discovery--not the extent to which that discovery depended on the "privileged material" represented by the tartrates. (Geison 1995, p. 80).[11]

Perhaps Pasteur became convinced because he had good evidence that he could separate the two forms of the crystal. Would Geison, or those who agree with him, deny that Kepler discovered his first law, that planets move in ellipses with the sun at one focus, because, to a very good approximation, the planets do move in ellipses with the sun at one focus?[12]

Geison goes even further. He claims Pasteur, and by implication other scientists, produce evidence in support of their preconceptions."We will become aware of his ingenious capacity for producing empirical evidence in support of positions he held *a priori*. In other words, one aim of this book is to show the extent to which nature can be rendered pliable in the hands of a scientist of Pasteur's skill, artistry, and ingenuity. But it will also suggest that not even Pasteur's prodigious talent always sufficed to twist the lion's tail in the direction he sought. Nature is open to a rich diversity of interpretations, but it will not yield to all" (p. 16). Although Geison does allow constraints by the material world, he does not seem to believe those constraints are very powerful.[13] Some constructivists are not even willing to go as far as Geison does in allowing a role for the physical world. "Reality will tolerate alternative descriptions without protest. We may say what we will of it, and it will not disagree. Sociologists of knowledge rightly reject realist epistemologies that *empower* reality" (Barnes 1991, p. 331).

The critics of science, both postmodern and constructivist, also ignore the pragmatic efficacy of science. (Recall Dawkins' comment cited earlier).[14] It is not just the successful practice of science, which is, after all, decided by scientists themselves, but rather evidence from the "real" world that underlies the judgment that science provides us with reliable knowledge about the world. A light comes on when a switch is thrown, objects fall down rather than up, rockets are launched toward , and reach, the moon, and synchrotrons work. Numerous examples of this kind provide grounds for believing that science is actually telling us something reliable about the world. It seems odd that the world would behave so that it fits the interests of scientists or their preconceptions.

The reason the constructivist view is a serious challenge is that their work looks at the actual practice of science. To readers without an adequate science background or knowledge of the particular episodes discussed, the accounts offered by social constructivists may appear persuasive and convincing.[15] Sandra Harding has criticized those of us who believe that science is a reasonable enterprise for not dealing with such episodes. "It is significant that the Right's objections virtually never get into the nitty-gritty of historical or ethnographic detail to contest the accuracy of social studies of science accounts. Such objections remain at the level of rhetorical flourishes and ridicule" (Harding 1996, p. 15).[16] In a sense, Harding is correct, but the blame should be equally distributed. There are very few episodes from the history of science which are discussed from both constructivist and evidence persepectives.[17]

This collection of essays is, in part, an effort to provide my part of the discussion of two cases in which the same historical episode is discussed from these two different points of view. "How to Avoid the Experimenters' Regress" is a discussion of the early attempts to detect gravity waves and has also been discussed from a contructivist position by Harry Collins (1985). "Discovery, Pursuit, and Justification," includes a discussion of the early experiments on atomic parity violation and their relation to the Weinberg-Salam unified theory of electroweak interactions, an episode discussed from a constructivist view by Andrew Pickering (1984a). (This episode is also discussed in "The Resolution of Discordant Results").

The other essays also argue that science is a reasonable enterprise based on experimental evidence, criticism, and reasoned discussion. "The Appearance and Disappearance of the 17-keV Neutrino" and "Instrumental Loyalty and the Recycling of Expertise" present detailed histories of two episodes from the recent history of physics. The former deals with how scientists decided that a proposed new elementary particle, the 17-keV neutrino, did not exist. The latter discusses a sequence of experiments on K^+ meson decay, showing the normal practice of science and how scientists make use of what they have learned from previous experiments to further investigate a phenomenon. A detailed account of the origin of the fifth-force hypothesis, a proposed modification of Newtons law of universal gravitation is presented in "The Rise of the Fifth Force."

The remaining essays use these case studies to examine particular issues in the practice of science and to illustrate the different roles that experiment plays in science.[18] "The Resolution of Discordant Results" shows how, in four different episodes: gravity waves; atomic parity violation, the fifth force, and the 17-keV neutrino, the discord between different experimental results was resolved. I argue that the resolution of the discord was based on epistemological and methodological grounds, that the decision was indeed based on experimental evidence, criticism, and reasoned discussion. "Calibration" examines a particular method by which scientists argue for the validity of their experimental results, namely, the use of a surrogate signal to demonstrate that an apparatus is working properly. Collins (1985) has argued that this procedure cannot be used to provide grounds for belief in an experimental result. I argue that it not only can, but does.

Figure 1. The simultaneous release of three steel balls from the eleventh floor of Gamow Tower, University of Colorado. The position of the one inch ball is shown by the arrow. Courtesy of John Taylor.

Figure 2. The position of the three balls after they have fallen approximately one hundred feet. The position of the one inch ball is indicated by the arrow. It is approximately one foot higher than the other two balls. Courtesy of John Taylor.

Experimental results can also be part of an enabling theory in the design of experiments, as illustrated in "Laws and Experiment." Yet another role for experiment is that of providing grounds for belief in the truth of physical theories and in the existence of entities involved in those theories. In "There Are No Antirealists in the Laboratory" I argue that experimental practice supports this view.[19]

These accounts will be technical. The history of physics involves physics. If I wish to argue that physics is a reasonable enterprise then I must examine the actual practice of physics, not simplified accounts. Mythological accounts of Galileo's experiment dropping unequal masses from the top of the Leaning Tower of Pisa do not tell us about science. The cartoon in the frontispiece of this book suggests several interesting questions we might ask about that experiment. Although Galileo probably didn't perform the experiment, it is unlikely, had he done so, that he would have been on the wrong side of the tower. A more interesting question is how he managed to release the masses at the same time. One might also ask what the balls were made of, how heavy they were, and how one detected the equality of fall.

My colleague John Taylor has recently replicated the Galileo experiment. He dropped three steel balls weighing 16 pounds, 8 pounds, and 2 ounces, respectively, from the top of Gamow Tower on the University of Colorado at Boulder campus, and photographed them as they fell a distance of approximately 100 feet. The method of release used was to place the balls at the edge of a hinged platform. The platform was released and allowed to rotate freely. A simple calculation shows that the acceleration of the edge of the platform is 3/2 g, where g is the acceleration due to gravity. Because the balls will accelerate at g, the platform falls out from under them, releasing them at the same time. The successful simultaneous release is shown in Figure 1. I note, however, that the platform has the bottom of the three spheres at the same height. Because the spheres were of different sizes (5 inches, 4 inches, and 1 inch in diameter, respectively), the centers of mass of the spheres were at *different* distances from the center of the earth, and thus, the gravitational force on each sphere was slightly different. The largest difference is approximately one part in 10^{16}, which is negligible in comparison with the approximately 1% effect observed in the final result, shown in Figure 2. After a fall of approximately 100 feet the heaviest ball is about one foot ahead of the lightest one (shown by the arrow), and one inch ahead of the intermediate ball.

Does this result refute Newton's laws that predict that all objects will fall at the same rate, regardless of mass? No. This prediction holds only if there are no resistive forces acting on the balls. In this experiment there is air resistance.[20] The force of air resistance depends on the size of the object. Because these spheres had both different sizes and different masses (they did have the same density), the acceleration of each ball will be different. Let me demonstrate this.

The net force acting on one of the balls is $F_{NET} = mg - Kv = ma$, where mg is the gravitational force on the ball and Kv is the resistive force, m is the mass of the object, and a is its acceleration. (I am assuming here that the resistive force is proportional to the velocity).[21] This gives $a = g - Kv/m$, where $K = 6\pi r\mu$ (Stokes' Law), r is the radius of the sphere, and μ is the viscosity of air. The mass of the ball, $m = 4/3\pi r^3 \rho$, where ρ is the density of the ball, and r is its radius. Thus $a = g - 9/2\ (\mu/\rho)(1/r^2)v$. The predicted acceleration is largest for the ball with the largest radius, which is what we observe. Even what appears to be a simple experiment involves complex experimental design and the experiment-theory comparison needs care and calculation.[22]

I encourage the reader without an extensive background in physics to follow the histories as stories. One can understand the significance of the episodes presented without necessarily understanding all of the technical details, just as one can understand the significance of the Galileo experiment without understanding the technical details presented above. I believe, however, that these details must be present if we are to distinguish and decide between different accounts of science such as those provided by the critics of science and myself.

My intent in this book is to argue positively that science is a reasonable enterprise by presenting evidence from detailed case studies from the history of contemporary physics. The reader will note immediately that my own methodology mirrors what I believe are the methods of science.[23] This is intentional. I plan to present evidence along with reasoned and critical discussion to persuade the reader that this same process constitutes the practice of science. If it does, then science can provide us with knowledge.

I challenge the critics of science, both postmodern and constuctivist, to provide convincing alternative explanations of the episodes and issues discussed...

NOTES

[1] I regard knowledge, in a rough and ready way, as justified belief. I distinguish between knowledge and truth. Thus, I believe that Newton's laws of motion and his law of universal gravitation were knowledge, certainly for the 18th and 19th centuries, although we currently regard them as false. Even today, Newton's laws are good enough for many purposes..

[2] This is a continuation of the work I began in *The Neglect of Experiment* (1986), *Experiment, Right or Wrong* (1990), and *The Rise and Fall of the Fifth Force: Discovery, Pursuit, and Justification in Modern Physics* (1993a). All of the examples presented will be from contemporary physics. I am primarily interested in science as it exists today. I also believe that, although there may be some minor differences, the methodology of physics is the same as that of other sciences.

[3] George Levine suggests that Ross is not as antiscience as he appears. "Ross is deliberately playful and theatrical throughout the book -- rather too much so, I think. The opening statement seems impossible not to read as a provocation. 'This book is dedicated to all the science teachers I never had. It could only have been written without them.' That's cute, of course. And it places Ross rather aggressively, it would seem, in the antiscience camp" (Levine 1996b, p.123). Levine suggests that Ross is playing with language and that "Since one of Ross' main points is that in order to act meaningfully in the politics of contemporary culture, we need to become *more* literate in science and technology, this opening aggression suggests a more humble and science-favoring reading than it seems at first to allow" (pp. 123-124). I believe that Levine is too sympathetic to Ross. Ross was, after all, the editor of the "Science Wars" volume of *Social Text*, a volume devoted solely to the critics of science.

[4] Not all of the nonsensical statements were written by Sokal. He cites various statements by postmodern critics of science as well. Consider the following statement by Derrida on Einstein's theory of relativity. "The Einsteinian constant is not a constant, is not a center. It is the very concept of variability -- it is, finally, the concept of the game. In other words, it is not the concept of some*thing* -- of a center starting from which an observer could master the field-- but the very concept of the game" (Derrida, cited in Sokal 1996a, p. 221). This comment makes no sense.

[5] One should also distinguish between the content of science and its technological applications.

[6] Constructivist criticism and postmodern criticism are often lumped together. There are, I believe, significant differences between them. Constructivist criticism actually deals with the details of science. Most often, postmodern criticism does not. Except for Sokal's article, the "Science Wars," issue of *Social Text* contains virtually no science.

[7] The view of scientific knowledge by social constructivists is not totally negative. In an earlier work Collins states that, "For all its fallibility, science is the best institution for generating knowledge about the natural world that we have" (Collins 1985, p. 165). I suppose he means it.

[8] Other scholars have suggested that social, class, religious, or political interests also play a role in science.

[9] Pickering's later views (1987) suggest that the material procedure of the experiment, including the experimental apparatus; the theoretical model of the apparatus; and the theoretical model of the phenomena are plastic resources that the investigator brings into a relationship of mutual support. Pickering's (1995) even more recent post-humanist view (his term) includes the "dance of agency," which I gloss as the interaction between experimenters and the physical world. Pickering views the "dance" as the dialectic between accommodation and resistance provided by human and material agency. Although the terminology has changed and Pickering seems to allow more of a role for the material world, his applications of his new views to questions of experiment are not very different from his earlier explanation, based on the interests of scientists.

[10] This is discussed below in detail in "Discovery, Pursuit, and Justfication."

[11] It is difficult to make sense of what Geison means by "privileged material." He attributes the phrase to Salomon-Bayet without any further explanation. If it is the fact that the kind of optical activity Pasteur was investigating occurs only for certain substances one wonders why he thinks this is privileged. After all, different substances often behave in different ways.

[12] The story of how Kepler came to his first law is actually quite complex, but this does not, I believe, challenge my view that the way the world is determined what Kepler found.

[13] Geison actually accuses Pasteur of fraud. "Here we deal not with mere acquiescence in the formulaic genre of scientific papers and the associated 'inductivist' image of science, but with discrepancies between Pasteur's public and private science in cases where the word 'deception' no longer seems so inappropriate, and even 'fraud' does not seem entirely our of line in the case of one or two major episodes" (p.16). For an incisive rebuttal and criticism of Geison's view of Pasteur see Perutz (1995).

[14] To be fair, not all such critics ignore this. "Of course, we would be fools to behave as though there is no knowledge of the natural world to be had and that science has no better shot at it than any other professionals, or nonprofessional"(Levine 1996b, p. 124).

[15] In their recent book *The Golem (1993)*, intended for a popular audience, Collins and Pinch present brief accounts of seven episodes from science intended to show that science is indeed unreliable. One of these is the gravity wave episode, discussed in detail below, in which I believe their account is wrong. The other accounts are also questionable..

[16] I note here that Harding, and others, associate defending science as a rational activity with political conservatism. This *ad hominem* name-calling is untrue. Defenders of science come in all political persuasions. For a detailed answer to Harding's objection see Koertge (1997).

[17] Some exceptions are: 1) the discovery of weak-nuetral currents, Galison (1987) and Pickering (1984b); 2) early attempts to detect gravity waves, Collins (1985, 1994) and Franklin (1994); 3) the solar neutrino problem, Pinch (1986) and Shapere (1982); and 4) atomic parity violation experiments, Pickering (1984a) and Franklin (1990, Chapter 8).

[18] I have left all of these essays in their original published form. This means that the later essays, which make use of the detailed case studies to illustrate the various issues, will contain repetitons, for which I apologize.

[19] As discussed below my view can be called "conjectural" realism. Although we may have good reasons for belief in the truth of scientific laws and in the existence of the entities involved in such laws, we may be wrong.

[20] Galileo was quite aware that because of air resistance the actual experimental result would differ from the result of an ideal (no air resistance) experiment. He remarked, "But I, Simplicio, who have made the test can assure you that a cannon ball weighing one or two hundred pounds, or even more, will not reach the ground by as much as a span ahead of a musket ball weighing only half a pound, provided both are dropped from a height of 200 cubits" (Galileo 1954, p. 62). Galileo's estimate of the size of the effect is reasonably accurate.

[21] Nothing in this discussion depends on the velocity dependence of the resistive force.

[22] Does the fact that the theory under test was used in the design of the apparatus prevent us from testing that theory? I think not. Although theory tells us that the release of the balls is simultaneous we have an independent check on that simultaneity (see Figure 1). If the test confirms the theory then all of the assumptions, including those used in the experimental design, are confirmed. The fact that the theory is also implicated in the design of the experiment may make the discussion more complex, but it does not make a solution impossible.

[23] This is not a circular argument. My method could show that these are not the methods of science.

CHAPTER 1

HOW TO AVOID THE EXPERIMENTERS' REGRESS

1. COLLINS AND THE EXPERIMENTERS' REGRESS

Harry Collins is well known for both his skepticism concerning experimental results and evidence and for what he calls the "experimenters' regress," the view that a correct outcome is one obtained with a good experimental apparatus, whereas a good experimental apparatus is one that gives the correct outcome. He has expressed this view at length in *Changing Order* (Collins 1985).

He illustrates these views with his history of the early attempts to detect gravitational radiation, or gravity waves. He argues that the decision between the claimed observation of gravitational waves by Weber and the failure to detect them in six other experiments could not be made on reasonable or rational grounds. This results from the fact that one can't legitimately regard the subsequent experiments as replications[1] and that one cannot provide independent reasons for belief in either result. He argues that we can't be sure that we can actually build a gravity wave detector and that we might have been fooled into thinking we had the recipe for constructing one, and that "we will have no idea whether we can do it until we try to see if we obtain the correct outcome. *But what is the correct outcome ?"*

> What the correct outcome is depends upon whether or not there are gravity waves hitting the Earth in detectable fluxes. To find this out we must build a good gravity wave detector and have a look. But we won't know if we have built a good detector until we have tried it and obtained the correct outcome! But we don't know what the correct outcome is until...and so on ad infinitum.
>
> The existence of this circle, which I call the 'experimenters' regress,' comprises the central argument of this book. Experimental work can only be used as a test if some way is found to break into the circle. The experimenters' regress did not make itself apparent in the last chapter because in the case of the TEA-laser the circle was readily broken. The ability of the laser to vaporize concrete, or whatever, comprised a universally agreed criterion of experiment quality. There was never any doubt that the laser ought to be able to work and never any doubt about when one was working and when it was not. Where such a clear criterion is not available, the experimenters' regress can only be avoided by finding some other means of defining the quality of an experiment; a criterion must be found which is independent of the experiment itself. (Collins 1985, p. 84).

More succinctly, "Proper working of the apparatus, parts of the apparatus and the experimenter are defined by the ability to take part in producing the proper experimental outcome. Other indicators cannot be found (p. 74)."

Collins argues that there are no formal criteria that one can apply to decide whether or not an experimental apparatus is working properly. In particular, Collins argues that calibration of an experimental apparatus cannot provide such a criterion.

> Calibration is the use of a surrogate signal to standardize an instrument. The use of calibration depends on the assumption of near identity of effect between the surrogate signal and the unknown signal that is to be measured (detected) with the instrument. Usually this assumption is too trivial to be noticed. In controversial cases, where calibration is used to determine relative sensitivities of competing instruments, the assumption may be brought into question. Calibration can only be performed provided this assumption is not questioned too deeply (p. 105).

In Collins' view the regress is broken by negotiation within the appropriate scientific community, which does not involve what we might call epistemological criteria, or reasoned judgment. Thus, the regress raises serious questions concerning both experimental evidence and its use in the evaluation of scientific hypotheses and theories. If no way out of the regress can be found then he has a point.

In this paper I will examine Collins' account of the first attempts to detect gravitational radiation. I will then present my own account of the episode, which differs substantially from his, and argue that his account is misleading and provides no grounds for belief in the experimenters' regress. I will show that calibration, although an important component of the decision, was not decisive in this case precisely because the experiments used a new type of apparatus to try to detect a hitherto unobserved phenomenon, and that the case of gravity wave detection is not at all typical of scientific experiments. I will also argue that the regress was broken by reasoned argument.

Before I begin, I would like to address an important methodological difference between Collins' account and my own. Collins bases his account of the episode almost entirely on interviews with some of the scientists involved. They are not named and are identified only by letter. My own account is based on the published literature. A supporter of Collins might argue that the published record gives a sanitized version of the actual history,[2] and that what scientists actually believed is contained in the interviews. I suggest that the interviews do not, in fact, show the scientists' consideration of the issues raised by the discordant results, and that these considerations are contained in the published record. In this particular episode, we have a published discussion among the participants, in which they explicitly addressed the issues as well as each others' arguments. I see no reason to give priority to off-the-cuff comments made to an interviewer, and to reject the accounts that scientists wished to have made as part of the permanent record.[3] There is no reason to assume that because arguments are presented publicly that they are not valid, or that the scientists did not actually believe them. There are, in fact, good reasons to believe that these are the arguments believed by the scientists. After all, a scientist's reputation for good work is based primarily on the published record, and it seems reasonable that they would present their strongest arguments there.[4] In addition, although Collins presents evidence that the various arguments were weighted differently by different scientists, the arguments presented were, in fact, the same as those given in publications. Neither does Collins' account demonstrate that the decision was based on anything other than the combined evidential weight of these arguments. As we shall see, there was considerable interchange between Weber and his critics, and that criticisms were offered by others, answered by Weber, and these answers were themselves evaluated. The published record indicates that the decision was based on a reasoned evaluation of the evidence.

Let us now consider in detail Collins' discussion of gravity wave detectors.

2. COLLINS'S ACCOUNT OF GRAVITY WAVE DETECTORS

Collins illustrates the experimenters' regress and his skepticism concerning experimental results with the early history of gravity wave detectors.[5] He begins with a discussion of the original, and later to become a standard, apparatus developed by Joseph Weber (Figure 1). Weber used a massive aluminum alloy bar,[6] or antenna, which was supposed to oscillate when struck by gravitational radiation.[7] The oscillation was to be detected by observing the amplified signal from piezo-electric crystals attached to the antenna. The expected signals were quite small (the gravitational force is quite weak in comparison to electromagnetic force) and the bar had to be insulated from other sources of noise such as electrical, magnetic, thermal, acoustic, and seismic forces. Because the bar was at a temperature different from absolute zero, thermal noise could not be avoided, and to minimize its effect Weber set a threshold for pulse acceptance. Weber claimed to have observed above-threshold pulses, in excess of those expected from thermal noise.[8] In 1969, Weber claimed to have detected approximately seven pulses/day due to gravitational radiation.

The problem was that Weber's reported rate was far greater than that expected from calculations of cosmic events (by a factor of more than 1000), and his early claims were met with skepticism. During the late 1960s and early 1970s, however, Weber introduced several modifications and improvements that increased the credibility of his results. He claimed that above-threshold peaks had been observed simultaneously in two detectors separated by one thousand miles. Such coincidences were extremely unlikely if they were due to random thermal fluctuations. In addition, he reported a 24 hour periodicity in his peaks, the sidereal correlation, that indicated a single source for the radiation, perhaps near the center of our galaxy. These results increased the plausibility of his claims sufficiently so that by 1972 three other experimental groups had not only built detectors, but had also reported results. None was in agreement with Weber.

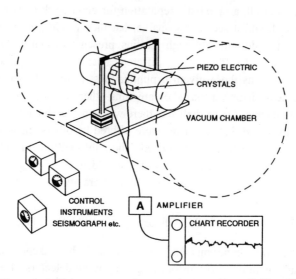

Figure 1. A Weber-type gravity wave detector. From Collins, 1985.

At this point Collins invokes the experimenters' regress cited earlier. He argues that if the regress is a real problem in science then scientists should disagree about what constitutes a good detector, and that this is what his fieldwork shows. He presents several excerpts from interviews with scientists working in the field that show differing opinions on the quality of detectors.[9] There were also different reasons offered for scientists's belief in Weber's claims. These included the coincidences between two separated detectors, the fact that the coincidence disappeared when one detector signal was delayed relative to the other, and Weber's use of the computer for analysis.[10] Not everyone agreed. Collins argues that these differing opinions demonstrate the lack of any consensus over formal criteria for the validity of gravitational wave detectors. According to Collins, the decision as to what counts as a competently performed experiment is coextensive with the debate about what the proper outcome of the experiment is.

Collins notes that after 1972 Weber's claims were less and less favored. During 1973 three different experimental groups reported negative results and subsequently these groups, as well as three others, reported further negative results. No corroboration of Weber's results was reported during this period. Although in 1972 approximately a dozen groups were involved in experiments aimed at checking Weber's findings, by 1975 no one, except Weber himself, was still working on that particular problem. Weber's results were regarded as incorrect. There were, however, at least six groups working on experiments of much greater sensitivity, designed to detect the theoretically predicted flux of gravitational radiation.

The reasons offered by different scientists for their rejection of Weber's claims were varied, and not all of the scientists engaged in the pursuit agreed about their importance.

During the period 1972-1975 it was discovered that Weber had made several serious errors in his analysis. His computer program for analyzing the data contained an error and his statistical analysis of residual peaks and background was questioned and thought to be inadequate. Weber also claimed to find coincidences between his detector and another distant detector when, in fact, the tapes used to provide the coincidences were actually recorded more than four hours apart. Weber had found a positive result where even he would not expect one. Others cited the failure of Weber's signal to noise ratio to improve, despite his "improvements" to his apparatus. In addition, the sidereal correlation disappeared.

Perhaps most important were the uniformly negative results obtained by six other groups. Collins points out that only one of these experimental apparatuses was not criticized by other groups, and that all of these experiments were regarded as inadequate by Weber.

> Under these circumstances it is not obvious how the credibility of the high flux case fell so low. In fact, it was not the single uncriticized experiment that was decisive;...Obviously the sheer weight of negative opinion was a factor, but given the tractability, as it were, of all the negative evidence, it did not *have* to add up so decisively. There was a way of assembling the evidence, noting the flaws in each grain, such that outright rejection of the high flux claim was not the necessary inference (Collins 1985, p. 91).

If Collins is correct in arguing that the negative evidence provided by the replications of Weber's experiment, the application of what we might call epistemological criteria, combined with Weber's acknowledged errors is insufficient to explain the rejection of Weber's results then he must provide another explanation. Collins offers instead the impact of the negative evidence provided by scientist Q.[11] Collins argues that it was not so much the power of Q's experimental result, but rather the forceful and persuasive presentation of that result and his careful analysis of thermal noise in an antenna that turned the tide. Q was also quite aggressive in pointing out Weber's mistakes. After Q's second negative result, no further positive results were reported.[12]

Actually, no positive results, other than Weber's, were reported before Q's publication. In fact, I have found no reports of positive results with a Weber bar detector by anyone other than Weber and his collaborators. Collins regards Q's work as the explanation of how the experimenters' regress was solved in this case. "The growing weight of negative reports, all of which were indecisive in themselves, were crystallized, as it were, by Q. Henceforward, only experiments yielding negative results were included in the envelope of serious contributions to the debate (p. 95)."

Collins concludes, "Thus, Q acted as though he did not think that the simple presentation of results with only a low key comment would be sufficient to destroy the credibility of Weber's results. In other words, he acted as one might expect a scientist to act who realized that evidence and arguments alone are insufficient to settle unambiguously the existential status of a phenomenon (p. 95)."

Scientists did offer other explanations of the discordant results of Weber and his critics. These included possible differences in the detectors; i.e. piezo-electric crystals or other strain detectors, the antenna material, and the electronics; different statistical analysis of the data, the pulse length of the radiation, and calibrations of the apparatus. These last three figure prominently in the subsequent history. Finally there was the invocation of a new, "fifth force," the possibility that the gravity wave findings were the result of mistakes, deliberate lies, or self-deception, and the explanation by psychic forces. Collins notes that by 1975 all of these alternative explanations, except for the accepted view that Weber had made an error, had disappeared from the scientific discussions. "This is exactly the sort of change we would expect to take place as the field reached consensus (p. 99)." Collins suggests that this was not a necessary conclusion, and that scientists might reasonably investigate these more radical possibilities.

Finally, Collins deals with the attempt to break the experimenters' regress by the use of experimental calibration. (See the earlier discussion of calibration). Experimenters calibrated their gravity wave detectors by injecting a pulse of known electrical energy at one end of their antenna and measuring the output of their detector. This served to demonstrate that the apparatus could detect energy pulses and also provided a measure of the sensitivity of the apparatus. One might, however, object that the electrostatic pulses were not an exact analogue of gravity waves. Another experimenter did use a different method of calibration. He used a local, rotating laboratory mass to more closely mimic gravity waves.[13]

According to Collins, Weber was initially reluctant to calibrate his own antenna electrostatically, but did eventually do so. His observations included, however, a quite different method of analyzing the output pulses. He used a non-linear, energy algorithm, whereas his critics used a linear, amplitude algorithm. (For a discussion of this difference see Appendix 1). The critics argued that one could show quite rigorously, and mathematically, that the linear algorithm was superior in detecting pulses. The issues of the calibration of the apparatus and the method of analysis used were inextricably tied together. When the calibration was done on Weber's apparatus, it was found that the linear algorithm was twenty times better at detecting the calibration signal than was Weber's non-linear algorithm. For the critics, this established the superiority of their detectors. Weber did not agree. He argued that the analysis and calibration applied only to short pulses, those expected theoretically and used in the calibration, while the signal he was detecting had a length and shape that made his method superior.

Collins regards Weber's agreement to the calibration procedure as a mistake. He had, by agreeing to it, also accepted two assumptions. The first was that gravitational radiation interacted with the antenna in the same way as electrostatic forces. Second, he accepted that the localized insertion of an energy pulse at the end of the antenna had a similar effect to that of a gravity wave that interacted with the entire antenna from a great distance.

Collins concludes,

The anomalous outcome of Weber's experiments could have led toward a variety of heterodox interpretations with widespread consequences for physics. They could have led to a schism in the scientific community or even a discontinuity in the progress of science. Making Weber calibrate his apparatus with the electrostatic pulses was one way in which his critics ensured that gravitational radiation remained a force that could be understood within the ambit of physics as we know it. They ensured physics' continuity--the maintenance of links between past and future. Calibration is not simply a technical procedure for closing debate by providing an external criterion of competence. In so far as it does work this way, it does so by controlling interpretive freedom. It is the control on interpretation which breaks the circle of the experimenters' regress, not the 'test of a test' itself (Collins 1985, pp. 105-6).

Collins states that the purpose of his argument is to demonstrate that science is uncertain. He concludes, however, "For all it's fallibility, science is the best institution for generating knowledge about the natural world that we have (p. 165)."

3. DISCUSSION

Although I agree with Collins concerning the fallibility of science and on its status as "the best institution for generating knowledge about the natural world we have," I believe there are serious problems with his argument. These are particularly important because the argument, despite Collins's disclaimer, really seems to cast doubt on experimental evidence and on its use in science, and therefore on the status of science as knowledge.

Collins's argument can be briefly summarized as follows. There are no other rigorous independent criteria for either a valid result or for a good experimental apparatus, independent of the outcome of the experiment. This leads to the experimenters' regress in which a good detector can only be defined by its obtaining the correct outcome, whereas a correct outcome is one obtained using a good detector. This is illustrated by the discussion of gravity wave detectors. In practice the regress is broken by negotiation within the scientific community, but the decision is not based on anything that one might call epistemological criteria. This casts doubt on not only the certainty of experimental evidence, but on its very validity. Thus, experimental evidence cannot provide grounds for scientific knowledge.

A) GRAVITY WAVE DETECTION[14]

Collins might correctly argue that the case of gravity wave detectors is a special case, one in which a new type of apparatus was being used to try to detect a hitherto unobserved quantity. I agree.[15] I do not, however, agree that one could not present arguments concerning the validity of the results, or that one could not evaluate the relative merits of two results, independent of the outcome of the two experiments. The regress can be broken by reasonable argument. I will also demonstrate that the published record gives the details of that reasoned argument. Collins's view that there were no formal criteria applied to deciding between Weber and his critics may be correct. But, the fact that the procedure was not rule-governed, or algorithmic, does not imply that the decision was unreasonable. (See discussion in Galison (1987), pp. 76-7).

Let us now examine the early history of attempts to observe gravity waves. As we shall see, it was not a question of what constituted a good gravity wave detector, but rather a question of whether or not the detector was operating properly and whether or not the data were being analyzed correctly. There is a distinction between data and results, or phenomena, as Bogen and Woodward (1988) have pointed out. All of the experiments did, in fact, use variants of the Weber antenna, and, with the exception of Weber, similar analysis procedures. The discordant results reported by Weber and his critics are not unusual occurrences in the history of physics, particularly at the beginning of an experimental investigation of a phenomenon.[16]

There was a clear claim by Weber that gravity waves had been observed. There were several other results of experiments to detect such waves that were negative. In addition, there were admitted errors made by Weber and serious questions raised concerning Weber's analysis and calibration procedures. To be fair, not everyone working in the field, particularly Weber, agreed about the importance of these problems. Collins expresses some surprise that the credibility of Weber's results fell so low. "...given the tractability, as it were, of all the negative evidence, it did not have to add up so decisively (Collins 1985, p. 91)." I am not surprised. I believe that Collins has seriously overstated the tractability of the negative results and understated the weight of the evidence against Weber's results. The fact that Weber's critics might have disagreed about the force of particular arguments does not mean that they did not agree that Weber was wrong. To decide the question we must look at the history of the episode as given in published papers, conference proceedings, and public letters. I believe that the picture these give is one of overwhelming evidence against Weber's result, and that the decision, although not rule governed, was reasonable, and based on epistemological criteria.

I begin with the issue of calibration and Weber's analysis procedure. The question of determining whether or not there is a signal in a gravitational wave detector, or whether or not two such detectors have fired simultaneously is not easy to answer. There are several problems. One is that there are energy fluctuations in the bar due to thermal, acoustic, electrical, magnetic, and seismic noise, etc. When a gravity wave strikes the antenna its energy is added to the existing energy. This may change either the amplitude or the phase, or both, of the signal emerging from the bar. It is not just a simple case of observing a larger signal from the antenna after a gravitational wave strikes it. This difficulty informs the discussion of which was the best analysis procedure to use.

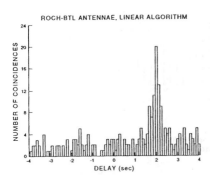 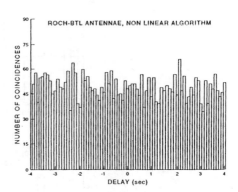

Figure 2. A plot showing the calibration pulses for the Rochester-Bell Laboratory collaboration. The peak due to the calibration pulses is clearly seen. From Shaviv and Rosen 1975.

Figure 3. A time-delay plot for the Rochester-Bell Laboratory collaboration, using the non-linear algorithm. No sign of any zero-delay peak is seen. From (Shaviv and Rosen 1975).

The non-linear, or energy, algorithm preferred by Weber was sensitive only to changes in the amplitude of the signal. The linear algorithm, preferred by everyone else, was sensitive to changes in both the amplitude and the phase of the signal. (See discussion in Appendix 1. Weber preferred the non-linear procedure because it resulted in proliferation, several pulses exceeding threshold for each input pulse to his detector. "We believe that this kind of cascading may result in observation of a larger number of two-detector coincidences for algorithm (6) [non-linear] than for (7) [linear],at certain energies (Weber, p. 246)."[17] Weber admitted, however, that the linear algorithm, preferred by his critics, was more efficient at detecting calibration pulses. He stated, "It is found for pulses which increase the energy of the normal mode from zero to kT that algorithm (7) [linear] gives a larger amount of response pulses exceeding thresholds, than algorithm (6) [non-linear]. Perhaps this is the reason that algorithm (7) is preferred by a number of groups (Weber, p. 247)." (I note here that Weber's earlier statement indicated that more than one pulse was detected for a single input pulse using the non-linear algorithm. His second statement refers to the efficiency of detecting individual calibration pulses. The language is somewhat confusing). Similar results on the superiority of the linear algorithm for detecting calibration pulses were reported by both Kafka (pp. 258-9) and Tyson (pp. 281-2). Tyson's results for calibration pulse detection are shown for the linear algorithm in Figure 2, and for the non-linear algorithm in Figure 3. There is a clear peak for the linear algorithm, whereas no such peak is apparent for the non-linear procedure. (The calibration pulses were inserted periodically during data taking runs. The peak was displaced by two seconds by the

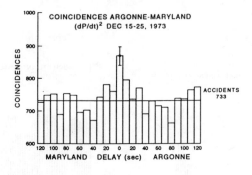
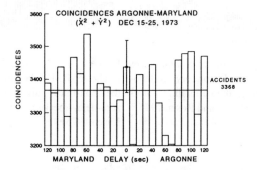

Figure 4. Weber's time delay-data for the Maryland-Argonne collaboration for the period Dec. 15-25, 1973. The top graph uses the non-linear algorithm, whereas the bottom uses the linear algorithm. The zero-delay peak is seen only with the non-linear algorithm. From Shaviv and Rosen 1975.

insertion of a time delay, so that the calibration pulses would not mask any possible real signal, which was expected at zero time delay).

Nevertheless, Weber preferred the non-linear algorithm. His reason for this was that this procedure gave a more significant signal than did the linear one. This is illustrated in Figure 4, in which the data analyzed with the non-linear algorithm is presented in (a) and for the linear procedure in (b). "Clearly these results are inconsistent with the generally accepted idea that $\dot{x}^2 + \dot{y}^2$ [the linear algorithm] should be the better algorithm (Weber, pp. 251-2)." Weber was, in fact, using the positive result to decide which was the better analysis procedure. If anyone was "regressing," it was Weber.

Weber's failure to calibrate his apparatus was criticized by others. "Finally, Weber has not published any results in calibrating his system by the impulsive introduction of known amounts of mechanical energy into the bar, followed by the observation of the results either on the single detectors or in coincidence (Levine and Garwin 1973, p. 177)."

His critics did, however, analyze their own data using both algorithms. If it was the case that, unlike the calibration pulses where the linear algorithm was superior, using the linear algorithm either masked or failed to detect a real signal, then using the

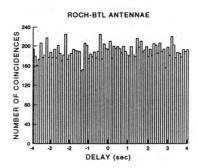

Figure 5. A time-delay plot for the Rochester-Bell Laboratory collaboration, using the linear algorithm. No sign of a zero-delay peak is seen. From Shaviv and Rosen 1975.

non-linear algorithm on their data should produce a clear signal. None appeared. Typical results are shown in Figures 3 and 5. Figure 3, which is Tyson's data analyzed with the non-linear algorithm, not only shows no calibration peak, but it does not show a signal peak at zero time delay. It is quite similar to the data analyzed with the linear algorithm shown in Figure 5. (I note that for this data run no calibration pulses were inserted).[18] Kafka also reported the same result, no difference in signal between the linear and the non-linear analysis.

Weber had an answer. He suggested that although the linear algorithm was better for detecting calibration pulses, which were short, the real signal of gravitational waves was a longer pulse than most investigators thought. He argued that the non-linear algorithm that he used was better at detecting these longer pulses. The critics did think that gravitational radiation would be produced in short bursts. For example, Douglass and others (1975) remarked that, "the raw data are filtered in a manner optimum for short pulses (p. 480)." "The filter was chosen and optimized on the basis of optimal filter theory and the assumption that bursts of gravitational radiation would be much shorter in duration then the 0.1 sec response time of the detector electronics (pp. 480-1)."

Still, if the signal was longer, one would have expected it to show up when the critics's data was processed with the non-linear algorithm. It didn't. (See Figure 3). Tyson remarked,

> I would merely like to comment that all the experiments of the Weber type, where you have an integrated calorimeter which asks the question: 'Did the energy increase or decrease in the last tenth of a second?'-- all those experiments, of which my own, Weber's, and Kafka's are an example -- would respond in a similar manner to a given pulse shape in the metric given the same algorithm. I think it must be something which only your (Weber) detector is sensitive to and not ours (Tyson, p. 288).

Drever also reported that he had looked at the sensitivity of his apparatus with arbitrary waveforms and pulse lengths. Although he found a reduced sensitivity for

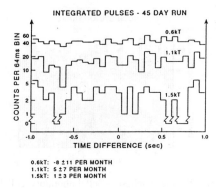

Figure 6. Drever's data looking for long pulses. No zero-delay peak is seen. From Shaviv and Rosen 1975.

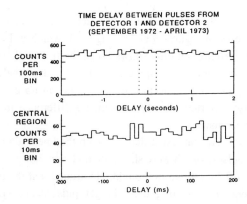

Figure 7. Drever's time delay plot. No sign of a peak at zero-delay is seen. From Shaviv and Rosen 1975.

longer pulses, he did analyze his data to explicitly look for such pulses. He found no effect (Figure 6). He also found no evidence for gravity waves using the short pulse (linear) analysis (Figure 7).

Drever summarized the situation in June 1974 as follows.

> Perhaps I might just express a personal opinion on the situation because you have heard about Joseph Weber's experiments getting positive results, you have heard about three other experiments getting negative results and there are others too getting negative results, and what does this all mean? Now, at its face value there is obviously a strong discrepancy but I think it is worth trying hard to see if there is any way to fit all of these apparently discordant results together. I have thought about this very hard, and my conclusion is that in any one of these experiments relating to Joe's one, there is always a loophole. It is a different loophole

> from one experiment to the next. In the case of our own experiments, for example, they are not very sensitive for long pulses.
>
> In the case of the experiments described by Peter Kafka and Tony Tyson, they used a slightly different algorithm which you would expect to be the most sensitive, but it is only the most sensitive for a certain kind of waveform. In fact, the most probable waveforms. But you can, if you try very hard, invent artificial waveforms for which this algorithm is not quite so sensitive.[19] So it is not beyond the bounds of possibility that the gravitational waves have that particular kind of waveform. However, our own experiment would detect that type of waveform; in fact, as efficiently as it would the more usually expected ones, so I think we close that loophole. I think that when you put all these different experiments together, because they are different, most loopholes are closed. It becomes rather difficult now, I think, to try and find a consistent answer. But still not impossible, in my opinion. One cannot reach a really definite conclusion, but it is rather difficult, I think to understand how all the experimental data can fit together (Drever, pp. 287-8).

There was considerable cooperation among the various groups. They exchanged both data tapes and analysis programs. "There has been a great deal of intercommunication here. Much of the data has been analyzed by other people. Several of us have analyzed each other's data using either our own algorithm or each other's algorithms (Tyson, p. 293)." This led to the first of several questions about possible serious errors in Weber's analysis of his data. Douglass first pointed out that there was an error in one of Weber's computer programs.

> The nature of the error was such that any above-threshold event in antenna A that occurred in the last or the first 0.1 sec time bin of a 1000 bin record is erroneously taken by the computer program as in coincidence with the next above-threshold event in channel B, and is ascribed to the time of the later event. Douglass showed that in a four-day tape available to him and included in the data of (Weber 1973), nearly all of the so-called 'real' coincidences of 1-5 June (within the 22 April to 5 June 1973 data) were created individually by this simple programming error. Thus not only some phenomenon besides gravity waves *could*, but in fact *did* cause the zero-delay excess coincidence rate (Garwin 1974, p. 9).

Weber admitted the error, but did not agree with the conclusion.

> This histogram is for the very controversial tape 217. A copy of this tape was sent to Professor David Douglass at the University of Rochester. Douglass discovered a program error and incorrect values in the unpublished list of coincidences. Without further processing of the tape, he (Douglass) reached the incorrect conclusion that the zero delay excess

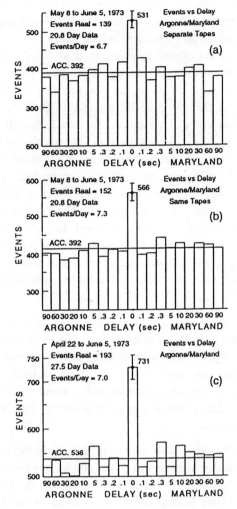

Figure 8. Weber's results. The peak at zero time delay is clearly seen. From (Weber et al. 1973).

was one per day. This incorrect information was widely disseminated by him and Dr. R.L. Garwin of the IBM Thomas J. Watson Research Laboratory. After all corrections are applied, the zero delay excess is 8 per day. Subsequently, Douglass reported a zero delay excess of 6 per day for that tape (Weber, p. 247).

Although Weber reported that his corrected result had been confirmed by scientists at other laboratories and that copies of the documents had been sent to editors and workers in the field I can find no corroboration of any of Weber's claims in the published literature. At the very least, this error raised doubts about the correctness of Weber's results (shown in Figure 8).

Another serious question was raised concerning Weber's analysis of his data. This was the question of selectivity and possible bias. Tyson characterized the difference between Weber's methods and those of his critics.

> I should point out that there is a very important difference in essence in the way in which many of us approach this subject and the way Weber approaches it. We have taken the attitude that, since these are integrating calorimeter type experiments which are not too sensitive to the nature of pulses put in, we simply maximize the sensitivity and use the algorithms which we found maximized the signal to noise ratio, as I showed you. Whereas Weber's approach is, he says, as follows. He really does not know what is happening, and therefore he or his programmer is twisting all the adjustments in the experiment more or less continuously, at every instant in time locally maximizing the excess at zero time delay. I want to point out that there is a potentially serious possibility for error in this approach. No longer can you just speak about Poisson statistics. You are biasing yourself to zero time delay, by continuously modifying the experiment on as short a time scale as possible (about four days), to maximize the number of events detected at zero time delay. We are taking the opposite approach, which is to calibrate the antennas with all possible known sources of excitation, see what the result is, and maximize our probability of detection. Then we go through all of the data with that one algorithm and integrate all of them. Weber made the following comment before and I quote out of context: "Results pile up." I agree with Joe (Weber). But I think you have to analyze all of the data with one well-understood algorithm (Tyson, p. 293).

A similar criticism was offered by Garwin, who also presented evidence from a computer simulation to demonstrate that a selection procedure such as Weber's could indeed produce his positive result.

> Second, in view of the fact that Weber at CCR-5 [a conference on General Relativity held in Cambridge][20] explained that when the Maryland group failed to find a positive coincidence excess "we try harder," and since in any case there has clearly been selection by the Maryland group (with the publication of data showing positive coincidence excesses but with no publication of data that does not show such excesses),[21] James L. Levine has considered an extreme example of such selections. In Figure [9] is shown the combined histogram of "coincidences" between two independent streams of random computer-generated data. This "delay histogram" was obtained by partitioning the data into 40 segments. For each segment, "single events" were defined in each "channel" by assuming one of three thresholds a, b, or c. That combination of thresholds was chosen for each segment which

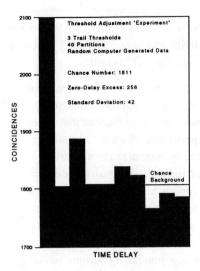

Figure 9. The result of selecting thresholds that maximized the zero-delay signal, for Levine's computer simulation. From Garwin 1974.

gave the maximum "zero delay coincidence" rate for that segment, The result was 40 segments selected from one of nine "experiments." The 40 segments are summarized in Figure [9], which shows a "six-standard-deviation" zero-delay excess (Garwin 1974, pp. 9-10).

Weber denied both charges.

> It is not true that we turn our knobs continuously. I have been full time at the University of California at Irvine for the last six months, and have not been turning the knobs by remote control from California (Weber's group and one of his antennas was located at the University of Maryland). In fact, the parameters have not been changed for almost a year. What we do is write the two algorithms on a tape continuously. The computer varies the thresholds to get a computer printout which is for 31 different thresholds. The data shown are not the results of looking over a lot of possibilities and selecting the most attractive ones. We obtain a result that is more than three standard deviations for an extended period for a wide range of thresholds. I think it is very important to take the point of view that the histogram itself is the final judge of what the sensitivity is (Weber, pp. 293-4).

Weber did not, however, specify his method of data selection for his histogram. In particular, he did not state that all of the results presented in a particular histogram had the same threshold.

Interestingly, Weber cited evidence provided by Kafka as supporting a positive gravity wave result. Kafka did not agree. This was because the evidence resulted from

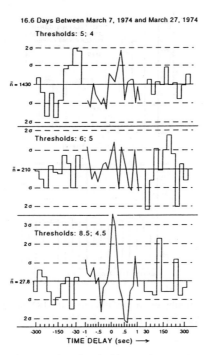

Figure 10. Kafka's results using varying thresholds. A clear peak is seen at zero-delay. From Shaviv and Rosen 1975.

performing an analysis using different data segments and different thresholds. Only one showed a positive result, indicating that such selectivity could produce a positive result. Kafka's results are shown in Figure 10. Note that the positive effect is seen in only the bottom graph. "The very last picture (Figure [10]) is the one in which Joe Weber thinks we have discovered something, too. This is for 16 days out of 150. There is a 3.6 σ [standard deviation] peak at zero time delay, but you must not be too impressed by that. It is one out of 13 pieces for which the evaluation was done, and I looked at least at 7 pairs of thresholds. Taking into account selection we can estimate the probability to find such a peak accidentally to be of the order of 1% (Kafka, p. 265)."

There was also a rather odd result reported by Weber.

> First, Weber has revealed at international meetings (Warsaw, 1973. etc.) that he had detected a 2.6-standard deviation excess in coincidence rate between a Maryland antenna [Weber's apparatus] and the antenna of David Douglass at the University of Rochester. Coincidence excess was located not at zero time delay but at "1.2 seconds," corresponding to a 1-sec intentional offset in the Rochester clock and a 150-millisecond clock error. At CCR-5, Douglass revealed, and Weber agreed, that the Maryland Group had mistakenly assumed that the two antennas used the same time reference, whereas one was on Eastern Daylight Time and the

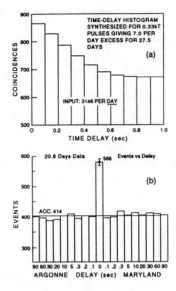

Figure 11. (a) Computer simulation result obtained by Levine for signals passing through Weber's electronics. (b) Weber's reported result. The difference is clear. From Levine and Garwin 1974.

other on Greenwich Mean Time. Therefore, the "significant" 2.6 standard deviation excess referred to gravity waves that took four hours, zero minutes and 1.2 seconds to travel between Maryland and Rochester (Garwin 1974, p. 9).

Weber answered that he had never claimed that the 2.6 standard-deviation effect he had reported was a positive result, By producing a positive result where none was expected, Weber had, however, certainly cast doubt on his analysis procedures.

Levine and Garwin (1974) and Garwin (1974) raised yet another doubt about Weber's results. This was the question of whether or not Weber's apparatus could have produced his claimed positive results. Here again, the evidence came from a computer simulation.

> Figure [11(b)] shows the 'real coincidences' confined to a single 0.1 sec bin in the time delay histogram. James L. Levine and I observed that the Maryland Group used a 1.6 Hz bandwidth "two-stage Butterworth filter." We suspected that mechanical excitations of the antenna (whether caused by gravity waves or not) as a consequence of the 1.6 Hz bandwidth would not produce coincident events limited to a single 0.1 sec time bin. Levine has simulated the Maryland apparatus and computer algorithms to the best of the information available in (Weber and others 1973) and has shown that the time-delay histogram for coincident pulses giving each

antenna 0.3 kT is by no means confined to a single bin, but has the shape shown in Figure [11(a)] (Garwin 1974, p. 9).

Let us summarize the evidential situation concerning gravity waves at the beginning of 1975. There were discordant results. Weber had reported positive results on gravitational radiation, whereas six other groups had reported no evidence for such radiation. The critics's results were not only more numerous, but had also been carefully cross-checked. The groups had exchanged both data and analysis programs and confirmed their results. The critics had also investigated whether or not their analysis procedure, the use of a linear algorithm, could account for their failure to observe Weber's reported results. They had used Weber's preferred procedure, a non-linear algorithm, to analyze their data, and still found no sign of an effect. They had also calibrated their experimental apparatuses by inserting electrostatic pulses of known energy and finding that they could detect a signal. Weber, on the other hand, as well as his critics using his analysis procedure, could not detect such calibration pulses.

There were, in addition, several other serious questions raised about Weber's analysis procedures. These included an admitted programming error that generated spurious coincidences between Weber's two detectors, possible selection bias by Weber, Weber's report of coincidences between two detectors when the data had been taken four hours apart, and whether or not Weber's experimental apparatus could produce the narrow coincidences claimed.

It seems clear that the critics's results were far more credible than Weber's. They had checked their results by independent confirmation, which included the sharing of data and analysis programs. They had also eliminated a plausible source of error, that of the pulses being longer than expected, by analyzing their results using the non-linear algorithm and by looking for such long pulses. They had also calibrated their apparatuses by injecting known pulses of energy and observing the output.

In addition, Weber's reported result failed several tests. Weber had not eliminated the plausible error of a mistake in his computer program. It was, in fact, shown that this error could account for his result. It was also argued that Weber's analysis procedure, which varied the threshold accepted, could also have produced his result. Having increased the credibility of his result when he showed that it disappeared when the signal from one of the two detectors was delayed, he then undermined his result by obtaining a positive result when he thought two detectors were simultaneous, when, in fact, one of them had been delayed by four hours. As Garwin also argued, Weber's result itself argued against its credibility. The coincidence in the time delay graph was too narrow to have been produced by Weber's apparatus. Weber's analysis procedure also failed to detect calibration pulses.

Contrary to Collins, I believe that the scientific community made a reasoned judgment and rejected Weber's results and accepted those of his critics. Although no formal rules were applied, i.e. if you make four errors, rather than three, your results lack credibility; or if there are five, but not six, conflicting results, your work is still credible; the procedure was reasonable.

I also question Collins's account of Garwin's role (Scientist Q). Although Garwin did present strong and forceful arguments against Weber's result, the same arguments were being made at the time by other scientists, albeit in a somewhat less aggressive manner. Collins's point that Garwin behaved as if he thought that a reasoned argument would not be sufficient to destroy the credibility of Weber's result also seems questionable. Garwin's behavior could also be that of a scientist who believed that Weber's results were wrong, and that valuable time and resources were being devoted to the investigation of an incorrect result, and who thought that Weber's adherence to his incorrect result was casting doubt on all of the good work being done in the field.[22] It might also just be the case that Garwin is a forceful and powerful polemicist.

I also question the role of Garwin as the crystallizer of the opposition to Weber. As we have seen, other scientists were presenting similar arguments against Weber. At GR7, Garwin's experiment was mentioned only briefly, and although the arguments about Weber's errors and analysis were made, they were not attributed to the absent Garwin.[23]

For those who prefer a theory-first view of science, I note that although disagreement with theoretical predictions may have played a role in the skepticism about Weber's initial results, it played no major role in the later dispute. Once Weber had established the credibility of his results by varying the time delay and seeing the effect disappear and by observing the sidereal correlation the argument became almost solely experimental. Was Weber really observing gravitational radiation?

B) CALIBRATION

A point that should be emphasized is that although calibration, and its success or failure, played a significant role in the dispute, it was not decisive, as Collins correctly points out. Other arguments were needed. In most cases, failure to detect a calibration signal would be a decisive reason for rejecting an experimental result. In this case it was not. The reason for this was precisely because the scientists involved seriously considered the question of whether or not an injected electrostatic energy pulse was an adequate surrogate for a gravity wave. It was doubts as to its adequacy that led to the variation in analysis procedures and to the search for long pulses.

The detection of gravitational radiation is not a typical physics experiment. Although experiments may find new phenomena it is not usual to have an experiment in which a new type of apparatus is used to search for a hitherto unobserved phenomenon. In a typical physics experiment there is usually little question as to whether or not the calibration signal is an adequate surrogate for the signal one wishes to detect. It is usually the case that calibration of the apparatus is independent of the phenomenon one wants to observe. A few illustrative cases will help.

Consider the problem I faced as an undergraduate assistant in a research laboratory. I was asked to determine the chemical composition of the gas in a discharge tube and I was given an optical spectroscope. The procedure followed was to use the spectroscope to measure the known spectral lines from various sources such as hydrogen, sodium, and mercury. The fact that I could measure these known lines accurately showed that the apparatus was working properly.[24] In addition to providing a check on whether or not I could measure spectral lines of optical wavelengths, this procedure also provided a calibration of my apparatus. I could determine small corrections to my results as a function of wavelength. I then proceeded to determine the composition of the gas in the discharge tube by measuring the spectral lines emitted and comparing them with known spectra. There was no doubt that the calibration procedure was adequate. The calibration lines measured spanned the same wavelength region as the ones I used to determine the composition.

Let us consider a more complex experiment, that of the Princeton group (Christenson *et al.* 1964) that observed the decay $K^0_L \rightarrow 2\pi$ and established the violation of CP symmetry (combined particle-antiparticle and space-reflection symmetry). The decay was detected by measuring the momenta of the two charged decay particles and reconstructing their invariant mass, assuming the decay particles were pions, and reconstructing the direction of the decaying particle relative to the beam. If it was a K^0_L decay into two pions the mass should be the mass of the K^0_L and the angle should be zero. An excess of events was indeed found at the K^0_L mass and at zero angle to the beam. In order to demonstrate that the apparatus was functioning properly and that it could detect such decays, it was checked by looking at the known phenomenon of the regeneration of K^0_S mesons, followed by their decay into two pions. If it was operating properly the distributions in mass and angle in the case of both the K^0_S decays and the proposed K^0_L decays should have been identical. They were. (For details of this experiment see Franklin (1986, Chapters 3, 6, 7). Here too there was no doubt that the surrogate and the phenomenon were sufficiently similar. Both detected two particle decays of particles which had the K^0 mass, and which were travelling parallel to the beam.

A somewhat different example is provided by an experiment to measure the K^+_{e2} branching ratio (Bowen *et al.* 1967). (For further discussion of this experiment see Franklin (1990, Chapter 6). In this case the decay positron resulting from the decay was to be identified by its momentum, its range in matter, and by its counting in a Cerenkov counter set to detect positrons. The proper operation of the apparatus was shown, in part, by the results themselves. Because the K^+_{e2} decay was very rare (approximately 10^{-5}) compared to other known K^+ decay modes such decays in coincidence with noise in the Cerenkov counter would be detected. In particular, the muon from $K^+_{\mu 2}$ decay, which had a known momentum of 236 MeV/c, was detected. A peak was observed at the predicted momentum, establishing that the apparatus could measure momentum accurately. In addition, the width of the peak determined the experimental momentum resolution, a quantity needed for the analysis of the experiment. The Cerenkov counter was checked, and its efficiency for positrons measured, by comparing it to a known positron detector in an independent experiment. The apparatus was also sensitive to K^+_{e3} decay. This decay produced high energy positrons with a maximum momentum of 227 MeV/c, which was quite close to the 246 MeV/c momentum expected for K^+_{e2} decay. High energy K^+_{e3} positrons were used to determine the range in matter expected for the K^+_{e2} positrons, and to demonstrate that the apparatus could indeed measure the range of positrons in that energy region. The approximately 10 percent difference in momentum was considered small enough, given the known behavior of positrons in this energy region. In this case, too, there was no doubt as to the adequacy of the calibration.

In all three cases the calibration of the apparatus did not depend on the outcome of the experiment in question. In these cases proper operation of the experimental apparatus was demonstrated independently of the composition of the gas discharge, whether or not the K^0_L actually decays into two pions, or what the K^+_{e2} branching ratio was. Clearly, three examples do not demonstrate that calibration always works, but they are, I believe, far more typical of the calibration procedures used in physics than is gravity wave detection. I also believe that in cases such as these they are legitimately more decisive. Had any of these calibration procedures failed, then the results of the experiments would have been rejected. In the case of gravity waves, as we have seen, calibration, while important, was not decisive. Scientists are quite good at the pragmatic epistemology of experiment.

Collins also claims that calibration is not a "test of a test," but rather breaks the circle of the experimenters' regress by its control of the interpretation of experimental results. He offers Weber's failed calibration as an explanation of why alternative explanations of the discordant results of Weber and his critics were not offered after 1975.

There is a simpler explanation for the lack of alternatives. Weber's result was reasonably regarded as wrong. There is no need to explain an incorrect result.

4. CONCLUSION

I have argued that Collins's argument for the experimenters' regress is wrong. He conflates the difficulty of getting an experiment to work with the problem of demonstrating that it is working properly. This leads him, particularly in the case of the TEA laser, to argue against the possibility of the replication of an experiment. (See discussion in note 1). The impossibility of replication, combined with what he claims is the lack of formal criteria for the proper operation of an experimental apparatus leads to the experimenters' regress. Gravity wave detection is then used to illustrate the regress.

I believe that I have shown that his account of gravity waves is incorrect. Epistemological criteria were reasonably applied to decide between Weber's result and those of his critics. I have also argued that although calibration was not decisive in the case of gravity wave detectors, nor should it have been, it is often a legitimate and important factor, and may even be decisive, in determining the validity of an experimental result.

Both the argument about the impossibility of replication and the lack of criteria in deciding the validity of experimental results fail. The history of gravity wave detectors does not establish what Collins claims it does. There are no grounds for belief in the experimenters' regress.

5. EPILOGUE

At the present time gravity waves have not been detected by either the use of Weber bar antennas or by the newer technique of using an interferometer, in which the gravitational radiation will have a differential effect on the two arms of the interferometer and thus change the observed interference pattern. The radiation has not been detected even though current detectors are several orders of magnitude more sensitive than those in use in 1975.[25]

Gravity waves have, however, been observed. They have been detected by measuring the change in orbital period of a binary pulsar. Such a binary system should emit gravitational radiation, thereby losing energy and decreasing the orbital period. This effect was initially measured using the two results of (Hulse and Taylor 1975), which provided the initial measurement of the period, and of (Weisberg and Taylor 1984), which measured the period at a later time. The measured change in the period was $(-2.40 \pm 0.09) \times 10^{-12}$ s s^{-1}, in excellent agreement with the theoretical prediction of $(-2.403 \pm 0.002) \times 10^{-12}$ s s^{-1}. "As we have pointed out before most relativistic theories of gravity other than general relativity conflict strongly with our data, and would appear to be in serious trouble in this regard. It now seems inescapable that gravitational radiation exists as predicted by the general relativistic quadrupole formula (Weisberg and Taylor 1984, p. 1350)."[26] If General Relativity is correct, Weber should not have observed a positive result.

APPENDIX 1

"Let the output voltage of the gravitational radiation antenna amplifier be given by

$$A = F(t) \sin(\omega_0 t + \phi), \tag{1}$$

where ω_0 is the normal mode angular frequency The amplitude $F(t)$ and the phase ϕ have values characteristic of signals and noise. It is now common practice to obtain from (1) the amplitude and phase by combining (1) with local reference oscillator voltages $\sin \omega_0 t$ and $\cos \omega_0$ to obtain:

$$A \cos \omega_0 t = \tfrac{1}{2} F(t) [\sin(2\omega_0 t + \phi) + \sin \phi], \tag{2}$$
$$A \sin \omega_0 t = \tfrac{1}{2} F(t) [\cos \phi - \cos(2\omega_0 t + \phi)]. \tag{3}$$

After filtering with a time constant short compared with the antenna relaxation time, (2) and (3) become the averages

$$x = \langle F(t) \cos \phi/2 \rangle, \tag{4}$$
$$y = \langle F(t) \sin \phi/2 \rangle. \tag{5}$$

An incoming signal may change phase and amplitude of the detector voltage, depending on the initial noise-induced phase relations. The detector output voltage includes narrow band noise of the normal mode of the antenna V_{ANT} and relatively wide band noise V_N from transducers and electronics. To search for sudden changes in amplitude we may observe a function of the derivative of the power P which for convenience is taken as the (positive) quantity:

$$(dP/dt)^2 = [\Delta(x^2+y^2)/\tau]^2 = [\Delta[V_{ANT}+V_N]^2/\tau]^2 \rightarrow [2\Delta(V_{ANT}V_N)/\tau]^2. \quad (6)$$

(6) is independent of the phase. Incoming signals which change only the phase would therefore be missed and to include such cases we may search for sudden changes in the quantity

$$(dx/dt)^2 + (dy/dt)^2 = [[\Delta V_{ANT}+ \Delta V_N]_x^2 + [\Delta V_{ANT} + \Delta V_N]_y^2]/\tau^2 \quad (7)$$

Suppose we insert a sequence of calibration test pulses with the short duration Δt at times t_1, t_2, t_3 ...t_n and search for the single pulse detector response only at times $t_1 + \Delta t$, $t_2 + \Delta t$, $t_3 + \Delta t$,...$t_n + \Delta t$. It is found for pulses which would increase the energy of the normal mode from zero to kT that algorithm (7) gives a larger amount of response pulses exceeding thresholds, than algorithm (6) . Perhaps this is the reason that algorithm (7) is preferred by a number of groups.

However, a study of chart records shows that algorithm (7) produces single response pulses for each test pulse while algorithm (6) may produce a sequence with more than 20 pulses following insertion of a single test pulse, many of them large enough to cross thresholds. This is a consequence of occurrence of the term $\Delta(V_{ANT}V_N)$ in (6). The single pulse excites the antenna and V_{ANT} remains large for the antenna relaxation time. The rapidly varying wide band noise V_N then produces the sequence of large pulses. This does not occur in (7) because ΔV_{ANT} instead of V_{ANT} is combined with ΔV_N. For very weak signals the term $2V_{ANT}\Delta V_{ANT}$ may be important for (6).

In one series of observations 50 single kT pulses were introduced at two-minute intervals. One hundred and ninety-two response pulses exceeding threshold set at five per minute were emitted by the receiver for algorithm (6) in consequence of the proliferation process. (Weber in (Shaviv and Rosen 1975), pp. 245-6)."

NOTES

[1] Collins offers two arguments concerning the difficulty, if not the virtual impossibility of replication. The first is philosophical. What does it mean to replicate an experiment? In what way is the replication similar to the original experiment? A rough and ready answer is that the replication measures the same physical quantity. Whether or not it, in fact, does so can, I believe, be argued for on reasonable grounds, as discussed below.

Collins' second argument is pragmatic. This is the fact that in practice it is often difficult to get an experimental apparatus, even one known to be similar to another, to work properly. Collins illustrates this with his account of Harrison's attempts to construct two versions of a TEA leaser (Transverse Excited Atmospheric) (Collins 1985, pp. 51-78). Despite the fact that Harrison had previous experience with such lasers, and had excellent contacts with experts in the field, he had great difficulty in building the lasers. Hence the difficulty of replication.

Ultimately Harrison found errors in his apparatus and once these were corrected the lasers operated properly. As Collins admits, "...in the case of the TEA laser the circle was readily broken. The ability of the laser to vaporize concrete, or whatever, comprised a universally agreed criterion of experimental quality. There was never any doubt that the laser ought to be able to work and never any doubt about when one was working and when it was not (Collins 1985, p. 84)."

Although Collins seems to regard Harrison's problems with replication as casting light on the episode of gravity waves, as support for the experimenters' regress, and as casting doubt on experimental evidence in general, it really doesn't work. As Collins admits (see quote in last paragraph), the replication was clearly demonstrable. One may wonder what role Collins thinks this episode plays in his argument.

[2] Trevor Pinch recently remarked that an account based only on publications was "bloodless (private communication)."

[3] Michael Lynch (1991) has, in a somewhat different case, argued that what scientists said when they were recording their data has more importance in evaluating their experimental claims than is their published considerations. This conflates data and experimental results. For a discussion of the general issue see (Bogen and Woodward 1988) and for discussion of this specific case see (Franklin 1993b).

[4] Someone might object that the scientist is merely putting their best foot forward, and that the public arguments are not those they actually believed. I don't believe this to be the case, and Collins has certainly not presented any evidence to support this view. I have presented evidence that, at least in one case, the arguments offered in private were the same as those offered publicly. In the case of the Fifth Force, a modification of the law of gravity, I have examined the private E-mail correspondence between the proposers of the hypothesis, and compared it with the published record. There is no difference in the arguments offered. See Franklin (1993a, pp. 35-48).

[5] As discussed earlier, one cannot examine Collins's sources in any detail. Collins uses interviews almost exclusively, and to maintain anonymity he refers to scientists by a letter only. In addition there are no references given to any of the published scientific papers involved, not even to those of Weber.

[6] This device is often referred to as a Weber bar.

[7] Gravitational radiation is produced when a mass is accelerated.

[8] Given any such threshold there is a finite probability that a noise pulse will be larger than that threshold. The point is to show that there are pulses in excess of those expected statistically.

[9] This might also be expected when a new detector is first proposed and there has been little experience in its use. Although one may think about sources of background in advance, it is the actual experience with the apparatus that often tells scientists which of them are present and important.

[10] Weber originally analyzed the data using his own observation of the output tapes.

[11] Any reader of the literature will easily identify Q as Richard Garwin.

[12] Collins does not imply that there was anything wrong with the behavior of Q and his group. "There is no reason to believe that they had anything but the best motives for these actions but they pursued their aim in an unusually vigorous manner (1985, p. 95)."

[13] A local oscillating mass is also not an exact analog. Although it produces tidal gravitational forces in the antenna, it does not produce gravity waves. Only a distant source could do that. Such a mass would, however, have a gravitational coupling to the antenna, rather than an electromechanical one.

[14] I will rely, primarily, on a panel discussion on gravitational waves that took place at the Seventh International Conference on General Relativity and Gravitation (GR7), Tel-Aviv University, June 23-28, 1974. The panel included Weber and three of his critics, Tyson, Kafka, and Drever, and included not only papers presented by the four scientists, but also included discussion, criticism, and questions. It includes almost all of the important and relevant arguments concerning the discordant results. The proceedings were published as Shaviv and Rosen (1975). Unless otherwise indicated all quotation in this section are from Shaviv and Rosen (1975). I shall give the author and the page numbers in the text.

[15] One might then wonder why he uses such an atypical example as his illustration of the experimenters' regress.

[16] For a discussion of other similar episodes, that of experiments on atomic parity violation and on the Fifth Force in gravity see Franklin (1990, 1993a, 1993b).

[17] One might worry that this cascading effect would give rise to spurious coincidences.

[18] Collins does not discuss the fact that Weber's critics exchanged both data and analysis programs, and that they analyzed their own data with Weber's preferred non-linear analysis algorithm and failed to find a signal. This fact, as documented in the published record, would seem to argue for the use of epistemological criteria in the evaluation of the discordant experimental results.

[19] Weber did, in fact, report such a waveform (1975).

[20] I have been unable to find a published proceedings of this conference. Richard Garwin (private comminication) has informed me that these proceedings were never published.

[21] As Weber answered, the Maryland group had presented data showing no positive coincidence excess at GR7. Garwin was not, however, at that meeting, and the proceedings were not published until after Garwin's 1974 letter appeared.

[22] Several scientists working on gravitational radiation mentioned that they thought Weber had, at least to some extent, discredited work in the field (private communication).

[23] The panel discussion on gravitational waves covers 56 pages, 243-298, in Shaviv and Rosen (1975). Tyson's discussion of Garwin's experiment occupies one short paragraph (approximately one quarter of a page) on p. 290.

[24] It also showed the experimenters that I was working properly.

[25] An account of an experiment using such a detector appears in (Astone and others 1993). Using a very sensitive cryogenic antenna they set a limit of no more than 0.5 events/day, in contrast to Weber's claim of approximately seven events/day.

[26] More recent measurements and theoretical calculations give $(2.427 \pm 0.026) \times 10^{-12}$ ss^{-1} (Measured) (Taylor and Weisberg 1989) and $(2.402576 \pm 0.000069) \times 10^{-12}$ s s^{-1} (Theory) (Damour and Taylor 1991).

CHAPTER 2

THE APPEARANCE AND DISAPPEARANCE OF THE 17-KEV NEUTRINO

It is a fact of life in empirical science that experiments often give discordant results. This is nowhere better illustrated than in the recent history of experiments concerning the existence of a heavy, 17-keV neutrino.[1] What makes this episode so intriguing is that both the original positive claim, as well as all subsequent positive claims, were obtained in experiments using one type of apparatus, namely those incorporating a solid-state detector, whereas the initial negative evidence resulted from experiments using another type of detector, a magnetic spectrometer.[2] This is an illustration of discordant results obtained using different types of apparatus. One might worry that the discord was due to some crucial difference between the types of apparatus or to different sources of background that might mimic or mask the signal.

The 17-keV neutrino was first "discovered" by Simpson in 1985. The initial replications of the experiment all gave negative results, and suggestions were made that attempted to explain Simpson's result using accepted physics, without the need for a heavy neutrino. Subsequent positive results by Simpson and others led to further investigation. Several of these later experiments found evidence supporting that claim, whereas others found no evidence for such a particle. Some theorists attempted to explain away the result, and others tried to explain it and to incorporate it within existing theory without the need for a new particle, or to look for the further implications of such a particle, or to propose a new theory which would incorporate the new particle.[3] The question of the existence of such a heavy neutrino remained unanswered for several years. Recently, doubt has been cast on the two most convincing positive experimental results, and errors found in those experiments. In addition, recent, extremely sensitive experiments have found no evidence for the 17-keV neutrino. The consensus is that it does not exist. The discord has been resolved by a combination of finding errors in one set of experiments and a preponderance of evidence.

I. THE APPEARANCE

A. "THE DISCOVERY"

The 17-keV neutrino was first reported in 1985 by Simpson (1985).[4] He had searched for a heavy neutrino by looking for a kink in the energy spectrum, or in the Kurie plot,[5] at an energy equal to the maximum allowed decay energy minus the mass of the heavy neutrino, in energy units. The fractional deviation in the Kurie plot value $\Delta K/K \sim R[1 - M_2^2/(Q - E)^2]^{1/2}$, where M_2 is the mass of the heavy neutrino, R is the intensity of the second neutrino branch, Q is the total energy available for the transition, and E is the

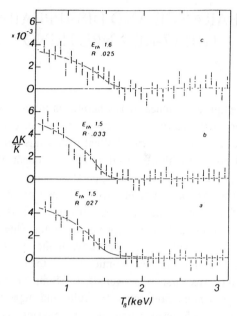

Figure 1. The data of three runs presented as ΔK/K (the fractional change in the Kurie plot) as a function of the kinetic energy of the β particles. E_{th} is the threshold energy, the difference between the endpoint energy and the mass of the heavy neutrino. A kink is clearly seen at E_{th} = 1.5 keV, or at a mass of 17.1 keV. Run a included active pileup rejection, whereas runs b and c did not. c was the same as b except that the detector was housed in a soundproof box. No difference is apparent. From (Simpson 1985).

energy of the electron.[6] Simpson's result is shown in Fig. 1. A kink is clearly seen at an energy of 1.5 keV, corresponding to a 17 keV neutrino. "In summary, the β spectrum of tritium recorded in the present experiment is consistent with the emission of a heavy neutrino of mass about 17.1 keV and a mixing probability of about 3%" (Simpson 1985, p. 1893).

Simpson had been using the apparatus for some time.[7] In 1981 he had attempted to measure, or to set an upper limit on, the mass of the neutrino (to be correct, the mass of the electron antineutrino) by a precise measurement of the end-point energy of the beta-decay spectrum of tritium.[8] If the neutrino had mass then the measured endpoint energy would be lower than that predicted, by an amount equal to the mass of the neutrino. In addition, the shape of the energy spectrum near the endpoint was sensitive to the mass of the neutrino. "The precision measurement of the β spectrum of tritium near its endpoint seems to offer the best chance of determining, or putting a useful limit on the mass m_v of the electron antineutrino" (Simpson 1981a, p. 649). Earlier measurements on tritium had been made with magnetic spectrometers, whereas Simpson used a different type of experimental apparatus, in which the tritium was implanted in a Si(Li) x-ray detector, a solid-state device. Although such an apparatus had worse energy

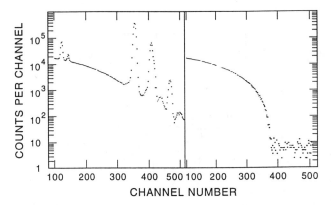

Figure 2. Logarithmic display of a typical spectrum in the multichannel analyzer. The x rays shown on the left side are those of Cu, Mo, and Ag. The ability of the chopper system to eliminate the x rays is clear. From Simpson 1981a.

resolution than did the magnetic spectrometers (300 eV as opposed to 50 eV), Simpson felt that that disadvantage could be circumvented to a large extent. In addition, source effects and final-state interactions would be different in the two types of experiment. "Clearly, it would be nice to have an experiment different enough from the above [magnetic spectrometers], yet accurate enough to check on the present upper limit on m_ν," (Simpson 1981a, p. 649).

Simpson devoted considerable effort to both the calibration of the apparatus and the details of data recording and analysis.[9] Two of the key elements of the measurement were the energy calibration and the energy resolution. The energy was calibrated using x rays of known energy from copper, molybdenum, and silver. The calibration, as well as the stability of the entire recording apparatus, was constantly monitored. Beta-decay spectrum data, as well as those data plus calibration data were recorded with the use of a slotted wheel, an x-ray chopper. This allowed x rays from the copper-molybdenum calibration source to strike the detector when the slots were open. When the slots were closed, the calibration x rays were excluded. The signal from the detector was routed to different halves of the same multichannel analyzer, depending on whether or not the slots were open. Thus, one should observe only the beta-decay spectrum when the slots were closed, and that spectrum with the x-ray calibration lines superimposed, when the slots were open. This is seen in Fig. 2. The energy resolution was determined at the same time using both copper and molybdenum x rays, and in separate experiments using x rays from iron and silver.

In Simpson's earlier low-mass neutrino search the energy resolution and calibration near the endpoint energy of 18.6 keV had been crucial. In the heavy-neutrino search, one had to worry about these factors at low energy, approximately 1.5 keV. "Because of the difficulty of energy calibrating an x-ray detector below about 6 keV the calibration was established in the following way. The x-rays from Cu and Br, and the Mo K_α were

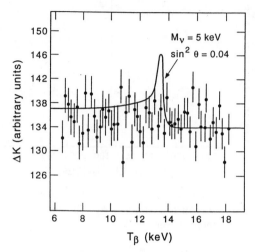

Figure 3. The magnitude of the difference of adjacent points of the Kurie plot for ^{3}H as a function of the kinetic energy of the b particles. The smooth curve is theoretically expected for a heavy neutrino with a mass of 5 keV and a mixing strength of four percent. From Simpson 1981b.

used to determine a linear calibration (with a typical rms deviation of 6 eV). The precision pulser was then used to measure the pulse-height response over the whole ADC [analog to digital converter] range. This was combined with the x-ray calibration to determine a calibration over the whole energy range" (Simpson 1985, p. 1891).

Another possible problem was pileup, a spectral distortion due to the chance occurrence of two nearly simultaneous β decays. "In one run a pile-up rejection signal from the amplifier was used to veto piled up pulses, and in two others this was not done in order to check that the rejection process did not create an artifact in the spectrum" (Simpson 1985, p. 1891). (See Fig. 1. For further details of the experiment and its analysis see Simpson (1981a)). The results of his first search were, "The measurement implies a mass < 65 eV with 95% confidence and a best value of 20 eV which is however only 0.2 standard deviations from zero mass" (Simpson 1981a, p. 649).

Simpson subsequently became aware of theoretical work (McKellar 1980; Shrock 1980) that showed that endpoint measurements were sensitive to neutrino mass only if it were the dominant decay mode. "There is considerable interest in whether the neutrino (or antineutrino) emitted in weak interactions is a mass eigenstate or a linear superposition of primitive neutrinos of definite mass. If the latter is the case, then energy spectra of β particles will show kinks associated with the emission of energetically allowed neutrinos of different mass. An examination of β spectra can therefore be used to look for massive neutrinos and, if observed, to determine the mixing amplitudes" (Simpson 1981b, p. 2971). Simpson, using the same apparatus that

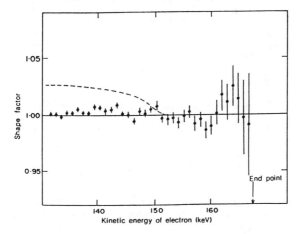

Figure 4. The shape factor of the β-spectrum of ^{35}S. The dotted line is the shape expected for a 17-keV neutrino with a mixing strength of three percent. From Datar *et al.* 1985.

he had used in his earlier experiment, searched for a neutrino with a mass between 100 eV and 10 keV. He found no evidence for such a neutrino (Fig. 3).

During the period 1981-1985 there had been, and continues to be, interest in whether or not there are massive neutrinos. This was due, in part, to reports by a Soviet group (Lubimov *et al.* 1980) that gave limits on the mass of the neutrino of $14 \leq m_v \leq 46$ eV, at the 99% confidence level.[10] Schreckenbach *et al.* (1983) had also searched for a massive neutrino and reported, "To conclude, we have found no evidence for a massive neutrino in the nuclear beta decay of ^{64}Cu for the range $m_v = 30\text{-}460$ keV. Limits below 1% were achieved" (p. 208). Boehm and Vogel reviewed the subject of neutrino mass in 1984 and concluded, "To date there has been no confirmed evidence that neutrinos have finite mass. A reported deviation in the beta decay endpoint in ^{3}H [tritium], if confirmed, may yet indicate a mass in the range 20-30 eV [a reference to the result reported by Lubimov *et al.*]" (Boehm and Vogel 1984, p. 131). This was where matters stood when Simpson reported the existence of the 17-keV neutrino.

B. THE INITIAL REACTION

1. Experimental
Simpson's positive result for the 17-keV neutrino was published in April, 1985. By the end of the year the results of five other experimental searches for the particle had appeared in the published literature (Altzitzoglou *et al.* 1985; Apalikov *et al.* 1985; Datar *et al.* 1985; Markey and Boehm 1985; Ohi *et al.* 1985). All of them were negative. The experiments set limits of less than one percent for a 17-keV branch of the decay, in contrast to Simpson's value of three percent (See Table I). Typical results are

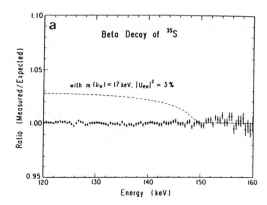

Figure 5. The ratio of the measured ^{35}S beta-ray spectrum to the theoretical spectrum. A three percent mixing of a 17-keV neutrino should distort the spectrum as indicated by the dashed curve. From (Ohi et al. 1985).

shown in Figs. 4 and 5 and should be compared to Simpson's result shown in Fig. 1. No kink of any kind is apparent.

Each of the experiments examined the beta-decay spectrum of ^{35}S, and searched for a kink at an energy of 150 keV, 17 keV below the endpoint energy of 167 keV. Three of the experiments, those of Altzitzoglou et al., of Apalikov et al., and of Markey and Boehm, used magnetic spectrometers. Those of Datar et al. and of Ohi et al. used Si(Li) detectors, the same type used by Simpson. In the latter two cases, however, the source was not implanted in the detector, as Simpson had done, but was separated from it. Such an arrangement would change the atomic physics corrections to the spectrum. In addition, as noted above, the experiments used a ^{35}S beta-decay source, which had a higher endpoint energy than did the tritium used by Simpson (167 keV in contrast to 18.6 keV). As discussed below, this higher endpoint energy made particular corrections to the beta-decay spectrum less important.

2. Theoretical

Questions were also raised concerning the theoretical model used by Simpson to analyze his data. In order to demonstrate that a kink existed in the beta-decay spectrum, one had to compare the measured spectrum with that predicted theoretically. This involved a rather complex calculation, which included various atomic physics effects, particularly screening by atomic electrons, and it was Simpson's calculation of these effects that was questioned. Haxton noted, "A number of conventional approximations in treating final-state Coulomb effects should fail for small β energies [Simpson's kink had been observed at very low energy]... A particular class of the neglected atomic effects, those corresponding to exchange terms in the sudden approximation are shown to generate corrections of order η^4 [a parameter related to the electron energy] to the standard Coulomb function, producing a distortion in the β spectrum qualitatively similar to that observed by Simpson. Similarly, the standard treatment of screening corrections becomes unreliable whenever η is not small. Thus it is possible that a

Table I: Summary of Results (Hime 1992)

Experiment	Isotope	$(\sin^2\theta) \times 100$	M_2 (keV)	Reference
Solid State				
Guelph	^{3}H in Si(Li)	2-3	17.1	Simpson (1985)
INS Tokyo	^{35}S	<0.15 (90% CL)	17	Ohi (1985)
Bombay	^{35}S	<0.60 (90% CL)	17	Datar (1985)
Guelph	^{3}H in Si(Li)	1.10 ± 0.30	17.07 ± 0.09	Hime (1989)
	^{3}H in HPGe	1.11 ± 0.14	16.93 ± 0.07	Hime (1989)
	^{35}S	0.73 ± 0.11	16.9 ± 0.4	Simpson (1989)
Oxford	^{35}S	0.78 ± 0.09	16.95 ± 0.35	Hime (1991)
	^{63}Ni	0.99 ± 0.22	16.75 ± 0.36	Oxford Report
LBL	^{14}C in HPGe	1.2 ± 0.3	17.1 ± 0.6	Sur (1991)
IBEC Studies				
CERN/Isolde	^{125}I	<2.0 (98% CL)	17	Borge (1986)
Zagreb	^{55}Fe	<1.6 (95% CL)	15-45	Zlimen (1988, 1990)
	^{71}Ge	1.6 ± 0.8	17.1 ± 1.3	Zlimen (1991)
LBL	^{55}Fe	0.85 ± 0.45	21 ± 2	Norman (1991)
Buenos Aires	^{71}Ge	0.80 ± 0.25	13.8 ± 1.8	TANDAR Preprint
Mag. Spec.				
Princeton	^{35}S	<0.40 (99% CL)	17	Altzitzoglou (1985)
ITEP	^{35}S	<0.17 (90% CL)	17	Apalikov (1985)
Caltech	^{35}S	<0.25 (90% CL)	17	Markey (1985)
	^{63}Ni	<0.25 (90% CL)	17	Wark (1986)
Chalk River	^{63}Ni	<0.28 (90% CL)	17	Hetherington (1987)
Caltech	^{35}S	<0.60 (90% CL)	17	Becker (1991)
Munich	^{177}Lu	<0.80 (83% CL)	17	Conf. Report

complete treatment of atomic effects will provide a conventional explanation of the observed distortion" (Haxton 1985, p. 807). Haxton's own calculation indicated that "Exchange corrections are shown to produce a distortion in the tritium beta spectrum similar in shape to that for heavy neutrino emission, though significantly smaller" (p. 807). See Fig. 6.

A similar point was made by Eman and Tadic (Eman and Tadic 1986).

The recent observation of a distortion in the β decay of tritium for electron kinetic energies T < 1.5 keV depends on the choice of the Fermi function F(Z,W). This function enters into the Kurie plot in which the expression $K = [N_\beta(Z,W)/pWF(Z,W)]^{1/2}$ is plotted vs. T. Here $N_\beta(Z,W)$ is the measured number of β particles at an energy W and a momentum p, and Z is the charge of the daughter nuclei. In principle, the Fermi function F(Z,W) includes all known effects, such as finite size, screening, radiation, exchange, and higher multipoles. Screening corrections will be discussed in the next section. These corrections lower the value of the

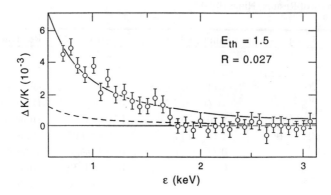

Figure 6. Simpson's data compared to the theoretical deviation in $\Delta K/K$ attributable to the neglect of screening corrections in the standard treatment of Coulomb distortions. The dashed line is the theoretical calculation. The solid line is the theoretical result multiplied by six. From (Haxton 1985).

Fermi function $F(Z,W)$ for β particles of low kinetic energy T. Hence the value of K increases at low T, in comparison to the Fermi function $F_o(Z,W)$ calculated for the Coulomb potential.

The main aim of this paper is to study the screening corrections. Should these turn out to be smaller than those used by Simpson, the value of K would increase at low T, so that the hump [kink] in the Kurie plot would disappear. In fact, our analysis indicates that this might very probably be the case, so that the observed distortion might have a more conventional origin. However, the uncertainties in the calculation of the Fermi function do not allow one to rule out heavy-neutrino emission completely (p. 2128).

The results of their calculation are shown in Fig. 7. The calculation, however, depended strongly on a parameter, D, whose value was not well determined. They also noted that experiments on ^{35}S involved higher kinetic energies, where screening effects were expected to be less important.

A further attempt to explain Simpson's result using accepted physics was made by Lindhard and Hansen (1986). They considered atomic physics corrections beyond those already discussed. "A detailed account of the decay energy and Coulomb-screening effects raises the theoretical curve in precisely the energy range [1.5 keV in the tritium beta-decay spectrum] so that little, if any, of the excess remains" (p. 965). Drukarev and Strikman (1986) also considered atomic effects in beta decay. They concluded, "The final-state interaction of a β electron with atomic electrons has been calculated to accuracy $(\alpha Z/v)^2$. It is shown that previous studies devoted to the final-state interaction have not taken into account all diagrams contributing in the first nonvanishing approximation. Correct allowances for the final-state interaction makes it impossible to

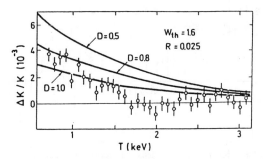

Figure 7. ΔK/K as a function of the kinetic energy of the β particles. The curves are not fits to the experimental data of Simpson (1985). From (Eman and Tadic 1986).

explain the discrepancy between the theory and the experimental results of Simpson by the emission of a neutrino" (p. 686).

A different criticism of Simpson's analysis was offered by Kalbfleisch and Milton (1985). They suggested that his result might be an artifact of systematic effects in his experiment. In particular, they noted that Simpson had used a piecewise treatment of the spectrum: 0.7 to 3.2 keV, 6.5 to 18 keV, and 9.5 to 17 keV. Simpson had also allowed the endpoint energy to vary considerably in each of the segments, from 18.7 to 19.3 keV. This was far larger than accepted variations. Simpson himself had remarked on this point. "In fitting Q, M_2, R, and an overall normalization were varied. While Q is now well determined to lie between about 18.57 and 18.61 keV, it was necessary to allow it to vary to achieve a good fit in the energy range of interest which is a long way from the endpoint. Incomplete pile-up rejection, *inadequacy of the screening correction to F(E,Z) [the Fermi function*, and any remaining inaccuracies of the energy calibration could account for obtaining a Q value different from the true one" (Simpson 1985, p. 1892, emphasis added). They also suggested that Simpson's result argued for a serious discrepancy between theory and experiment for the lifetime of tritium.

By the end of 1985, there were apparently well-confirmed experiments that disagreed with Simpson's claim of a 17-keV neutrino, albeit with a different source (^{35}S in contrast to ^{3}H) and, in some cases, with different types of experimental apparatus. There were also plausible suggestions that might explain his result using accepted physics, and which did not involve a heavy neutrino.[11] Work continued.

C. THE SEARCH GOES ON

Although Simpson's claim had been severely challenged, not everyone agreed that it had been conclusively refuted. The situation was more uncertain than it appeared in the published literature. In January 1986, Simpson presented a paper at the Moriond workshop on massive neutrinos (Simpson 1986b),[12] in which he presented supportive

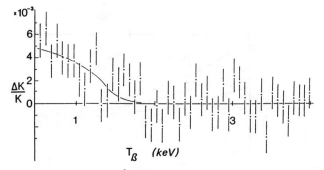

Figure 8. Distortion of the Kurie plot of ^{3}H as a function of electron kinetic energy, obtained with the low-dose detector. From (Simpson 1986b).

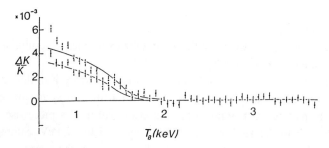

Figure 9. The effects of screening on the deviation of the Kurie plot. The upper curve used a screening potential of 99 eV, whereas the lower curve used 41 eV. The upper curve gives $\tan^2\theta$ = 0.028 (mixing probability) and threshold energy 1.57 keV. The lower curve gives a mixing probability of 0.022 and threshold energy 1.53 keV. From (Simpson 1986b).

results from an experiment which used a somewhat different apparatus. In this case the detector had been implanted with tritium at a different energy, and with a much lower concentration (about 1/40 that of the original detector). The results are shown in Fig. 8.[13] They are "consistent with the emission of a 17.1 keV neutrino, with a mixing probability between 2 and 3%. It would seem to be not accidental that two detectors by different manufacturers implanted quite differently with very different amounts of tritium should show the same distortion of the β-spectrum of tritium" (Simpson 1986b, p. 569).[14]

Simpson also discussed the question, raised by Haxton and by Eman and Tadic, of the adequacy of the exchange and screening corrections used in his theoretical model. He remarked that different corrections did produce changes in the β-spectrum and in ΔK/K, but found that they reduced the size of the kink by approximately 20%. This agreed with Haxton's estimate of the effect. The kink was, however, still clearly present when a different, and presumably better, calculation was used (Fig. 9). Simpson also questioned the negative results reported in the five experiments on ^{35}S. He argued that the type of analysis used, which fitted the beta-decay spectrum over a rather large

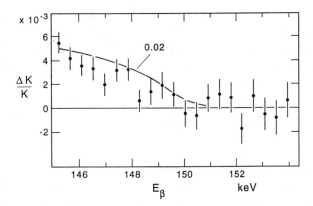

Figure 10. ΔK/K for the ^{35}S spectra of (Ohi and others 1985) as recalculated by Simpson. From (Simpson 1986a).

energy range, would tend to minimize the effect due to a heavy neutrino. He commented that 45 percent of the effect occurred within 2 keV of the neutrino threshold, and that "...in trying to fit a vary large portion of the β spectrum, the danger that slowly-varying distortions of a few percent could bury a threshold effect seems to have been disregarded. One cannot emphasize too strongly how delicate is the analysis when searching for a small branch of a heavy neutrino, and how sensitive the result may be to apparently innocuous assumptions" (Simpson 1986b, p. 576).[15] Simpson reanalyzed the results of each of the five experiments and argued that two of those (Apalikov *et al.* 1985; Ohi *et al.* 1985) showed statistically significant effects that agreed with his tritium results. His reanalysis of the result of Ohi *et al.* is shown in Fig. 10 (see also Simpson 1986a). He also stated that the result of Datar *et al.* was, in fact, consistent with his, but that because of statistical limitations nothing more could be concluded. For the last two experiments, those of Altzitzoglou *et al.* (1985) and Markey and Boehm (1985), he argued that the analysis was inadequate to decide whether or not there was a distortion in the β spectrum at 150 keV, the 17 keV neutrino threshold.

The situation seemed unresolved. Borge and collaborators (1986), after summarizing the uncertain evidence, which included Simpson's reanalysis, remarked, "Rather than entering into this controversy here, we provide our own independent piece to the puzzle" (p. 591).[16] Their experiment looked at a related, but somewhat different, phenomenon in beta decay, internal bremsstrahlung in electron capture (IBEC). In ordinary electron capture a nucleus of charge Z absorbs an atomic electron, transforming itself into a nucleus with charge Z-1, with the emission of a monoenergetic neutrino. In the process of capture, the electron may interact with the atomic electrons and produce a photon (usually in the x-ray energy region). This is internal bremsstrahlung electron capture. This latter process produces a continuous spectrum of x-rays, and is reduced relative to ordinary electron capture by a factor of α, the fine structure constant, approximately 1/137. Under certain favorable conditions, as shown by De Rujula (1981), when the energy available for the decay, the Q value, is resonant with electron binding energies in the atom, the rate of IBEC can be increased by several orders of magnitude. It can then be used as a sensitive alternative test for a massive neutrino.[17] This experiment also involved the detection of x-rays rather than the detection of electrons, which made it somewhat different.

In this experiment, too, much depended on the theoretical model used for comparison with the experimental result. Borge and his collaborators found that when they fitted their spectrum of ^{125}I with a six parameter curve, which included the mass and the mixing probability of the heavy neutrino as free parameters, "that the effect of the heavy neutrino to a large extent can be absorbed by other parameters....Thus, in comparing different hypotheses for m_2, c_2 [the heavy neutrino mass and mixing probability] it is essential each time to carry out an independent adjustment of the other free parameters. Analogous problems occur, of course, in the ^{35}S experiments. We feel, *in complete agreement with the opinions expressed by J.J. Simpson...that the limits on c_2 derived in [the experiments of Ohi et al. (1985) and of Datar et al. (1985)] are misleading as the parameters were not fitted again under the assumption of a heavy neutrino; instead the contribution from this was simply added.* The approach taken here and in also in Refs. [7] and [10] (Altzitzoglou et al. 1985; Markey and Boehm 1985), leads to much more conservative limits on c_2" (Borge *et al.* 1986, pp. 593-4, emphasis added). They concluded, however, that their result excluded a 17 keV neutrino with a mixing probability of 2-4%, at confidence levels of 98% and 99.9% for the ends of the interval and that "It supports the results of the ^{35}S measurements, which exclude the corresponding antineutrino" (p. 595).[18]

Negative evidence on the 17 keV neutrino continued to accumulate. Hetherington *et al.* (1987) reported no evidence for the heavy neutrino in their measurement of the beta-decay spectrum of ^{63}Ni, using a magnetic spectrometer. A preliminary negative result had been presented at the 1986 Osaka conference (Hetherington *et al.* 1986), "However, there was some concern about this conclusion because of the relatively strong absorption in the detector window and other possible instrumental effects. In this paper we present results from an entirely new set of data taken with a thinner window and with explicit evaluation of the impact of instrumental corrections" (Hetherington *et al.* 1987, p. 1504).[19]

There was evidence of continuing cooperation and collaboration within the beta-decay community. "Simpson drew our attention to the fact that a measurement of the shape of the ^{63}Ni beta spectrum could provide an ideal test of the existence of the 17 keV neutrino. This spectrum's endpoint (67 keV) is lower than that of ^{35}S [167 keV], offering better resolution and counting statistics, but high enough to avoid the very low energy problems associated with tritium. It has a single allowed branch with a half-life long enough (100 yr) to avoid normalization problems" (Hetherington *et al.* 1987, p. 1504).

The need for care in the performance of the experiment was also evident. In preliminary measurements, excess counts were found above the endpoint energy of the ^{63}Ni spectrum, which indicated the presence of background, most probably due to

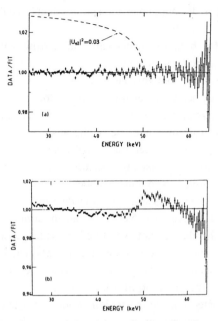

Figure 11. "Ratio of data to fit for wide scan spectrum. (a) $|U_{e2}|^2$ fixed to zero and the other parameters optimized. For comparison the dashed line shows the expected shape for $|U_{e2}|^2 = 3\%$. The best fit parameters are $E_o = 66.946$ keV, $\alpha = 0.00065$ keV^{-1}, and $\chi^2_v = 0.862$. (b) Best fit for $|U_{e2}|^2$ fixed at 3%. The other parameters are $E_o = 67.019$ keV, $\alpha = 0.0049$ keV^{-1}, and $\chi^2_v = 7.76$. The shape of the plot and the reduced χ^2 value clearly rule out this large a mixing fraction for the 17 keV neutrino" From (Hetherington et al. 1987, p. 1510).

scattering of the decay electrons. Extra antiscatter baffles were were added to the experimental apparatus, which solved the problem.[20] The group took data in both a broad energy range, 25-70 keV, with additional runs in the narrower energy range, 46-54 keV, in which effects of the 17 keV neutrino, if it existed, would appear. Thus, such effects could be searched for in both narrow and wide energy ranges. Recall Simpson's earlier comment about the possibility that using a wide energy range might hide a threshold effect due to the 17 keV neutrino.

There were also difficulties in calculating the expected spectrum shape that was to be compared with the experimental data. Despite the best efforts of the group, "it was found in the analysis that a shape 'correction' of the form $S = (1 + \alpha E)$ was required in order to obtain a good fit. This is probably caused by uncertainties in the instrumental corrections e.g. window absorption, penetration through the edges of the counter slits, electrostatic effects on transmission, etc....It should be noted that the inclusion of an unknown shape correction (α) does not bias the result obtained for $|U_{e2}|^2$ [the mixing probability] provided that both parameters are allowed to float simultaneously (as was the case in all results quoted here except where otherwise noted). This reflects the ability of the least squares technique to distinguish between a continuously varying effect in the data and a discontinuous threshold effect" (Hetherington *et al.* 1987, p. 1508).[21]

Their conclusions, for both the wide-scan and narrow-scan spectra, agreed (the results for the wide-scan spectrum are shown in Fig. 11). "The shape of the plot and the reduced χ^2 value clearly rule out this large a mixing fraction [3%] for the 17 keV neutrino" (p. 1510). They set an upper limit of 0.3% for the mixing probability of the 17 keV neutrino. They agreed with Simpson that the stricter limit of 0.15% set by Ohi *et al.* was probably not warranted because of the analysis procedure used. They did, however, offer a note of caution concerning Simpson's analysis. "It has been argued [by Simpson] that in order to avoid systematic errors, only a narrow portion of the beta spectrum should be employed in looking for the threshold effect produced by heavy neutrino mixing. If one accepts this argument, our data in the narrow scan region set an upper limit of 0.44%. However, we feel that concentrating on a narrow region and excluding the rest of the data is not warranted provided adequate care is taken to account for systematic errors. The rest of the spectrum plays an essential role in pinning down other parameters such as the endpoint. Furthermore, concentrating on too narrow a region can lead to misinterpretation of a local statistical anomaly as a more general trend which, if extrapolated outside the region, would diverge rapidly from the actual data (p. 1512)."[22] This experiment was generally regarded as the most complete magnetic spectrometer experiment done to that point (see Bonvicini 1993, p. 98).

Further evidence against the 17-keV neutrino was provided by Zlimen and collaborators (1988), using the internal bremsstrahlung technique on ^{55}Fe. They concluded, "We obtain a negative result and, at the 99.7% confidence level, our limit for the fraction of emitted neutrinos in the mass range 16.4 -> 17.4 keV is < 0.0074" (p. 539).

The group was quite concerned with the construction of a theoretical model to compare with their experimental result.

> The decay of ^{55}Fe (Q = 231.4 ± 0.7 keV) is an allowed transition and the theoretical understanding of such transitions is well developed... An experimental investigation has been made Berenyi *et al.*, with an accuracy comparable to that attained in the study of β-ray shape factors. The agreement between experiment and theory is better than 1% over a wide energy range. As our analysis is limited to a relatively narrow energy range we can be confident that the shape of the IBEC spectrum is known to a high degree of accuracy. However, it must be emphasized that our technique does not depend on there being an absolute accuracy of 1%. It is only necessary that the theory is sufficiently well-established that, in the absence of heavy neutrino emission, there are no kinks in the spectrum in the energy region used in our analysis. The recent careful investigation of Borge *et al.* has also shown that the shape of the IB spectrum is in excellent agreement with the theoretical predictions (p. 540).

The technique used was to look for kinks in the IBEC spectrum produced by the emission of a heavy neutrino. Internal bremsstrahlung can proceed from different atomic shells, which does, in fact, produce kinks in the spectrum. The kink due to a 17-keV neutrino would occur at an energy of $(Q - 17.1 - B_{1s})$, where Q is the energy available for decay and B_{1s} is the binding energy of the 1s state, the lowest energy atomic state, and the dominant decay mode. No kinks are expected below this energy and in order to have an energy range in which only one kink was expected they set an upper limit to their energy of $(Q - 17.1 - B_{2s})$, where B_{2s} is the binding energy of the 2s state.

At the end of 1988 the situation seemed much as it had been at the end of 1985. There seemed little reason to believe in the existence of a 17-keV neutrino. Aside from Simpson's original result, and his reanalysis of the negative results of others, no other evidence for such a particle had been presented. There had been nine negative experimental searches as well as plausible explanations that might explain his result using accepted physics.

D. THE TIDE STARTS TO TURN

In April 1989,[23] two new experimental results, obtained by Simpson and Hime, were published that supported the existence of the 17-keV neutrino (Hime and Simpson 1989; Simpson and Hime 1989). The effect of discussions and criticism within the research community on the performance and analysis of experiments, noted earlier, is clearly seen in these papers.

The first experiment was done on ^{3}H (tritium; Hime and Simpson 1989), the same substance used in Simpson's original experiment. Once again, the tritium was implanted in a solid-state detector, but in this experiment the detector was a hyperpure crystal of germanium, rather than a Si(Li) detector. "It was deemed important to check the earlier result by measuring the ^{3}H β spectrum in a different detector" (p. 1837). One problem with embedding the tritium in a germanium detector is that the embedding process may cause radiation damage, which causes pulse-height defects and will therefore result in an incorrect spectrum. It was known, however, that such damage could be removed by annealing at a temperature > 200°C, whereas the tritium remains bound in the germanium for temperatures up to 500°C.[24] The annealing was done in several steps. The crystal was first removed from the cryostat and allowed to warm to room temperature. Although this seemed to remove the pulse-height defect, a 0.45 mm dead layer remained in the detector. Further annealing took place *in situ* using heating coils to attain temperatures from 90°C to 135°C. A dead layer of 0.14 mm remained and further annealing, by heating to 180°C for about 10 hours, was done. This solved the problem completely. Possible experimental difficulties concerning whether or not the decay electron energy would be completely absorbed in the detector were solved by implanting the tritium in the center of the detector, to avoid edge effects, and by embedding it with sufficient energy so that its depth was approximately 0.3 mm, which was large compared to the mean absorption length in germanium of approximately 20

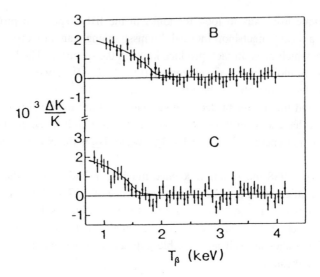

Figure 12. "The fractional deviations ΔK/K in the Kurie plots of spectrum B and spectrum C from a straight line using an effective screening potential of 42.6 eV. The smooth curves are as predicted in Eq. (5) for the emission of a heavy neutrino after accounting for resolution smearing. M_2 = 16.85 keV with $\sin^2\theta$ = 1.1% in the case of spectrum B with FWHM ≈ 405eV. M_2 = 17.00 with $\sin^2\theta$ = 1.3% in the case of spectrum C with FWHM = 310 eV" (Hime and Simpson 1989, p. 1846). These results are for tritium. M_2 is the mass of the heavy neutrino and $\sin^2\theta$ is the mixing probability. Run C had somewhat better energy resolution.

μm for 18.6-keV photons, and also large compared to the mean path length for 18.6 keV electrons (~ 1.6 μm), the maximum energy of the decay. (For further details see Hime and Simpson 1989, pp. 1839-1841).

Data were taken both after the *in situ* annealing, using two different detector electronics systems, and in two longer runs taken after the annealing process was completed (Runs B and C; Run C had improved energy resolution). The results from all four runs were consistent and the results for Runs B and C are shown in Fig. 12. Hime and Simpson concluded, "The excess of counts observed in the low-energy region of the tritium spectrum is best described by the emission of a 16.9 ± 0.1 keV neutrino and a mixing probability between 0.6 and 1.6% when allowance is made for uncertainty in the effective screening potential appropriate for tritium bound within a crystal lattice" (p. 1837).

Notice that the mixing probability has decreased by approximately a factor of three when compared to Simpson's original result. Recall, however, that questions had been raised concerning the theoretical corrections for screening and exchange. Hime and Simpson remarked that using the screening potential suggested by Lindhard and Hansen (1986) reduced the original 3% mixing probability to 1.6%. They also analyzed both their new germanium data and the original Si(Li) data, allowing the screening potential to vary, and looking for the best fit. They found, for the original data, best values of 38

± 10 eV and (1.1 ± 0.3)%, for the screening potential and mixing probability, respectively. This was in good agreement with the values 42.6 eV and (1.1 ± 0.2)% for the new germanium data.[25]

The second experiment done by Simpson and Hime (1989) was on ^{35}S, the element whose spectrum had provided considerable evidence against the existence of the 17-keV neutrino. This experiment used two different ^{35}S sources, 0.5 μCi and 5.0 μCi, and a Si(Li) detector. This detector also runs at liquid nitrogen temperature (-196°C) and a problem was found with the buildup of water vapor on the detector. "In addition, a copper cryopanel of ~300 cm^2 surface area surrounds the silicon detector and is cooled to liquid nitrogen temperature through a copper cold finger. This provides a large cold surface that freezes out residual water vapor in the chamber that can otherwise freeze on the detector surface. Without this cold surface a continuous build-up on the detector took place as observed by a continuous energy shift of the internal conversion electron lines of ^{57}Co. The centroid positions and shape of these lines remain sufficiently stable for periods of 4-5 days with the copper cryopanel in place" (Simpson and Hime 1989, p. 1826). An experimental problem had been identified and solved.

The main source of possible distortion of the β spectrum in their spectrometer was backscatter or back diffusion of the decay electrons. Simpson and Hime claimed that experiments by others had shown that the fraction of electrons backscattered was approximately 32 percent, and that it was essentially independent of the electron energy (See Simpson and Hime, 1989, for references).

The two runs with different sources, B and C, gave similar results and were combined to give a final result $M_2 = 16.9 ± 0.4$ keV and $\sin^2\theta = (0.73 ± 0.09)\%$, for the mass and mixing probability of the heavy neutrino, respectively (Fig. 13). "A threshold anomaly 17 keV from the end point in the measured ^{35}S spectra from that expected from theory for the emission of a single-component massless neutrino is the only distortion observed in the spectrum over the energy interval ranging from 110 to 166 keV. The agreement between theory and experiment below this anomaly indicates that the systematic effects associated with the technique of the measurement including detector response function and background, are well understood. It is very unlikely that systematic uncertainties would affect the shape of the spectrum in only an isolated region and not continuously over the entire β spectrum. It must be emphasized that *no arbitrary* shape factor has been required in analyzing the ^{35}S spectra to achieve a good fit, reinforcing confidence in the knowledge of the systematic features governing the shape of the measured spectrum" (p. 1833).

They regarded the two experiments, on tritium and on ^{35}S, as providing increased support for the existence of the 17-keV neutrino. "The present result [on ^{35}S] is remarkably similar to the results of the measurement of the β spectrum of tritium. Since the β energy and the experimental technique are so different in the ^{3}H and ^{35}S measurements it would have to be a remarkable coincidence for extraneous experimental effects to produce the similar results" (p. 1835).[26]

Simpson and Hime also discussed the previous negative results and concluded, "The present results are in disagreement with the claims of previous groups measuring β

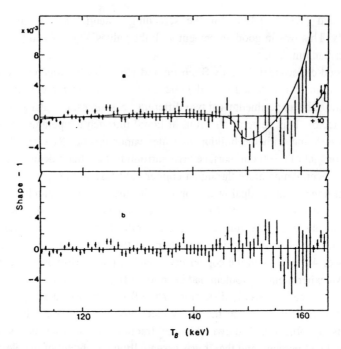

Figure 13. The deviation of the shape function from a constant for the combined data of runs B and C (Two different ^{35}S sources, 0.5 µCi and 5.0µCi). In (a) the theoretical spectrum has $\sin^2\theta = 0$. $\chi^2/\nu = 2.0$. The smooth curve shows the shape expected for $M_2 = 17$ keV and $\sin^2\theta = 0.008$. In (b) the experimental data are divided by a theoretical fit with $M_2 = 17$ keV and $\sin^2\theta = 0.0075$. $\chi^2/\nu = 1.0$" (Simpson and Hime 1989, p. 1830).

spectra of ^{35}S and ^{63}Ni. In the present experiment all important systematic effects are understood and accounted for. This is not generally the case in the other experiments,[27] and it can be argued that, with the possible exception of two of the previous ^{35}S experiments (Apalikov et al. 1985; Ohi et al. 1985), these results are more correctly described as providing no support for a 17-keV neutrino at the 0.7% level rather than ruling it out. The two exceptions perhaps give weak confirmation of the 17-keV neutrino" (p. 1835). Not everyone working in the field agreed.

The continuing experimental work by Simpson and Hime encouraged further theoretical work on corrections to the tritium beta-decay spectrum.[28] Weisnagel and Law (1989) included internal bremsstrahlung effects, which had not been previously considered, as well as other atomic physics effects to produce what they considered to be the most complete theoretical model of the spectrum. Using Simpson's original data and their model, they reported " a best fit for the neutrino mass $m_\nu = 17.2$ keV and a mixing probability R = 2.5% (p. 904)," in agreement with Simpson's original result. (See Fig. 14).

Weisnagel and Law also suggested that the previous theoretical work of Eman and Tadic (1986) and of Lindhard and Hansen (1986), which had attempted to explain Simpson's result on the basis of atomic physics effects, had overestimated the size of

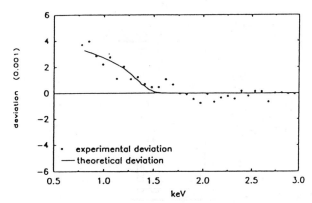

Figure 14. The deviation spectrum for the tritium experimental data of Simpson as a function of electron kinetic energy for the sum of all the effects studied by Weisnagle and Law (1989) in the energy region up to 3 keV.

these effects because of differences in spectrum normalization procedures. Their own calculation indicated that the effect suggested by Lindhard and Hansen could explain only one-third of the electron surplus seen by Simpson.

The positive reports by Simpson and Hime encouraged Zlimen and collaborators (1990) to extend their analysis of the internal bremsstrahlung spectrum of ^{55}Fe to the mass range 15-45 keV, in comparison with their original result for the range 16.4-17.4 keV. They concluded, "We have no evidence for the existence of a heavy neutrino with a mass larger than 20 keV. Although our results confirm that any possible heavy neutrino in the 15-20 keV region have relative intensities well below the value of 3% of (Simpson 1985a) they do not exclude the new results of Simpson and Hime. New detailed measurements are needed in this energy range" (p. 426). This was the first report by anyone other than Simpson and his collaborators, or from Simpson's own reanalysis of other experimenters' data, of a result that was consistent with the existence of a 17-keV neutrino. More would follow.

Even before these results were available in the published literature they had been presented at workshops and conferences and communicated privately to others.[29] For example, Sheldon Glashow, a Nobel Prize-winning theoretical physicist, cited both the published work of Simpson and Hime (1989) as well as private communications from Hime and from Norman at Berkeley, both of which would soon be published, as evidence for the existence of the 17-keV neutrino. He then incorporated the heavy neutrino in a model that accounted for the solar neutrino deficit while remaining "in accord with known constraints from particle physics and cosmological theory" (Glashow 1991, p. 255). He was, in fact, quite enthusiastic about Simpson's work. "Simpson's extraordinary finding proves that Nature's bag of tricks is not empty and demonstrates the virtue of consulting her, not her prophets. That a simple extension of

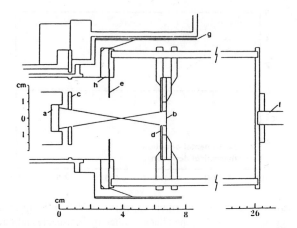

Figure 15. The experimental apparatus of (Hime and Jelley 1991). (a) Si(li) detector, (b) Source substrate, (c) Al detector aperture, (d) Cu source aperture, (e) Al anti-scatter baffle, (f) linear motion feed-through (g) liquid nitrogen cryo-panel, (h) teflon centering ring, (i) vacuum chamber.

the standard model seems to work on earth and in the stars shows she is not malicious" (p. 257).[30]

Glashow was not alone. In an article in the May 1991 *Physics Today*, Schwarzschild (1991) wrote an article entitled "Four of Five New Experiments Claim Evidence for 17-keV Neutrinos." Schwarzschild cited the positive results already published by Hime and Jelley, as well as the positive results presented at the 1990 Bratislava conference on nuclear physics by Norman (^{14}C and ^{55}Fe) and by Ljubocic. (All of these experiments will be discussed in detail below). The only new negative result cited had been presented at the 1991 Moriond Workshop by Becker and collaborators (1991). Schwarzschild cited Glashow's enthusiastic response, quoted above, along with Shrock's more cautious "To the theorists who say the 17-keV neutrino can't be right, *and* to those who offer a nice model purporting to explain it....our present theories don't even explain the well known fermion masses" (Shrock, quoted in Schwarzschild 1991, p. 19). Schwarzschild concluded, "On one thing everyone seems to agree. After six years, the experimenters must begin to resolve the stubborn discrepancy between the two different styles of beta-decay experiment [solid-state detectors and magnetic spectrometers]" (p. 19).

The first of the results to appear in print were those of Hime and Jelley. "After his apprenticeship with Simpson, Hime is now at Oxford, where he and Nick Jelley have recently completed a new measurement of the beta-decay spectrum of ^{35}S (Hime and Jelley 1991). This high-statistics extension of the Simpson technique with an improved instrumental geometry is, by consensus, the most compelling of the experimental results that claim to see the 17-keV neutrino" (Schwarzschild 1991, p. 17).[31]

The experimental apparatus used by Hime and Jelley is shown in Fig. 15.[32] It included both source and detector apertures as well as an aluminum anti-scatter baffle (d, c, and e in the figure, respectively). "The aim was to provide a well defined geometry in which electrons are normally incident on silicon, an improvement on the scheme used at the University of Guelph (Simpson and Hime 1989) where no form of collimation was used" (Hime 1993, p. 166). This was to ensure that electrons would not penetrate the edges of the detector and to guard against electrons scattering from the walls of the vacuum chamber into the detector, and thus possibly distort the energy spectrum.[33]

During the operation of the experiment it was found that some electrons were, in fact, losing additional energy by penetrating the edge of the detector and so the geometry of the apparatus was changed to reduce this effect (Runs 1 and 2). During both experimental runs the calibration and stability of the apparatus was monitored in two different ways: 1) daily calibration runs using gamma rays from a ^{57}Co source, and 2) using monoenergetic electrons from internal conversion sources at the beginning, middle, and end of each run.

The results of both runs were consistent with each other and combining results from both sets of data gave $M_2 = (17.0 \pm 0.4)$ keV and $\sin^2\theta = 0.0084 \pm 0.0006 \pm 0.0005$, for the mass of the heavy neutrino and the mixing probability, respectively (Fig. 16). (The two uncertainties on the mixing probability are statistical and an estimate of systematic uncertainty, respectively). They concluded, "The data strongly support the claim that the electron neutrino couples to a heavy mass eigenstate in agreement with the measurement of the ^{35}S spectrum at Guelph" (p. 448).

They noted that by restricting their data to the energy region above 120 keV they had made their result less sensitive to fine details of the electron response function. "Consequently, if the distortion observed in the beta spectrum arises from some unknown systematic feature associated with this method of measurement [solid-state detectors] then it appears to be much more subtle than a misunderstanding of the electron response function" (p. 448). They also cited a preliminary result from Berkeley on the ^{14}C spectrum (to be discussed in detail later) as well as the earlier positive results on tritium and ^{35}S as supporting the existence of the 17-keV neutrino. In addition, they remarked on the criticism of the earlier negative results. "To date no response has appeared concerning these criticisms nor has any new result been reported in the literature from the authors of the work" (p. 441).

The Berkeley results on ^{14}C were published at approximately the same time as Schwarzschild's article (Sur et al. 1991).[34] The technique was similar to that used by Simpson in which the radioactive source had been embedded in the solid-state detector. In this case the detector was a germanium crystal in which ^{14}C was dissolved. A novel feature of the detector was that the electrode was "divided by a 1-mm-wide circular groove into a 'center region,' 3.2 cm in diameter, and an outer 'guard ring.' By operating the guard ring in anticoincidence mode, we can reject events occurring near the boundary, which are not fully contained within the center region" (p. 2444). Such events would give an incorrect energy, and thus distort the spectrum. Their results are

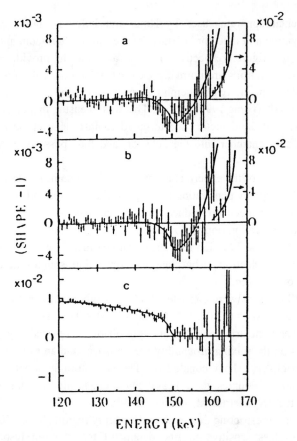

Figure 16. Shape factors for (a) run 1 and (b) run 2 obtained by dividing the experimental spectra by the best least squares fit to the region 120-160 keV when no heavy neutrino mixing is allowed. (c) Shape factor for the combined runs when normalizing to a single component over the region above 150 keV. The smooth curves in each case indicate the expected deviaition for the emission of a 17-keV neutrino with $\sin^2\theta = 0.009$. (Hime and Jelley 1991). In run 2 the source aperture was made smaller to reduce the fraction of decay electrons striking the edge of the detector aperture.

shown in Fig. 17, and give a value of 17 ± 2 keV and (1.40 ± 0.45 ± 0.14)% for the mass of the heavy neutrino and its mixing probability, respectively, "which supports the claim by Simpson that there is a 17-keV neutrino emitted with ~1% probability in β decay" (p. 2447).[35] They also claimed to rule out the null hypothesis (no heavy neutrino) at the 99% confidence level.

The Berkeley group included new results on the internal bremsstrahlung spectrum of ^{55}Fe as well as their ^{14}C results in (Norman et al. 1991). The ^{55}Fe experiment used a germanium detector (a solid-state device) and also made use of a sodium-iodide anticoincidence shield to veto both Compton scattered gamma rays as well as external background. They used the last 55 keV of the ^{55}Fe spectrum and obtained a best fit for

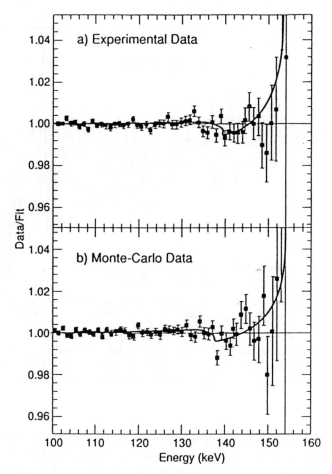

Figure 17. (a) The ratio of the data to a theoretical fit assuming no 17-keV neutrino. (b) Monte Carlo-generated data which contain a 1% fraction of a 17-keV neutrino. The curves illustrate the shape expected from the best fits to the data, which includes a 17-keV neutrino. (Sur *et al.* 1991).

M_2 = 21 ± 2 keV, a somewhat different value for the mass than had been found in all the other experiments, and a mixing probability of (0.85 ± 0.45)%, where both uncertainties are one standard deviation. They noted, however, that their fit was not as good as that obtained in their ^{14}C experiment, which they attributed to their lack of precise knowledge of their detector response function. They also found that the position of the "kink" (the mass of the heavy neutrino) could be moved by varying the energy dependence of the detector. They concluded, "Thus, these results are suggestive that there is a feature ~17 keV below the endpoint of the ^{55}Fe IB [internal bremsstrahlung] spectrum, but further study of this system is clearly necessary" (p. S298).

The results that Ljubicic had presented at Bratislava were published shortly thereafter (Zlimen *et al.*, 1991). The technique used was quite similar to that used in their earlier internal bremsstrahlung experiments (described earlier) though this experiment used a

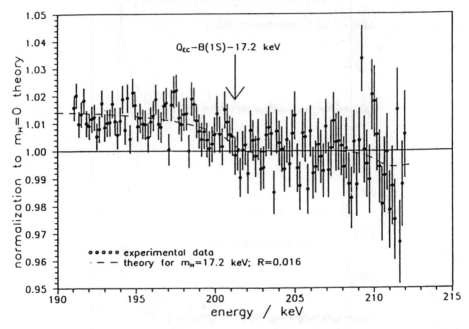

Figure 18. The experimental corrected IB spectrum divided by the theoretical spectrum which assumes no heavy neutrino, normalized to the region above the expected kink. The dashed line represents the theory for a 17.2 keV neutrino with a probability of 1.6%. (Zlimen et al. 1991).

^{71}Ge source. "The evidence for a small kink is not apparent in a visual inspection of the IB spectrum. However, if we normalize the spectrum to a spectrum which assumes $m_H = 0$ [no heavy neutrino] the kink becomes visible..." (pp. 562, 563; see also Fig. 18). Their values were $17.2^{+1.3}_{-1.1}$ keV and (1.6 ± 0.79)%, for the mass of the heavy neutrino and the mixing probability, respectively. Both of these results were at the 95% confidence level, and were in agreement with Simpson's results.

The one new negative result presented at this time was by Becker et al. (1991). The experiment used a magnetic spectrometer to measure the beta-decay spectrum of ^{35}S. They, too, allowed a varying shape factor in fitting their data. They fit their data in two energy ranges: a wide scan, 100-165 keV, and a narrow scan, 132-163 keV. In both cases the fit clearly favored no 17-keV neutrino and they ruled out such a neutrino with a mixing probability of 0.8% (the value found by Hime), with a confidence level greater than 99%. They admitted, however, that their wide-scan fit for no 17-keV neutrino, while considerably better than that which included such a neutrino, had an excessive χ^2 (53 for 35 degrees of freedom, with a probability of approximately 0.05, which they regarded as unlikely). They attributed this to possible systematic errors. The fit for the narrow scan was considerably better (χ^2 of 13.8 for 20 degrees of freedom, with probability of greater than 0.8).

Table II: Experimental Evidence for a 17-keV Neutrino (Simpson 1991)

Isotope	ν Mass (keV)	Mixing Angle θ	Reference
^{3}H (Si(Li))	17.1 ± 0.2	0.105 ± 0.015	Hime (1989), Simpson (1985)
^{3}H in Ge	16.9 ± 0.1	0.105 ± 0.015	Hime (1989)
^{35}S	16.9 ± 0.4	0.082 ± 0.008	Hime (1989)
	16.95 ± 0.35	0.088 ± 0.005	Simpson (1989)
^{14}C in Ge	17.0 ± 0.5	0.114 ± 0.015	Sur (1991)
^{63}Ni	16.75 ± 0.38	0.101 ± 0.011	Hime, Oxford Report (OUNP-91-20)

[a] The mixing probability is essentially the square of the mixing angle.

II. THE DISAPPEARANCE

This time it vanished quite slowly, beginning with the end of the tail, and ending with the grin, which remained some time after the rest of it had gone.
— The Cheshire Cat in Lewis Carrol's *Alice in Wonderland*.

A. THE TIDE EBBS

The year 1991 was the high point in the life of the 17-keV neutrino. Although the evidence for its existence was far from conclusive, its existence had been buttressed by the recent results of Simpson and Hime, of Hime and Jelley, of the Berkeley group (Norman, Sur, and others), and by Zlimen and others. From this point on, however, the evidence would be almost exclusively against it. Not only would there be high-statistics, extremely persuasive, negative results, but serious questions would also be raised about its strongest support.

The Europhysics Conference on High Energy Physics held in July 1991 illustrates the uncertain status of the 17-keV neutrino. Simpson (1992) offered a summary of the evidence favoring its existence, whereas Morrison (1991) offered a rather critical and negative review. Simpson summarized the recent positive evidence for the 17-keV neutrino, which, in the light of his previous criticism, he regarded as having more evidential weight than the negative results discussed earlier. (See Table II for the positive evidence cited by Simpson).

Morrison's summary was rather negative.[36] He began with Koonin's (1991) soon to be published calculation on tritium. Recall that questions had been raised earlier concerning atomic physics effects in tritium decay (Drukarev and Strikman 1986; Eman and Tadic 1986; Haxton 1985; Lindhard and Hansen 1986; Weisnagel and Law 1989). The calculated effects had all been rather smooth, and although one could argue about their size, there was a question as to whether or not such a smooth effect could account for a "kink" in the beta-decay spectrum. Koonin proposed the BEFS (beta environment fine structure), which gave rise to an oscillatory structure in the spectrum and depended on the embedding of the tritium in a crystal structure (Fig. 19). "At the 10^{-3} level of accuracy currently of interest, it will be important only for those sources in which tritium is embedded in a host material" (p. 469). This calculation seemed to cast doubt on the positive results obtained with tritium, all of which had used such a source. Morrison concluded, "The conclusion is that tritium, or another beta source with a low end-point, should not be used to look for heavy neutrinos because of uncertainty in the expected spectrum shape" (Morrison 1991, p. 600).[37]

Morrison also discussed Simpson's previous criticism of the negative searches. In particular, he examined Simpson's reanalysis of Ohi's data. "The question then is, How could the apparently negative evidence of Figure [5] become the positive evidence of Figure [10]? The explanation is given in Figure [20], where a part of the spectrum near 150 keV is enlarged. Dr. Simpson only considered the region 150 keV ± 4 keV (or more exactly + 4.1 and -4.9 keV). The procedure was to fit a straight line, shown solid, through the points in the 4 keV interval above 150 keV, and then to make this the baseline by rotating it down through about 20° to make it horizontal. This had the effect of making the points in the interval 4 keV below 150 keV appear above the extrapolated dotted line. This, however, creates some problems, as it appears that a small statistical fluctuation between 151 and 154 keV is being used: the neighboring points between 154 and 167, and below 145 keV, are being neglected although they are many standard deviations away from the fitted line. Furthermore, it is important, when analysing any data, to make sure that the fitted curve passes through the end-point of about 167 keV, which it clearly does not" (p. 600).[38]

Morrison also noted that the shape-correction factors needed in magnetic spectrometer experiments were smooth and unlikely to obscure a kink due to a heavy neutrino, and remarked that there were problems due to backscattering of electrons in the positive experiments of Simpson and Hime and of Hime and Jelley. In looking at the experimental situation, Morrison cited the most recent positive results along with new negative results on ^{35}S from Caltech (discussed below), from Grenoble on ^{177}Lu, and the tritium result of Bahran and Kalbfleisch. His summary of results is given in Table III.

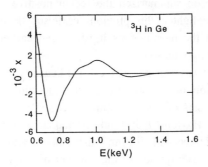

Figure 19. Beta environmental fine structure (BEFS) for tritium in a germanium crystal at a temperature of 80K. (Koonin 1991).

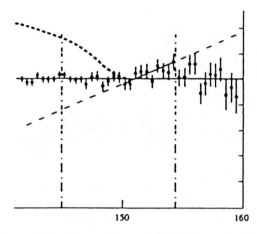

Figure 20. Morrison's reanalysis of Simpson's reanalysis of Ohi's result. (Morrison 1991).

The results of Bahran and Kalbfleisch (1992a) were presented at this conference.[39] Their experiment investigated the spectrum of tritium using a gas proportional chamber.[40] Their results are shown in Fig. 21, along with those of Simpson (1985) and of Hime and Simpson (1989). No excess of events is seen and their 99% confidence

Table III: Summary of Experimental Results (Morrison 1991)

Laboratory	First Author	Year	Source	Technique	Mass (keV)	% Mixing (CL %)
		(CL %)				
Guelph	Simpson	1989	^{35}S	Si(Li)	16.9±0.4	0.73±0.11
Oxford	Hime	1991	^{35}S	Si(Li)	17.0±0.4	0.84±0.08
Berkeley	Sur	1991	^{14}C	Ge	17 ± 2	1.40±0.5
Zagreb	Zlimen	1990	^{71}Ge	Int. Brems.	17.2	1.6(2sd)
Tokyo	Ohi	1985	^{35}S	Si(Li)	No evid.	<0.3(90%)
Princeton	Altzitzoglou	1985	^{35}S	Mag. sp.	No evid.	<0.4(99%)
ITEP	Apalikov	1985	^{35}S	Mag. sp.	No evid.	<0.17(90%)
BARC/TIFR	Datar	1985	^{35}S	Mag. sp.	No evid.	<0.6(90%)
Caltech	Markey	1985	^{35}S	Mag. sp.	No evid.	<0.25(90%)
Caltech	Becker	1991	^{35}S	Mag. sp.	No evid.	<0.6(90%)
Chalk River	Hetherington	1986	^{63}Ni	Mag. sp.	No evid.	<0.3(90%)
Caltech	Wark	1986	^{63}Ni	Mag. sp.	No evid.	<0.25(90%)
CERN-ISOLDE	Borge	1986	^{125}I	Int. brems.	No evid.	<0.9(90%)
ILL Grenoble	Schreckenbach	1991	^{177}Lu	Mag. sp.	No evid.	<0.2(90%)
Guelph	Simpson	1985	^{3}H	Si(Li)	~17.1	About 3
Guelph	Hime	1989	^{3}H	Ge	16.9±0.1	0.6-1.6
Oklahoma	Bahran	1991	^{3}H	Prop. ctr.	No evid.	<0.4(90%)

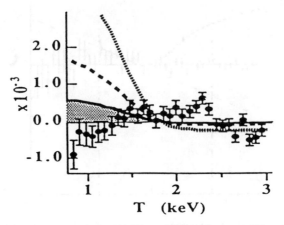

Figure 21. ΔK/K with upper limit of 0.4% is compared with the positive claims of Simpson 1985 (2.7%, dotted) and of Hime and Simpson 1988 (1.1%, dashed). (Bahran and Kalbfleisch 1991).

level upper limit was 0.4% for 17-keV neutrino mixing.[41]

In the published version of his paper, Simpson responded to Morrison's view that the early negative results should be taken seriously in evaluating the evidence.[42] He noted that the magnetic spectrometer experiments all needed a shape correction factor that was of the order of several percent in the energy region of interest. "It is therefore difficult to see how these experiments can rule out a 1% effect, which requires an accuracy of perhaps 0.2% over the analyzed region" (Simpson 1991, p. 598).[43]

Simpson's view of the early negative magnetic spectrometer results was strongly supported by Bonvicini's work [published first as a 1992 CERN report (CERN-PPE/92-54) and later as Bonvicini (1993)].[44] In this work Bonvicini discussed the question of whether or not a kink in the energy spectrum due to an admixture of a 17-keV neutrino could be masked by the presence of unknown distortions, such as the shape correction factors used in magnetic spectrometer experiments. "Most urgent in this discussion is why experiments where the β⁻ energy is measured calorimetrically tend to see the effect, and those which use spectrometers do not. My analysis ... shows that *large continuous distortions in the spectrum can indeed mask or fake a discontinuous kink* (emphasis added). In the process I point to some deep inconsistencies in all the spectrometer experiments considered here" (Bonvicini 1993, p. 97). He performed a detailed analysis and Monte Carlo simulation of what were then generally regarded as best experiments on either side of the 17-keV neutrino issue: the positive result from ^{35}S by Hime and Jelley (1991), and the negative result from ^{63}Ni by Hetherington and others (1987). He also analyzed several other experiments.[45]

Bonvicini concluded that the positive Hime and Jelley result was statistically sound. He cautioned, however, that the electron response function in this experiment had been only partially measured, and that this might be a possible problem.[46] As the subsequent history shows, this statement was prescient. Bonvicini's analysis of the experiment of Hetherington et al. concluded that although their use of a 2.5% shape correction factor was certainly acceptable when searching for a 3% kink, when one looked for a 0.8% kink more work was needed. His summary of the overall situation was as follows. "A look at the published data seems to indicate that the statistical criteria listed above would eliminate all the negative experiments considered here, but it is left to the authors to look at their data" (p. 114). As far as the positive experiments were concerned, he rejected the Hime-Jelley result on ^{63}Ni on the grounds of poor statistics and large correlation between kink and distortions. The Berkeley result had too large (6%) a shape correction factor. "The positive experiments do not offer definitive proof of a total ERF [electron response function] measurement, but having found the kink in the same apparatus with two different nuclei [the Hime-Jelley results on ^{35}S and ^{63}Ni] seems to eliminate the possibility of one common missing distortion as the source of the kink" (p. 114). "The ^{35}S result of Hime and Jelley is statistically sound, as they have run the checks suggested in this paper, while the earlier Simpson and Hime result[s] have not been analyzed. Thus there is only one experiment at this time and in my knowledge where one could say that a kink is certainly there" (p. 116).[47]

Bonvicini disagreed quite strongly with Morrison's very negative view of the situation. "The conclusions of this review differ from those of [a reference to a CERN report by Morrison]. I do not share the same enthusiasm for anything sporting a decent χ^2, or the same belief in infinitely small systematic errors. Conclusions based on 'experiment counting' (one counts experiments and the majority wins) is most definitely not the way to assess this controversy" (p. 114).

Bonvicini also suggested the need for more good experiments and gave criteria for what would constitute such a good experiment. These included: 1) direct measurement of the electron response function at more than one point across the fitted spectrum; 2) cross checks of the fitted correlation coefficients; 3) use of at most one small linear shape factor; 4) no narrow energy scans - the scan should include a range at least twice the neutrino mass; 5) an experiment should also show the results of a fit with a shape factor with one order more.

Bonvicini's work argued quite strongly that the negative results of the previous magnetic spectrometer experiments were inconclusive and suggested the design of experiments which either used no shape correction factor or had such overwhelming statistical accuracy that a kink would always be visible. As we shall see below, experiments of this type were, in fact, performed and were decisive in answering the question as to whether or not the 17-keV neutrino existed.[48]

The new Caltech result Morrison had referred to appeared in Radcliffe et al. (1992). This experiment also looked at the ^{35}S spectrum with a magnetic spectrometer. They took data in two different runs: a wide energy range, 130-167 keV; and a narrow scan of 10 keV around the kink expected at 150 keV for the 17-keV neutrino. Both runs were consistent with no heavy neutrino and excluded a 17-keV neutrino with a 0.85% mixing probability at the 99.3% confidence level and the 99.9% confidence level for

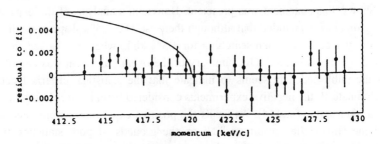

Figure 22. Data for Run B, a high statistics study of the beta spectrum near 420 keV, normalized above the kink. The curve shows the expected spectrum with a 0.85% admixture of a 17-keV neutrino. (Radcliffe *et al.* 1992).

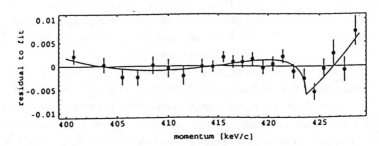

Figure 23. Synthetic kink induced in the beta spectrum of ^{35}S by a 17 μm aluminum foil. The solid curve is the spectrum expected with a 2.5% admixture of a 15.6-keV neutrino. (Radcliffe *et al.* 1992).

the wide and narrow scan runs, respectively. Their result for the narrow scan run is shown in Fig. 22. No kink is seen.

An interesting feature of this experiment was their simulation of a kink in the spectrum. All of the previous searches for a heavy neutrino with magnetic spectrometers had been negative and a question had been raised as to whether or not this type of apparatus was, in fact, capable of detecting such a kink. The experimenters shielded 10% of their detector with a 17 micron aluminum foil. The electrons would lose energy in passing through the foil and they expected this energy loss to produce a kink in the spectrum that would simulate a heavy neutrino with a 1% admixture. Their results with the foil in place are shown in Fig. 23. A kink is clearly visible and gave a best fit for a mass of 15.6 keV with a mixing factor of 2.5%, thus demonstrating that a magnetic spectrometer experiment was sensitive enough to detect a 17-keV neutrino, at least at that level. One might legitimately wonder (See Fig. 23) whether or not the apparatus was sensitive enough to detect a heavy neutrino with 1% mixing. The shape

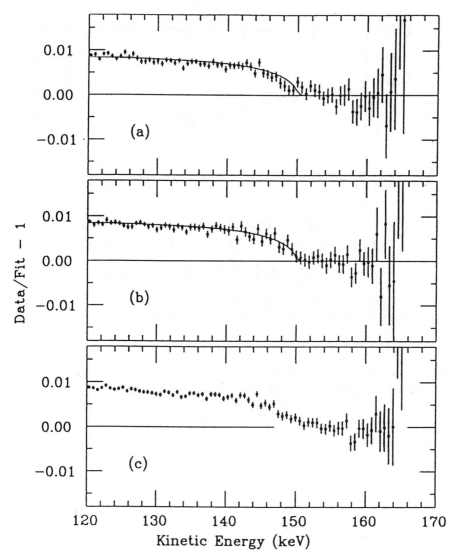

Figure 24. Relative residuals for the data of Hime and Jelley (1991) for (a) run 1, (b) run 2, and (c) combined. The solid curves are for a 17-keV neutrino with 0.9% mixing probability. (Piilonen and Abashian 1992).

of the spectrum distortion produced was also different from that expected for a heavy neutrino.

Further argument against the existence of the 17-keV neutrino was provided by the theoretical reanalysis of the data of Hime and Jelley (1991) by Piilonen and Abashian (1992). They used the published data[49] and constructed the relative deviations (DATA-FIT)/FIT, shown in Fig. 24. "While the combined data [c] is certainly consistent with

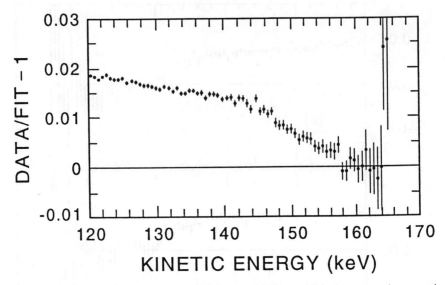

Figure 25. Relative residuals for the data of Hime and Jelley compared to a massless-neutrino spectrum calculated with the response function of Piilonen and Abashian (1992).

the 17 kev hypothesis there is simply not enough statistical precision near 150 keV to see an unmistakable kink (p. 226)." Piilonen and Abashian examined various alternative explanations for the results of Hime and Jelly. These included ambient background, unrejected pileup, the theoretical spectrum, and radiative corrections. None of these explained the measurements. They also performed a detailed simulation of the experimental apparatus to try to get "an accurate determination of all the contributions to the electron response function as well as their dependence on energy (p. 229)." Their simulation of electron scattering and energy loss was more complex than that used by Hime and Jelley. Their result is shown in Fig. 25 and was still in disagreement with a massless-neutrino beta spectrum. They concluded, "We agree with Hime and Jelley that there is a serious distortion in their ^{35}S data, though we cannot pinpoint any definite cause for it. We believe that if the original data is reanalyzed by Hime and Jelley with a more realistic electron response function such as we have derived in our simulation, then the consistency of this distortion with a two-component neutrino hypothesis (with $m_2 = 17$ keV) will disappear" (p. 233).

The most detailed summary of the evidential situation, and a moderate position, was provided by Hime in early 1992 (Hime 1992).[50] Although Hime was an active participant in the controversy, and one of those who provided persuasive evidence in favor of the 17-keV neutrino, his summary seems quite fair and judicious. He provided a reasonably complete history of the experiments and their results and devoted considerable attention to possible experimental problems or difficulties.

He first considered the issue of the atomic physics corrections to the tritium results and noted that taking account of the criticism reduced the size of Simpson's original result from 3% to ~1%. He observed that part of the difficulty with these calculations was that the experiments did not use free tritium, but rather tritium bound in a crystal lattice. He did not, however, mention Koonin's recent BEFS calculation. "The main point emerging from the analysis is that the *sudden* excess of counts in the tritium spectrum cannot be explained via atomic physics alone, unless effects are present that are yet to be contemplated" (Hime 1992, p. 1303).

Hime also discussed Simpson's reanalysis of the early negative results in ^{35}S, and remarked that they were based on a reanalysis of the data over only a narrow band of energy. "The difficulty remains, however, that an analysis using such a narrow region could mistake statistical fluctuations as a physical effect. The claim of positive effects in these cases [by Simpson] should be taken lightly without a more rigorous treatment of the data" (p. 1303).[51]

Hime also examined the issue of the uniformly negative results provided by magnetic spectrometer experiments. He noted that such experiments eliminated the problem of backscattering and energy loss that appeared in experiments using external sources and solid-state detectors. He also stated that whereas magnetic spectrometer experiments still required an extra shape-correction factor, "it is a point for debate whether or not sensitivity to a heavy neutrino is preserved. It is clear that the addition of extra degrees of freedom will reduce the sensitivity of the data but it remains difficult to see how a smooth correction would completely remove a 'kink'" (p. 1309). Hime observed that "given the obvious disagreement between magnetic spectrometer searches on the one hand and the positive results with solid state detectors on the other it is now generally agreed that insight into the discrepancy could be made if the sensitivity of a magnetic spectrometer to uncover a heavy neutrino signal could be experimentally demonstrated. Proposals include measurements with a mixed source (such as 99% ^{35}S + 1% ^{14}C), or artificially invoking energy loss in part of the spectrum at some predetermined level. This latter approach was suggested by the Caltech group [see earlier discussion of (Radcliffe and others 1992)] and has been implemented in their program" (p. 1310).

Hime also critically examined the positive evidence for the 17-keV neutrino, some of which he had himself provided. He argued that his Oxford results on ^{35}S and on ^{63}Ni had improved on the original Guelph results of Simpson, and of Simpson and Hime by changing the geometry of the experiment from a diffuse to a collimated source. He also argued that the dominant systematic uncertainty in these experiments, that due to uncertainty in the backscattering component of the electron response function had been adequately checked. "It seems that an alternative description of the Oxford data requires an effect that does not show up in direct measurements of the detector response to monoenergetic electrons" (p. 1305). He also noted that there were possible background and veto problems with the positive result on ^{14}C reported by the Berkeley group, and that an analysis of "unvetoed" data reduced their mixing probability for a heavy neutrino from 1.2% to 0.75%. He also questioned the TANDAR result. "Even more recently, an experiment at the TANDAR facility in Argentina measured the ^{71}Ge spectrum, yielding equally confusing values for a heavy neutrino mass. In particular, their data yield a best fit with $M_2 = 13.8 \pm 1.8$ keV and $\sin^2\theta = 0.80 \pm 0.25\%$. Fixing $M_2 = 17$ keV does not provide a better fit to the data than does a massless neutrino spectrum. There are, however, some unsettling aspects to this experiment. In the first place, in fitting the ^{71}Ge spectrum to a two-component neutrino, the chi-square distribution for the heavy neutrino mass exhibits two relative minima (one at about 9 keV and another, slightly deeper, at 13.8 keV). This is not expected for the emission of a single heavy neutrino and is inconsistent with other measurements finding evidence for a 17 keV neutrino. This suggests that, either the heavy neutrino hypothesis is not compatible with the data or that systematic effects are present which have not been properly accounted for" (pp. 1307-8).

Hime's summary of the situation seems quite reasonable (See Table I).

> The evidence accumulated both for and against the existence of a 17-keV neutrino presents an unresolved conundrum. On the positive side, a diverse range of isotopes have been studied in many experimental environments, all of which yield self-consistent results. While a working alternative for explaining these results has not been realized, potential hazards have not necessarily been exhausted.
>
> It remains an unsettled debate whether or not the null results obtained with magnetic spectrometers are weakened by unresolved systematic effects. In particular, a thorough analysis of the effects of polynomial shape corrections is desired. Results do exist, however, where systematic uncertainties associated with shape corrections have been properly analyzed, making it difficult to see how a 17 keV neutrino would not be revealed if it does indeed exist. A demonstration of the sensitivity to uncover a heavy neutrino signal in such an experiment could provide insight into this puzzle and potential schemes were outlined above. A useful alternative would be to determine the spectrometer response *a priori* via "complementary experiments."
>
> In the meantime a host of experimental efforts continue with the hope to elucidate the issue. All experiments share the difficult task of determining the shape of a continuous energy spectrum. Furthermore, the accuracy required is at the level of a few tenths of a percent. It is clear that the difficulties associated with low energy β decay measurements are predominantly of a systematic origin and that a resolution of the "17 keV conundrum" will require a careful and critical analysis of both positive and negative results (pp. 1312-13).

Thus, there were four differing summaries of the situation: one positive, by Simpson; one negative by Morrison; and two neutral, by Hime and by Bonvicini. The situation seemed unresolved. The experimental efforts under way, combined with the critical and careful analysis for which Hime had hoped, would decide the issue.

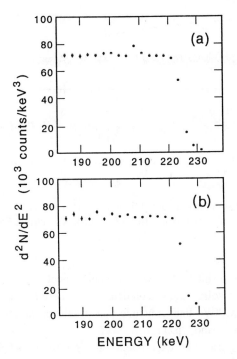

Figure 26. The second derivative of the inner bremsstrahlung spectrum of ^{55}Fe. (a) Monte Carlo data generated with a 1% admixture of a 17-keV neutrino. (d) experimental data after subtracting ^{59}Fe. From (Norman et al. 1992).

B. THE KINK IS DEAD

Support for the existence of the 17-keV neutrino began to erode shortly after the publication of Hime's review. In early August 1992, the Berkeley group presented a conference report which included a statistically improved result from ^{14}C of $M_2 = 17 \pm 1$ keV and a mixing probability of $(1.26 \pm 0.25)\%$ (Norman et al. 1993). They also reported "a high statistics measurement of the inner bremsstrahlung spectrum of ^{55}Fe and find no indication of the emission of a 17-keV neutrino" (p. 1123).[52] The analysis method used in the ^{55}Fe experiment was to examine the second derivative of the beta-decay spectrum for a kink. "In the present ^{55}Fe experiment, we have sufficiently high statistics that a true 'local' analysis could be performed [over a narrow energy region].[53] It is well known that taking the second derivative of a spectrum can sometimes reveal small peaks that might otherwise be missed. We have found that the second derivative technique is also a powerful way to reveal the distortion in a spectrum produced by the emission of a massive neutrino....The second derivative of the resulting spectrum is shown in Figure [26]. There is clearly no hint of a structure near 208 keV [the energy at which a kink due to a 17-keV neutrino would be expected]. Thus our ^{55}Fe experiment shows no evidence for the emission of a 17 keV neutrino" (p. 1125). They cited three

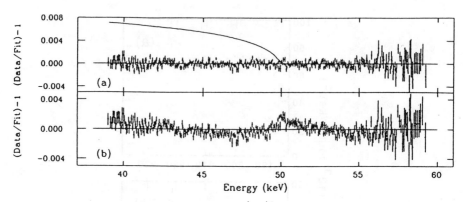

Figure 27. Deviations from the best global fit with $|U|^2$ free (a) and fixed to 1% (b). The curve in (a) indicates the size of a 1% mixing effect of the 17-keV neutrino. From (Ohshima 1992).

other recent negative results;[54] two with magnetic spectrometers on ^{35}S and ^{63}Ni, and one on ^{35}S, that used a solid-state silicon detector.[55] These results, along with their own, supported their view, "We thus conclude that, whatever causes the 'kink' in our ^{14}C spectrum, it is not a neutrino" (p. 1126).

The magnetic spectrometer experiment on ^{63}Ni by the Tokyo group was also presented at the conference (Ohshima 1993). There was also an earlier published result of the same experiment (Kawakami et al. 1992).[56] The experimenters noted some of the problems of experiments that used wide energy regions and commented that, "we have concentrated on performing a measurement of high statistical accuracy, in a narrow energy region, using very fine energy steps. Such a restricted energy scan ...also reduced the degree of energy-dependent corrections and other related systematic uncertainties" (Kawakami 1992, p. 45). The data were taken over three overlapping energy ranges; 41.2 - 46.3 keV, 45.7 - 51.1 keV, and 50.5 - 56.2 keV [the threshold for a 17-keV neutrino occurs at approximately 50 keV]. The results of their experiment are shown in Fig. 27, for (a) the mixing probability allowed to be a free parameter, and (b) with the probability fixed at 1%. The effect expected for a 17-keV neutrino with a 1% mixing probability is also shown in (a). No effect is seen. Their best value for the mixing probability of a 17-keV neutrino was [-0.011 ± 0.033 (statistical) ± 0.030 (systematic)]%, with an upper limit for the mixing probability of 0.073% at the 95% confidence level. This was the most stringent limit yet. "The result clearly excludes neutrinos with $|U|^2 \geq 0.1\%$ for the mass range 11 to 24 keV" (Ohshima 1992, p. 1128).[57]

Although the experiment's narrow energy range was designed to minimize the dependence of the result on the shape correction, the experimenters also checked on the sensitivity of their result to that correction. They normalized their data in the three energy regions using the counts in the overlapping regions, and divided their data into two parts: (A) below 50 keV, which would be sensitive to the presence of a 17-keV neutrino, and (B) above 50 keV, which would not. They then fit their data in (B) and extrapolated the fit to region (A). The resulting fit was far better than one that included a 1% mixture of the 17-kcV neutrino, demonstrating that the shape correction was not masking a possible effect of a heavy neutrino. Bonvicini noted that this experiment, with its very high statistics, had answered essentially all of his criticism of spectrometer experiments convincingly. "Thus, I conclude that this experiment could not possibly have missed the kink and obtain[ed] a good χ^2 at the same time, in the case of an unlucky misfit of the shape factor" (Bonvicini 1993, p. 115).

The 17-keV neutrino received another severe blow when Hime, following the suggestion of Piilonen and Abashian, extended his calculation of the electron response function of his detector to include electron scattering effects and found that he could fit the positive results of Hime and Jelley without the need for a 17-keV neutrino (Hime 1993). This seemed to remove one of the most persuasive pieces of evidence for the heavy neutrino. "It will be shown that scattering effects are sufficient to describe the Oxford β-decay measurements and that the model can be verified using existing calibration data. Surprisingly, the β spectra are very sensitive to the small corrections considered. Consequently, any reinterpretation of the data is reliable only if the scattering amplitudes can be computed or measured accurately, and *independent* of the β-decay measurements" (p. 166).

Hime briefly reviewed the evidence, noting that the major evidence against the existence of the 17-keV neutrino came from magnetic spectrometer experiments in which questions had been raised concerning the shape corrections. He commented that Bonvicini (in a CERN report, discussed earlier) had shown that non-linear distortions could mask the presence of a heavy neutrino signature and still be described by a smooth shape correction. He remarked, however, that "A measurement of the ^{63}Ni spectrum (Kawakami *et al.* 1992) has circumvented this difficulty. The sufficiently narrow energy interval studied, and the very high statistics accumulated in the region of interest, makes it very unlikely that a 17-keV threshold has been missed in this experiment" (p. 165). He also cited a new result from a group at Argonne National Laboratory (Mortara *et al.* 1993, discussed in detail below), that provided "convincing evidence against a 17-keV neutrino." In particular, the Argonne group had demonstrated the sensitivity of their magnetic spectrometer experiment to a possible 17-keV neutrino by admixing a small component of ^{14}C in their ^{35}S source and detecting the resulting kink in their composite spectrum. These negative results provided the impetus for Hime's reexamination of his result.

Hime's new Monte Carlo study included the effect of "electrons which enter the detector after scattering from the aluminum baffle (see Fig. 15), electrons which penetrate the edges of the apertures, and electrons which back-diffuse from the source substrate" (p. 167).[58] These scattering effects had not been included in the original analysis. The dominant effect in the experiment was the scattering from the aluminum baffle, which resulted in 1.2-1.4% of the electrons detected originating from scattering in the baffle, and which exhibited a peak in their energy distribution. The Oxford data were reanalyzed with these additional effects included.

Table IV: Reanalysis of Oxford data (Hime 1993)

Experiment	f_{baffle} (%)	$(\sin^2 \theta_{17}) \times 10^2$	Q (keV)	χ^2/ν
35S run #1	0	0	167.0169(33)	135.8/76
	0	0.752(100)	167.0626(47)	78.9/75
	1.544(175)[a]	0	167.0614(50)	75.2/75
	1.403(175)[b]	0	167.0655(50)	74.6/75
	1.335(175)[c]	0	167.0674(50)	74.4/75
35S run #2	0	0	167.0194(37)	127.3/76
	0	0.833(107)	167.0686(53)	68.6/75
	1.260(190)[a]	0	167.0549(56)	74.9/75
	1.187(190)[b]	0	167.0592(56)	74.7/75
	1.136(190)[c]	0	167.0601(56)	74.6/75
35S data combined	0	0	167.0182(24)	195.9/76
	0	0.816(75)	167.0679(34)	76.9/75
	1.414(131)[a]	0	167.0595(37)	78.3/75
	1.285(131)[b]	0	167.0622(37)	77.4/75
	1.224(131)[c]	0	167.0641(37)	77.2/75
63Ni	0	0	66.8218(39)	173.8/72
	0	1.018(99)	66.8654(56)	69.4/71
	2.730(265)[a]	0	66.8863(62)	71.2/71
	2.698(265)[b]	0	66.8816(62)	71.0/71
	1.676(265)	0	66.8792	68.9/72

[a] Antiscatter baffle (varied)
[b] Antiscatter baffle (varied) + apertures
[c] Antiscatter baffle (varied) + apertures + source backing

The results of this analysis are listed in Table [IV], where comparison is made between the present model and that of a 17-keV neutrino in the absence of intermediate scattering corrections. The inclusion of intermediate scattering effects describes the spectral distortions surprisingly well. Furthermore, the fraction of electrons in the additional LET [Low Energy Tail] component which has been fitted by the data is in very good agreement with that expected from the calculations presented above. The sensitivity of the data to the various effects also agrees with expectations. In the case of ^{35}S, for example, the data are only weakly sensitive to the effects of aperture penetration or to the presence of the source substrate. On the other hand, a fit to the ^{63}Ni data with only the aluminum baffle included yields a result that is inconsistent with both the ^{35}S data and calculations. The agreement is significantly improved,

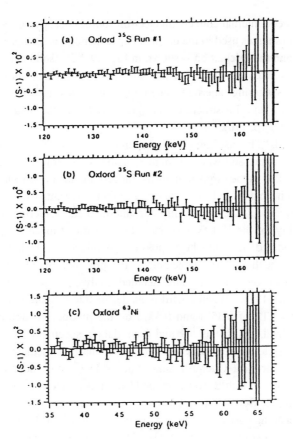

Figure 28. Shape factors extracted from Oxford, (a) ^{35}S run #1, (b) ^{35}S run #2, and (c) ^{63}Ni, data after implementing the best fit theoretical spectrum including intermediate scattering effects and assuming a single component. massless neutrino. From (Hime 1993).

however, after accounting for electron back-diffusion from the source substrate, and the effects of aperture penetration are marginal.

Residuals are presented in Fig. [28] in the form of shape factors derived from optimum fits to the data after including intermediate scattering effects and assuming a single-component massless neutrino. The ^{35}S shape factors (Figs. [28a] and [28b]) hint at spectral distortion beyond ~150 keV. While the intermediate scattering contributions cannot produce a "kink" per se, the chi-squared analysis (table [IV]) indicates that any difference between the two models considered [with and without a 17-keV neutrino] cannot be distinguished by the statistics of the data (p. 169).

Still, there remained a possibility that the new calculation was incorrect. Hime was able to independently confirm his model by measuring the electron response function using monoenergetic internal conversion electron sources occupying the same geometry as the beta-decay sources used in the original experiments. The comparison between the measurements and the calculation is shown in Fig. 29. "The solid curve drawn through these residuals is taken directly from the calculations presented above, including the effects of baffle-scattering, aperture penetration, and back-diffusion from the source substrate. The data reveal a structure that agrees well with the model, both in overall shape and intensity" (p. 170).[59]

Hime concluded, "The distortions observed in the ^{35}S and ^{63}Ni experiments at Oxford are significantly suppressed when account is made for intermediate scattering effects that were overlooked in the original analysis. Indeed, the heavy neutrino hypothesis can be replaced with that based on scattering effects. Essentially, there is a 100% correlation between f_{int}, the probability for intermediate scattering, and $\sin^2\theta_{17}$, the mixing probability for the 17-keV neutrino. Hence, without independent knowledge of the effects considered, it would be impossible to rule out a 17-keV neutrino based solely on fitting the β spectra. Nonetheless, the presence of intermediate scattering effects has been uncovered in a more detailed analysis of IC [internal conversion] electron spectra (p. 171)... When regard is made for intermediate scattering effects an upper limit (90% CL) of 0.35% and 0.53% on the mixing probability for a 17-keV neutrino is obtained, using the ^{35}S and ^{63}Ni data respectively" (p. 172). He also suggested that despite the very different geometries that intermediate scattering effects might explain the original Guelph results. Such effects could not, however, explain those results obtained with a souce embedded in the detector, such as Simpson's original result and the Berkeley result on ^{14}C.

Further evidence against the 17-keV neutrino was provided by the Argonne group headed by Freedman (Mortara *et al.* 1993). The experiment used a solid-state, Si(Li), detector, the same type used by Hime, an external ^{35}S source, and a solenoidal magnetic field to focus the decay electrons. The apparatus had a 2π sr solid angel (50% efficiency), which allowed a thin source which reduced scattering in the source, and still allowed a high counting rate. The solenoidal field was shaped so that the angle between the electron velocity and the solenoid axis decreased as the electron moved toward the detector. This helped to reduce backscattering from the detector, which is larger at glancing angles. In addition the field shape reflected some of the backscattered electrons to the detector by the magnetic mirror effect. The backscattering was reduced to less than 7% of the incident intensity, and a Monte Carlo simulation indicated that the fraction in the backscattering tail was nearly independent of energy. The apparatus also required no collimator. In additiom, the electron response function was measured at several points in the fitted spectrum. "The present experiment requires that we know the electron response function between 120 and 167 keV. Measurements of the conversion lines of ^{139}Ce at 127, 160, and 167 keV are the principal constraint on the model of the electron response function" (p. 395). Previous ^{35}S experiments had used an electron response function extrapolated from the lower energy ^{57}Co lines. Finally, and perhaps most importantly, this experiment required no arbitrary shape correction factor.

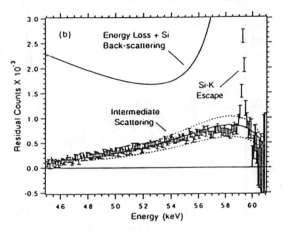

Figure 29. ^{109}Cd spectrum accumulated in Oxford geometry. Residuals extracted from the 61-keV K=IC tail when intermediate scattering efffects are neglected. The solid curve shows the effect calculated for intermediate scattering. From (Hime 1993).

The experimenters also demonstrated the sensitivity of their apparatus to a possible 17-keV neutrino.

> To assess the reliability of our procedure, we introduced a known distortion into the ^{35}S beta spectrum and attempted to detect it. A drop of ^{14}C-doped valine ($E_o - m_e \sim 156$ keV) was deposited on a carbon foil and a much stronger ^{35}S source was deposited over it. The data from the composite source were fitted using the ^{35}S theory, ignoring the ^{14}C contaminant. The residuals are shown in Figure [30]. The distribution is not flat; the solid curve shows the expected deviations from the single component spectrum with the measured amount of ^{14}C. The fraction of decays from ^{14}C determined from the fit to the beta spectrum is (1.4 ± 0.1)%. This agrees with the value of 1.34% inferred from measuring the total decay rate of the ^{14}C alone while the source was being prepared. This exercise demonstrates that our method is sensitive to a distortion at the level of the positive experiments. Indeed, the smoother distortion with the composite source is more difficult to detect than the discontinuity expected from the massive neutrino. (Mortara et al., p. 396).

Their final result, shown in Fig. 31, was $\sin^2\theta = -0.0004 \pm 0.0008$ (statistical) $\pm$ 0.0008 (systematic), for the mixing probability of the 17-keV neutrino. "In conclusion, we have performed a solid-state counter search for a 17 keV neutrino with an apparatus with demonstrated sensitivity. We find no evidence for a heavy neutrino, in serious conflict with some previous reports" (p. 396).

This experiment was clearly convincing. It met all the criteria previously suggested by Hime and Bonvicini along with a demonstrated ability to detect a kink in the spectrum had one been there. Along with the extremely high statistics Tokyo

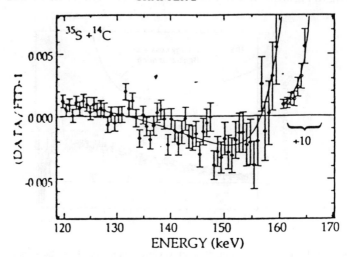

Figure 30. Residuals from fitting the beta spectrum of a mixed source of ^{14}C and ^{35}S with a pure ^{35}S shape; the reduced χ^2 of the data is 3.59. The solid curve indicates residuals expected from the known ^{14}C contamination. The best fit yields a mixing of $(1.4 \pm 0.1)\%$ and reduced χ^2 of 1.06. From (Mortara et al. 1993).

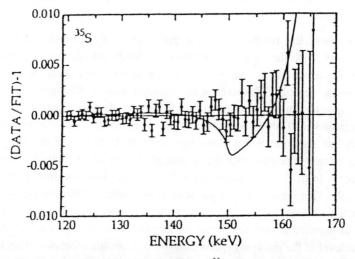

Figure 31. Residuals from a fit to the pile-up corrected ^{35}S data assuming no massive neutrino; the reduced χ^2 for the fit is 0.88. The solid curve represents the residuals expected for decay with a 17-keV neutrino and $\sin^2\theta = 0.85\%$; the reduced χ^2 of the data is 2.82. From (Mortara et al. 1993).

experiment discussed earlier, it provided very strong evidence against the existence of the 17-keV neutrino.

The Berkeley group published a later, higher statistics result for the internal bremsstrahlung spectrum of ^{55}Fe (Wietfeldt, 1993a). They discussed the question of whether or not a smooth distortion in the spectrum would affect their analysis. "Finally to verify that this analysis would not be affected by a smooth distortion in the spectrum, Monte Carlo spectra were generated with arbitrarily chosen linear and quadratic factors, with and without massive neutrinos; and these spectra were analyzed in the same way. The presence or absence of a neutrino kink was always correctly found....Thus, we conclude that the effect reported previously in lower statistics experiments, and interpreted to be the result of a 17 keV neutrino, is not in fact caused by a massive neutrino (p. 1762)." "In particular, a 17 keV neutrino with $\sin^2\theta = 0.008$ is excluded at the 7σ level" (p. 1759).[60]

Once again, Schwarzschild provided a summary of the situation in *Physics Today*, in an April 1993 article entitled, "In Old and New Experiments the 17-keV Neutrino Goes Away" (Schwarzschild 1993). He noted that the 17-keV neutrino had received five severe blows during the preceding twelve months. Among these were the high-statistics and persuasive results of Kawakami's group in Tokyo (Kawakami *et al.* 1992; Ohshima 1992; Ohshima *et al.* 1993) and those of the Argonne group (Mortara *et al.* 1993). He also quoted Hime on the question of electron scattering that led to his reanalysis. "I didn't pay too much attention to this critique [that of Piilonen and Abashian] at the time. We hadn't included scattering off the baffles because we knew that scattering at the detector was a much bigger source of electron energy degradation. That's something we *had* included in our fits and shown that it had no effect on our 17-keV signal. And besides, Simpson and I had seen the same 17-keV signal in a variety of earlier geometries at Guelph that had nothing to do with baffles. But after Freedman's result [Mortara *et al.* 1993] I knew I had to take a serious second look" (Hime, quoted in Schwarzschild 1993, p. 18). Schwarzschild remarked that Jelley had confirmed Hime's reanalysis of their results and had redesigned the apparatus to avoid the problems Jelley continued to take data and Schwarzschild reported that the 17-keV neutrino signal seemed to have gone away.

Schwarzschild also raised the issue of the positive results from ^{14}C obtained by the Berkeley group.

> After the report of Norman's initial four-month run, his Berkeley group found 17-keV signals in each of three additional runs of comparable statistics. Just as they were about to publish these new confirmatory results at the end of 1991, the group acquired a new data acquisition system that allowed them, for the first time, to discard events by off-line software veto. Because the betas have a range of about 100μm in germanium, one wants to discard decays that are too close to the edge of the crystal, lest they get out without depositing their full energy. To that end, the Berkeley group had surrounded the detector's fiducial volume with a Ge guard ring designed to veto events too close to the edge.
>
> All the 1990-91 running had been done with just an on-line hardware veto from the guard ring. But now, with the new software-veto capability, the group could take a closer look at the events the guard ring was discarding. And what they found was very disturbing; it called all their

previous results into question. Far too many events were being vetoed, and there was a peculiar correlation between the energies recorded in the central crystal and in the guard ring. Eventually it was found that the culprit was electronic cross talk between the central detector and the ring.

With the errant guard ring taken care out of commission, the Berkely group took new ^{14}C data throughout 1992 with a software fiducial veto and found that its 17-keV neutrino had vanished (Schwarzschild 1993, p. 18).

The Berkeley group continued to work on trying to find the reason for the artifact in their ^{14}C data. The cause, found in 1993, was quite subtle. The way in which the center detector was separated from the guard ring was by cutting a groove in the detector. "The n^+ is divided by a 1-mm-wide circular groove into a 'center region' 3.2cm in diameter, and an outer 'guard ring.' By operating the guard ring in anticoincidence mode, one can reject events occurring near the boundary which are not fully contained within the center region" (Sur et al. 1991 p. 2444). Such events would not give a full energy signal and would thus distort the observed spectrum.

What the Berkely group found was that ^{14}C decays occurring under the groove shared the energy between both regions without necessarily giving a veto signal, and thus gave an incorrect event energy, distorting the spectrum. They also found that, although their earlier tests had indicated that the ^{14}C was uniformly distributed in the detector, their new tests showed that between one third and one half of the ^{14}C was localized in grains. They also found that approximately 1% of the grains were located under the groove. Thus, the localization of the ^{14}C combined with the energy sharing gave rise to a distortion of the spectrum that simulated that expected from a 17-keV neutrino (Norman, private communication and Wietfeldt et al. (1993b, 1994)).

There was virtually no evidence left that supported the existence of the 17-keV neutrino. Simpson was not, however, totally convinced. Although he admitted that he owed Glashow a bottle of wine, the stake of a wager on the existence of the 17-keV neutrino, he remarked, "Still it's very peculiar that all these different experimental arrangements should have conspired to give the same spurious signal. At the moment it appears that only the Guelph results remain to be explained, so we're continuing our experiments" (Simpson, quoted in Schwarzschild 1993, p.18). As of November 1993 Simpson was still working on the problem. He agreed that the preponderance of evidence is against the existence of the 17-keV neutrino, but he hesitated to say that it is definitely gone. He noted that the presence or absence of the effect is quite sensitive to the method of data analysis used, although he believes that the later experiments seemed to avoid that problem by using both wide and narrow energy range analysis. He also remarked on the oddity that the very different experimental artifacts, those of Hime and of the Berkeley group, both gave effects at the same neutrino mass, an unlikely occurrence. These artifacts do not, however, explain his original positive results. He is currently searching for a possible error or artifact that might explain why his original result was incorrect (Simpson,1993).

The consensus of the physics community is, however, that the 17-keV neutrino does not exist.

III. DISCUSSION

This episode illustrates important points about the methodology of scientific practice. This methodology is particularly apparent in cases such as this in which discordant experimental evidence both supports and disconfirms an experimental result or a speculative hypothesis. Perhaps the most important point is that the decision that the 17-keV neutrino did not exist was a reasonable one, based on epistemological considerations. As we have seen, the discord between the experimental results was resolved by a combination of finding errors in one set of experiments with the accumulation of evidential weight in the other set.

Other commentators on science have questioned whether or not epistemological arguments enter into this type of decision.[61] For example, Harry Collins argues for what he calls the "experimenters' regress" (Collins 1985). In his discussion of the early experimental attempts to detect gravity waves Collins argues that we can't be sure that we can actually build a gravity wave detector, that we might have been fooled into thinking we had the recipe for constructing one, and that "we will have no idea whether we can do it until we try to see if we obtain the correct outcome. *But what is the correct outcome?*"

> What the correct outcome is depends upon whether or not there are gravity waves hitting the Earth in detectable fluxes. To find this out we must build a good gravity wave detector and have a look. But we won't know if we have built a good detector until we have tried it and obtained the correct outcome! But we don't know what the correct outcome is until...and so on ad infinitum.
>
> The existence of this circle, which I call the 'experimenters' regress,' comprises the central argument of this book. (Collins 1985), p. 84). [In these quotations one could easily substitute "17-keV neutrino" for "gravity waves" without changing the sense of Collins' statement].

More succinctly, "Proper working of the apparatus, parts of the apparatus and the experimenter are defined by the ability to take part in producing the proper experimental outcome. Other indicators cannot be found" (Collins 1985, p. 74). I have argued elsewhere that Collins' analysis of the gravity wave episode is incorrect (Franklin 1994). I also believe that the history of both gravity waves and of the 17-keV neutrino shows not only that such epistemological indicators can be found, but were.

What are these epistemological indicators or criteria? In previous work I have argued for an epistemology of experiment, a set of strategies that can be philosophically justified and used to argue for the validity of an experimental result. I have also shown that they are, in fact, used by practicing scientists. These include: 1) experimental checks and calibration, in which the experimental apparatus reproduces known phenomena; 2) reproduction of artifacts that are known in advance to be present; 3) intervention, in which the experimenter manipulates the object under observation; 4) independent confirmation using different experiments; 5) elimination of plausible sources of error and alternative explanations of the result (the Sherlock Holmes strategy); 6) use of the results themselves to argue for their validity; 7) use of an independently well-corroborated theory of the phenomena to explain the results; 8) use of an apparatus based on a well-corroborated theory; and 9) Use of statistical arguments (Franklin 1986; Franklin 1990).

The problem is that *all* of the experiments discussed here offered such strategies and arguments in support of their results.[62] The question was whether or not these epistemological strategies had been applied correctly.[63] How does one argue that such a strategy or a correction to an experimental result has been incorrectly applied? One possibility is to show that its use in a particular experiment generates a contradiction with accepted results. A second possibility is to show that some plausible source of error or an alternative explanation of the result has not been considered.[64] (What is considered plausible may change with time, as discussed below). One might also examine assumptions concerning the operation of the apparatus and demonstrate empirically that they are incorrect. One might also show that plausible explanations of results, suggested by others, are incorrect. All of these occurred in this episode.

Other criteria may also exist. In a particular experiment some epistemological strategies may have been applied successfully whereas others had failed. This is illustrated in the episode concerning experiments on atomic parity violation. In this case there was a conflict between the discordant results of the Washington and Oxford atomic parity violation experiments and the SLAC E122 experiment on electron scattering. The Oxford experiment had admitted systematic uncertainties that were the same size as the predicted effect. In addition, the Washington results were internally inconsistent. Both of these effects made their results less credible. The SLAC E122 experiment had no such failures and therefore had more evidential weight.[65]

Sometimes the failure to reproduce an observation, despite numerous attempts to do so, might be legitimately regarded as casting doubt on the original observation, even if no error has been found in that experiment. This would be a case of a preponderance of evidence.[66]

The history shows us that deciding on the correct answer to the question of the existence of the 17-keV neutrino involved not only numerous repetitions of the experiment, but also criticism and discussion of the experimental results, of the experimental apparatuses, and of the methods of analysis used. The history also shows that these criticisms and discussions were taken seriously and acted upon by the scientists involved. This was, in effect, applied epistemology.

Let us review in detail how the decision that the 17-keV neutrino did not exist was reached. What makes this process so interesting is that the original discordant results were obtained with two different, and seemingly reliable, types of experimental apparatuses. One might worry that it was a peculiarity of one of the types of apparatus that either created an artifact or that masked a real effect. As Schwarzschild remarked in

1991, "On one thing everyone seems to agree. After six years, the experimenters must begin to resolve the stubborn discrepancy between the two different styles of beta-decay experiments" (Schwarzschild 1991, p. 19). Both Simpson's original report of the 17-keV neutrino and the other positive results were obtained using solid-state detectors. Such detectors had been in wide use since the early 1960s and their use was well understood. The early negative results were obtained using magnetic spectrometers. This type of apparatus had been used in nuclear beta-decay experiments since the 1930s and both the problems and advantages of using this technique had been well studied. (See Franklin (1990, Chapter 1) for details of some early experiments).

Simpson's first report of the 17-keV neutrino was unexpected. It was not predicted, or even suggested, by any existing theory. Faced with such an unexpected result the physics community took a reasonable approach. Some scientists tried to explain the result within the context of accepted theory. They argued that a plausible alternative explanation of the result had not been considered. This involved the question of whether or not the theory used in the analysis of the data, and to compare the experimental result with the theory of the phenomenon, was correct. This is an important point. An experimental result is not immediately given by an examination of the raw data, but requires considerable analysis. In this case the analysis included atomic physics corrections, needed for the comparison of the theoretical spectrum and the experimental data. Everyone involved agreed that such corrections had to be made. There were, after all, large effects of this kind exhibited in the phenomenon of internal bremsstrahlung. The atomic physics corrections used by Simpson in his analysis, particularly the screening potential, were questioned by other scientists. All of these suggestions were aimed at accommodating the unexpected result. Several calculations indicated, at least qualitatively, that Simpson's result could be accommodated within accepted theory, and that there was no need for the suggestion of a new particle.

The physics community also tried to replicate Simpson's results. Within a year, five attempted replications of Simpson's experiment, using primarily, but not exclusively, magnetic spectrometers, all gave negative results. This was an attempt to provide independent confirmation of a result using different experiments. The apparatuses used in the attempts used a different decay source, ^{35}S as opposed to tritium, and magnetic spectrometer apparatuses as opposed to solid-state detectors. By using different sources one could check on whether or not Simpson's observed effect might be due to some atomic physics phenomena peculiar to his choice of decay source. Had positive results been found then one would have concluded that no such effects existed and the experiments would have provided more support for Simpson's original result than would have been the case if the experiments had used the same source.[67] The difficulty is that although greater support for a result is provided when different experiments agree, when such different experiments disagree we don't know which result is correct and will suspect that the different results are caused by some difference in the experimental apparatus, or in the analysis of the data.

In addition, Simpson offered several criticisms of those early negative results. These involved the analysis procedures used in those experiments. One feature of magnetic spectrometer experiments was the need for a smooth, energy-dependent "shape-correction factor" to obtain the decay spectrum. In ordinary experiments the use of such a factor was not crucial but, as Simpson pointed out, when one looked for effects of the order on one part in a thousand, then one had to be quite certain of one's analysis procedure. ($\Delta K/K$, the quantity of interest in the beta-decay spectrum, was approximately 10^{-3}. See Figs. 1 and 9). Simpson also questioned other aspects of the analysis. The first was the use of a wide energy range, rather than a narrow one, to calculate the expected spectrum and to fit the spectrum parameters. He noted that 45% of the effect of a heavy neutrino occurred within 2 keV of the neutrino threshold. He also criticized the procedure of merely adding the expected effect of a 17-keV neutrino to the best-fit spectrum, rather than incorporating the effect into the spectrum and then determining the best fit. He claimed that these procedures tended to minimize the effect of the proposed particle. In both of these criticisms, Simpson was questioning whether the Sherlock Holmes strategy of eliminating plausible sources or error or alternative explanations of the result had been correctly applied. The question was whether the analysis procedures used might mimic or mask the effect a 17-keV neutrino. Simpson was suggesting that they might. As we have seen, others agreed with Simpson.

Subsequent experiments acquired sufficient statistics so that a local analysis (a narrow energy range which minimized the effects of the shape correction factor) could be used; and other experiments used both a narrow and a wide energy range, so that any difference due to the energy range used in the analysis might be seen. In addition, the type of spectrum fitting suggested by Simpson was used in several of the later experiments. (See for example (Becker *et al.* 1991; Hetherington *et al.* 1987; Norman *et al.* 1993; Ohshima 1993; Radcliffe *et al.* 1992)). These experiments found no evidence for a 17-keV neutrino, and eliminated the analysis procedure as a possible explanation of their failure to find it. Simpson also reanalyzed the early negative results using his own preferred analysis procedure and argued that they did not, in fact, argue against the existence of the 17-keV neutrino.

Just as others took Simpson's criticisms seriously, so did Simpson react to the criticism of others. He reanalyzed his own data using the different atomic physics correction to the spectrum that his critics had suggested and found that his effect, although reduced in size, was still present.

Simpson had enough confidence in his own work to continue his investigation, despite both the criticism and the negative results,[68] and reported additional positive results (Hime and Simpson 1989; Simpson and Hime 1989). Recall that there was not much experimental work done on the 17-keV neutrino between the five negative results reported in 1986 and these new results. This further confirmation of the original Simpson result, using both a different source and a different solid-state detector, encouraged others to further investigate the phenomenon. Several of these experiments obtained positive results, with particularly persuasive results provided by Hime and Jelley (1991) and Sur *et al.* (1991).

At the same time, other experiments with both improved statistics and analysis procedures were finding increasingly persuasive negative results. Piilonen and

Abashian (1992) suggested that a background effect might be simulating the presence of a 17-keV neutrino in the Hime-Jelley experiment. The persuasive negative results encouraged Hime to consider the Piilonen-Abashian suggestion seriously[69] and to reanalyze his own result. He found, using an experimentally checked Monte Carlo calculation,[70] that scattering of the decay electrons in the experimental apparatus could explain his result, without the need for a 17-keV neutrino. At approximately the same time, the Berkeley group found an error in their own positive result on ^{14}C. Their attempt to guard against a spectrum distortion caused by decays close to the edge of their detector, which do not deposit their full energy, had not worked. It had, instead, caused a spectrum distortion that mimicked the effect of a 17-keV neutrino. Improvements in the apparatus had allowed them to examine the energy deposition in both the central detector and the guard ring for each event, whereas the previous set-up had not allowed this. They had assumed that there was no energy sharing and found that there was, an effect that distorted their spectrum. They had found, not merely a plausible source of error, but an actual error in their result.

The newer negative results were persuasive because of their improved statistical accuracy, and also because, in the case of the Argonne experiment, they were able to demonstrate that their experimental apparatus could detect a kink in the spectrum if one were present (Mortara *et al.* 1993). This was a direct experimental check that there were no effects present that would mask the presence of a heavy neutrino. These experiments met Hime's suggested criteria of a demonstrated ability to detect a kink combined with high statistics so that a local analysis of the spectrum could be done.[71] Ohshima (1993) had also shown that the shape correction factors used in their experiment did not mask any possible 17-keV neutrino effect. This combination of almost overwhelming and persuasive evidence against the existence of a 17-keV neutrino, combined with the demonstrated and admitted problems with the positive results, decided the issue. There was no 17-keV neutrino. It seems clear that this decision was based on experimental evidence, discussion, and criticism or, in other words, epistemological criteria. The process of designing a good "17-keV neutrino" detector was not simply a matter of deciding whether or not the particle existed, and then asserting that a good detector was one that gave the correct answer. The community decided which were the good detectors, based on epistemological criteria, and then decided that the particle did not exist.[72]

Another interesting aspect of this episode is that it was almost completely driven by experiment and observation. The existence of a heavy neutrino would have had important implications for theory, particularly for electroweak interactions, and for cosmology. In the early 1980s, for example, the well-established Weinberg-Salam unified theory of electroweak interactions might very well have accommodated a light, massive neutrino. A neutrino with mass in the keV range would, however, have made the revised theory "ugly."[73] A heavy neutrino would also have had important implications for astrophysics. During the 1980s astrophysicists were exploring the possibility that the "missing mass" in existing cosmological theories might be accounted for by heavy neutrinos.[74] Although these theoretical considerations might have had an

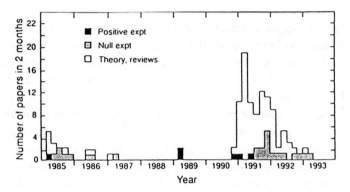

Figure 32. Preprints on the 17-keV neutrino received at CERN as a function of time. From (Morrison 1993).

effect on work on the 17-keV neutrino, they did not. Activity in the field was determined by experimental results and by the desire to answer the question of whether or not the 17-keV neutrino existed. These experiments had a life of their own.

This is clearly shown in Figure 32, which shows the number of preprints on the 17-keV neutrino received at CERN as a function of time (Morrison 1993). This is not a complete picture of the activity in the field because not everyone sent preprints of their work to CERN, but it does give an accurate relative picture of work in the field. The figure shows an initial spurt of activity, both experimental and theoretical, triggered by Simpson's original claim in 1985. Some of that early theoretical work consisted of the attempts to explain Simpson's result without the need for a heavy neutrino, discussed earlier. Within a year five negative experimental results were reported. These negative results, combined with the alternative explanations, had a chilling effect on work in the field. As we have seen, however, work continued, albeit at a low level of activity.

A second and much larger burst of activity began in late 1990. This coincided with the new positive results on the 17-keV neutrino reported by Hime and Jelley at Oxford and by Norman's group at Berkeley. One may speculate that the reason that these results had such a positive effect, whereas the positive results reported in 1989 by Simpson and Hime had no such effect, was that they were the first positive results reported by scientists other than Simpson. Physicists may have wondered whether or not it was some artifact produced by Simpson's experimental apparatus or some problem with his data analysis that was producing the effect.[75] The support for the 17-keV neutrino provided by other experimental groups using different experimental apparatuses and data analysis procedures was greater than that provided by Simpson's similar repetitions of his own experiment.[76] The numerous and persuasive negative results reported from 1991 to 1993 ended activity in the field. Experimental evidence has shown that the 17-keV neutrino did not exist.

One should also note the important and legitimate role that Monte Carlo calculations, computer simulations of experiments, played in this episode. It was Hime's Monte Carlo calculation of the effect of electron scattering in his experimental apparatus that convinced him, as well as the rest of the physics community, that his result supporting the existence of the 17-keV neutrino was incorrect. The Berkeley group also used Monte Carlo methods to check that their analysis procedure was not masking or creating the effect of the 17-keV neutrino. They deliberately inserted the effect of such a neutrino into some, but not all, of their simulations and found that their analysis procedure correctly identified the presence or absence of the neutrino in every case. Monte Carlo simulation was also important in Bonvicini's study.

Pickering has, however, questioned the use of such Monte Carlo calculations, and suggested that their use in experiments precludes the use of the results as evidential support (Pickering 1984). In discussing the use of such a simulation in the Gargamelle experiment, which reported the existence of weak neutral currents, Pickering noted that several of the inputs to the calculation could be questioned. These included the beam characteristics, the interaction of nucleons with atomic nuclei, neutron production, and idealized experimental geometry. "My object here is simply to demonstrate that assumptions were made which could be legitimately questioned: one can easily imagine a determined critic taking issue with some or all of these assumptions. Moreover, even if all of the assumptions were granted, it remained the case that they were input not to an analytic calculation, but to an extremely complex numerical simulation. The details of such simulations are enshrined in machine code and are therefore inherently unpublishable and not independently verifiable. Thus the sceptic could legitimately accept the input to the calculation but continue to doubt its output" (Pickering 1984, p. 96).

What Pickering overlooks is that considerable effort is devoted to checking the results of that calculation by comparison with experimental evidence that is independent of the result in question.[77] The results of this checking are, in fact, publicly available in the published work. Thus, Hime's Monte Carlo calculation had shown that intermediate scattering effects in his aluminum baffles could account for his data, just as well as did the assumption of a 17-keV neutrino. He checked his calculation by comparing it to data taken with the same experimental apparatus and geometry using a monoenergetic internal conversion electron source. The excellent fit between these measurements and his simulation argued for the correctness of his calculation (See Fig. 29).

Such checks are usually done. For example, in an experiment designed to measure the energy dependence of the form factor in K^+_{e3} decays, $K^+ \rightarrow e^+ + \pi^0 + \nu$, the way in which the energy dependent parameter λ was fixed was by comparing Monte Carlo generated spectra with different values of λ with the experimental data (Imlay et al. 1967). The Monte Carlo simulation was checked by comparing its results with a sample of background events.

> It was also necessary to know the energy distributions relating to background events. These distributions were obtained from the Monte-Carlo generated sample of spurious K^+_{e3} events. Indications of the validity of this calculation were obtained from the distributions of positron momentum, γ-ray energy, and π^0 energy for those *events which*

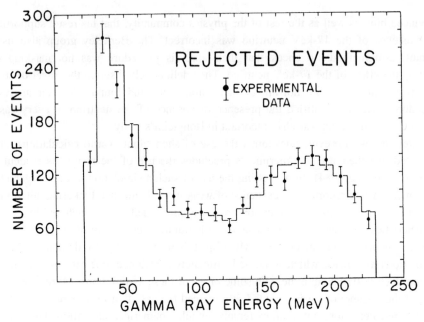

Figure 33. Comparison of Monte-Carlo-generated γ-ray spectrum with experimental data for rejected events. From (Imlay et al. 1967).

> *were rejected by selection criterion 3.* This criterion required that the counter behind each spark chamber give a pulse if the shower in the chamber contained sparks in either of its last two gaps. These rejected events should differ from the background events in the final sample of 1867 nominal K^+_{e3} events only with regard to selection criterion 3. Thus, when reconstructed as K^+_{e3} decays, the background events that passed and failed criterion 3 should have exactly the same distributions. These are shown in Fig. [33], along with the calculated distributions for Monte-Carlo generated spurious events. The good agreement provides strong support for the background calculation, particularly since these distributions differ substantially from the corresponding distributions for good events. (Imlay et al. 1967, p. 1209).

Pickering also overlooks the fact that the robustness of the results of a Monte Carlo calculation is checked against reasonable variations in the input parameters. This is because, as Pickering himself notes, these parameters are not exactly known. Typically, the results are not sensitive to such variations. If they are, then the results must be used with extreme care, and may not, in fact, be usable.

Determined critics or skeptics might question such Monte Carlo calculations, but they would have to discount the independent evidence provided.

In thinking about this episode, as well as other episodes, we should distinguish between the processes of pursuit and justification; between the further investigation of a phenomenon or a theory and the process by which they are accepted as knowledge by the scientific community (see note 68). Although both of these processes were going on simultaneously we should note that belief that a hypothesis or a result is correct is not a necessary prerequisite for working on it. The reasons for further investigating (pursuing) a hypothesis or result are not usually the reasons by which one justifies belief in them. With respect to the existence of the 17-keV neutrino the attitude of scientists working on the problem varied from belief to disbelief with various intermediate positions. Recall the differing summaries of the situation offered in 1991 and 1992. Simpson was quite positive, Morrison quite negative, and Hime adopted a moderate, agnostic position.

During the period 1985-1993 considerable theoretical work was done on the 17-keV neutrino.[78] These papers attempted to incorporate the particle into accepted particle theory, to include it in a new theory, or to look for further implications of such a particle. Not everyone was as positive as Glashow about the existence of the particle (see quotation earlier). More agnostic views were, "The possible discovery of a 17 keV neutrino in β-decay experiments is a challenge to both astrophysics and cosmology" (Altherr *et al.* 1991), p. 251). and "Recent experimental evidence for a 17 keV neutrino mass eigenstate with 0.8% mixing to v_e, while still disputed, has led to extensive theoretical investigations because it is very difficult to reconcile a particle with these properties with standard particle theories, not to mention cosmology and astrophysics" (Madsen 1992, p. 571).

Scientists may have other reasons than belief in the correctness of the theory or result for pursuing it further. As Madsen indicated above, an experimental result may call for a new theory because it is incompatible with accepted theory. One might also work on something because it fits in with an existing research program or because it looks like a fruitful, important, or interesting line of research. "The existence of massive neutrinos would have profound implications for both particle physics and astrophysics" (Norman *et al.* 1991, p. S291).

Experimenters may have additional experimental reasons for pursuit. These may include the fact that the experiment can be done with existing apparatus or with small modifications of it. The measurement may also fit in with an existing series of measurements in which the experimenters have expertise. We might call these instrumental loyalty and the recycling of expertise (see later essay). Simpson had been using a solid-state detector to search for massive neutrinos in β-decay experiments for several years before he reported the existence of the 17-keV neutrino, and the other groups had considerable experience in doing beta-decay experiments. Another reason for pursuit might be that the experimenters might have thought of a clever way to do the experiment. Thus, the Berkeley group remarked, "Moreover, we were aware of a unique detector... that was ideally suited for this experiment" (Sur *et al.* 1991, p. 2444).

How was the decision concerning the existence of the 17-keV neutrino made? I believe I have shown that the decision that it did not exist was made on the basis of valid experimental evidence. I have also argued that epistemological criteria were used in the evaluation of that evidence. The process also involved discussion and criticism that was taken seriously by everyone involved. Popper has characterized science as "critical rationality." That seems an apt description.

NOTES

[1] The units of mass are in keV/c^2, but physicists usually refer to the masses of particles in energy units such as keV. Physicists currently believe that the mass of the neutrino is zero, or very close to it.

[2] As we shall see below, two of the initial negative results were, in fact, obtained with solid state detectors. Simpson later argued that one of the experiments, (Ohi and et al. 1985), was incorrect, and that the other, (Datar et al. 1985), was inconclusive. There was also suggestive, although not conclusive, evidence from a third type of experiment, that detecting internal bremsstrahlung in electron capture (IBEC), a form of beta decay. This is also sometimes referred to as internal or inner bremsstrahlung. I note that not all of the IBEC experiments gave positive results. As discussed later, one of the experiments that convinced the physics community that the 17-keV neutrino did not exist did, in fact, use a solid-state detector (Mortara et al. 1993).

[3] In this paper I will not discuss the large amount of theoretical work on the 17-keV neutrino unless it impinges directly on the experiments or on the existence of such a particle.

[4] Although, as we shall see later, there is good reason to doubt the existence of the 17-keV neutrino, I shall speak of it as if it existed.

[5] In a normal beta-decay spectrum the quantity $K = (N(E)/[f(Z,E) (E^2 - 1)^{1/2}E])^{1/2}$ is a linear function of E, the energy of the electron. A plot of that quantity as a function of E, the energy of the decay electron, is called a Kurie plot.

[6] This neglects the effects of experimental energy resolution.

[7] Simpson reported, "The decay of tritium has been followed with this detector over a period of four years and the halflife has been determined to be 12.35 ± 0.03 yr, in very good agreement with published values (Simpson 1985, p. 1891)."

[8] Simpson was searching for a low mass neutrino with a mass of the order of tens of eV.

[9] Although I will discuss the details of Simpson's calibration and data analysis here, I will not, in general, discuss these issues for subsequent experiments unless questions have been raised concerning those details.

[10] The question of whether or not the neutrino has mass, or is a superposition of states which have mass, can be separated into two parts. The first is whether or not it is close to zero mass, but finite. The second is whether or not a heavy neutrino, with mass of order keV, exists. In this essay I will concentrate on the latter.

[11] There was considerable discussion among the active researchers in the field. Haxton, Eman and Tadic, and Lindhard and Hansen, all acknowledged helpful conversations with Simpson concerning both his experimental apparatus and his theoretical calculations. Although Eman's and Tadic's paper was not published until mid-1986, Simpson knew of it by private communication and made use of it in a calculation presented at the Moriond Workshop, 25 January-1 February, 1986.

[12] The Moriond Workshops play an extremely important role in speculative and/or controversial issues. They provide a forum for those working in the field to meet, present papers, and to have both formal and informal discussions and criticism. For a discussion of the role that the Moriond

Workshops played in another controversial episode, that of the Fifth Force, a proposed modification of Newton's law of gravity, see (Franklin 1993a).

[13] Details of the experimental apparatus are contained in Simpson (1985).

[14] Simpson is relying here, as he did earlier in his discussion of why he performed a search for a low mass neutrino with a solid-state detector, on the idea that "different" experiments provide more confirmation of a hypothesis or of an experimental result, than do repetitions of the "same" experiment. For a discussion of this see Franklin and Howson (1984).

[15] As we shall see below, others agreed with Simpson. Bonvicini (1993), in a very detailed analysis, showed that a smoothly varying shape correction factor could, in fact, either mask or mimic a kink in the spectrum. This will be discussed later. It was also noted that the method of analysis chosen might create a signal when one was not really present. This question of the energy range used in the analysis of the data will be quite important in the subsequent history.

[16] A preliminary report of this experiment appeared in Riisager (1986).

[17] As Borge et al. (1986) pointed out, IBEC is actually sensitive to the mass of the neutrino, whereas ordinary beta decay involves an antineutrino. This then made the experiment a test of CPT invariance, which requires that particles and antiparticles have identical masses.

[18] They also thanked Simpson for interesting discussions.

[19] At the same conference Wark and Boehm (1986) also presented negative results on the 17-keV neutrino.

[20] As discussed later, the presence of such antiscatter baffles themselves could be a source of problems.

[21] "The penalty paid for having an unknown shape correction is that its interdependence with $|U_{e2}|^2$ raises the error in that parameter (Hetherington et al. 1987, p. 1508)."

[22] The group reported a value for the endpoint energy E_o = 66.946 ± 0.020 keV, in disagreement with the accepted value of 65.92 ± 0.15 keV.

[23] The papers were received at the *Physical Review* on September 9, 1988 and the results were, no doubt, known to those working in the field well before publication.

[24] Such a detector is normally run at liquid-nitrogen temperature (-196°C).

[25] "...the goodness of fit is not a strong function of the screening potential used. However, it is important to emphasize that even when zero screening is used the excess of counts at low energy is not completely removed (p. 1846)." It was still consistent with a mixing probability of 0.5 percent.

[26] Recall the earlier discussion of the increased support by "different" experiments.

[27] Recall the earlier discussion of the shape factor needed in the experiment of (Hime and Simpson 1989).

[28] Although the work on Simpson and Hime may have encouraged the new work, it certainly did not initiate it. During the 1970s, Law, in collaboration with Campbell, published three papers on atomic physics corrections to nuclear beta decay. Law and Weisnagel were colleagues of Simpson at the University of Guelph and acknowledged discussions with Simpson and Hime.

[29] In general, unless specific references are made to private communications of to conference presentations, I shall use the published versions of the papers. The published versions usually have more details and are also when the physics community, rather than the group of specialists, becomes aware of the results. There are times. however, when I shall use these less formal presentations.

[30] Glashow was not an uncritical theorist who accepted experimental results merely because experimenters presented them. In an earlier episode, that of the Fifth Force, a modification of Newton's law of gravity, Glashow rejected both the speculation and the evidence it was based on. "Unconvincing and unconfirmed kaon data, a reanalysis of the Eötvös experiment depending of the contents of the Baron's wine cellar [an allusion to the importance of local mass inhomogeneities in the analysis], and a two-standard deviation geophysical anomaly! Fischbach and his friends offer a silk purse made out of three sows' ears and I'll not buy it (Quoted in (Schwarzschild 1986, p. 20)."

[31] Schwarzschild's comments appeared in *Physics Today* a semi-popular magazine that is distributed to all members of the American Physical Society.

[32] I include the details here because they will be important later in the story.

[33] "Improvements" may not make the experiment better, as we shall see later. For a case in which technological improvements to an apparatus precluded the replication of what ultimately were very important results see Franklin (1986, Ch. 2).

[34]. Although there are several references to the Berkeley results being presented at the Bratislava conference, no paper appears in the published proceedings.

[35] I note here that the Berkeley group also used a fitted "shape factor," something that had been criticized in the magnetic spectrometer experiments (see earlier discussion).

[36] A second summary appeared in Morrison (1992).

[37] Bonvicini (1993) also considered tritium experiments to be too limited statistically.

[38] The effect seen by Simpson was quite sensitive to the energy interval chosen. In general, an experimental result should be robust against such changes. Recall also the earlier comments of Hetherington and others concerning the danger of mistaking a statistical fluctuation for a physical effect.

[39] These results were published in Bahran and Kalbfleisch (1992).

[40] Using a gas source avoids problems associated with embedding the tritium in a crystal, but still requires atomic physics corrections.

[41] Bahran and Kalbfleisch note that a 1% anomaly had been seen in the tritium spectrum at approximately 1 keV, the kink energy for a 17-keV neutrino, in 1959 (Conway and Johnston 1959). This effect was attributed to a possible non-linearity in the energy response of their proportional chamber at low energies. They also noted that the experimental result of Hime and Jelley did not include the effect of radiative corrections to the spectrum. The results Simpson included in the published version of his talk included such corrections. The issue of tritium results is still unresolved.

[42] One of the difficulties of using papers in published conference reports is that they contain modifications made well after the conference. Thus, Simpson could respond to what Morrison had said. As seen below, Morrison responded to Simpson's criticism at a subsequent conference.

[43] Recall, however, that Ohi's experiment did not use such a device, but rather a solid-state detector.

[44] The major difference between the CERN report and the published paper is a detailed discussion of the negative Tokyo experiment (Oshima and others). This experiment is discussed in detail below. Quotations are from the 1993 published paper.

[45] Bonvicini ignored experiments on tritium on the grounds that the Coulomb correction factor in such experiments is quite large for low energy electrons (where the kink due to the 17-keV

neutrino would be seen) and difficult to calculate precisely. He also suggested for future work that experiments on tritium be avoided.

[46] The electron response function was measured at a single energy. Bonvicini suggested that it should be measured at several energies spanning the fitted energy spectrum.

[47] This last point appears only in the published paper (1993). The 1992 CERN report contains only the previous statement about the results with two different nuclei.

[48] I am grateful to an anonymous referee for emphasizing the importance of Bonvicini's work.

[49] They didn't have access to the raw data.

[50] Although the article was not published until 20 May 1992, it was received at *Modern Physics Letters on* 13 March 1992.

[51] Hetherington *et al.* (1987), Bonvicini, and Morrison also agreed on this point.

[52] The Berkeley group had earlier reported a result of $M_2 = 21 \pm 2$ keV with a mixing probability of $(0.85 \pm 0.45)\%$ for ^{55}Fe.

[53] High statistics avoids the problem of statistical fluctuations affecting the results. Recall Morrison's earlier discussion of Simpson's reanalysis of Ohi's data.

[54] Two of these results were also presented at the conference, and one had been communicated to the Berkeley group by preprint.

[55] Two of these experiments, those of Ohshima *et al.* and of Mortara *et al.*, will be discussed in detail below. The Caltech result, a preprint, has not appeared in the published literature.

[56] The published paper appeared in early August 1992, but had been received at the journal on 16 April 1992. The conference paper, presented in early August 1992, set a more stringent limit on the presence of the 17-keV neutrino. A more detailed account of the experiment appeared in (Ohshima and *et al.* 1993).

[57] The published value in (Kawakami *et al.* 1992) was $(0.018 \pm 0.033 \pm 0.033)\%$, with an upper limit of 0.095%. $|U|^2$ is the mixing probability.

[58] The Monte Carlo calculation included the best data then available, and, as we shall see below, was checked against an experimental result, independent of the beta-decay experiment. Pickering (1984) has argued that one can always question such Monte Carlo results. I shall discuss this point in detail later.

[59] Hime attributed the small peak at the high energy end to electron ionization of the silicon K-shell with the subsequent escape of silicon K x-rays. It does not cast any doubt on the confirmation of the model.

[60] The probability of a 7σ effect is 2.6×10^{-10}%.

[61] Collins and Pickering have, for example, argued that factors such as career interests, consistency with existing community commitments, recycling of expertise, and utility for future practice enter into such decisions. I believe the history shows no evidence of this, although they certainly enter into the question of pursuit, discussed below.

[62] Recall the effort that Simpson devoted to calibrating and checking his apparatus.

[63] Rasmussen (1993) has argued that these strategies are open to negotiation and dispute. That is certainly true in principle, but I do not agree that it happens in practice (Rasmussen 1993). I have previously presented case studies in which these strategies are explicitly used. See, for example, Franklin (1986, Ch. 7).

[64] In the case of the Fifth Force, a proposed modification of the law of gravity, the positive results reported by experiments measuring gravity on towers and in mineshafts were shown to have neglected the effects of local terrain. For details see Franklin (1993a).

[65] Not all experiments are equal. Some experiments are more equal than others. (With apologies to George Orwell). For details of this discussion see (Ackermann 1991; Franklin 1990; Franklin 1993b; Lynch 1991; Pickering 1991).

[66] In the case of the Fifth Force, no error has been found in Thieberger's positive result. There have, however, been numerous other experiments which have given negative results. The overwhelming weight of these negative results has persuaded the physics community, as well as Thieberger himself, that the original result is wrong.

[67] For a discussion of the support provided by the "same" and "different" experiments see (Franklin and Howson 1984).

[68] This is the question of pursuit, the further investigation of a phenomenon or of a theory, rather than justification, the process by which a result or theory becomes accepted as scientific knowledge. These are not always easy to separate, but it is quite clear from the history that a decision on the 17-keV neutrino had not yet been reached at this time. For further discussion see below and Franklin (1993b).

[69] New evidence may make an explanation more plausible.

[70] I will discuss the question of experimentally checking a Monte Carlo calculation later.

[71] In addition, Morrison showed that Simpson's most persuasive reanalysis of one of the early negative results was dependent on a statistical fluctuation. Hetherington *et al.* (1987) had also suggested that this might be a problem.

[72] These arguments provided good grounds for the belief that the 17-keV neutrino did not exist, but did not, of course, guarantee it.

[73] This was Steven Weinberg's description (private communication).

[74] Observations of galactic rotations also pointed to "dark matter" and missing mass.

[75] Glashow's positive comment on the 17-keV neutrino came only after he had learned of the Berkeley and Oxford results, although he did cite the 1989 Simpson-Hime results.

[76] This is an example of "different" experiments providing more support for a hypothesis than do repetitions of the "same" experiment. For a general discussion of this see (Franklin and Howson 1984). Recall, however, that the 1989 Simpson-Hime experiments did include significant differences from the original Simpson experiment. These included the use of a germanium, rather than a Si(Li), detector in the tritium experiment, and the use of a ^{35}S source with a Si(Li) detector in the other experiment. Simpson was certainly aware of possible problems in his first experiment, and also aware of the fact that different experiments would provide more support for his conclusion. See the earlier discussion of these 1989 experiments.

[77] In addition, the input parameters to the Monte Carlo calculations are the best and most reliable ones that the experimenters can find.

[78] A survey of papers and reprints received at the Stanford Linear Accelerator Center, a major research facility, for the period 1985-1992 shows approximately sixty theoretical papers on the 17-keV neutrino.

CHAPTER 3

INSTRUMENTAL LOYALTY AND THE RECYCLING OF EXPERTISE

In an ideal world, when experimental physicists are considering which experiment to do next they would ask themselves the question, "What is the best physics experiment that can be done?" In the real world the question often asked is, "What is the best physics experiment that I can do with an already existing apparatus, or with a minor modification of that apparatus?" This is not to say that physicists never construct a new apparatus to perform an experiment, they often do.[1] Rather, it is the recognition that physicists have areas of knowledge and of expertise and that there is a certain cognitive and economic efficiency in using an already existing apparatus. In such a case the experimenters already have substantial knowledge about how the apparatus works and what the experimental difficulties might be. They will be aware of backgrounds that might mask or mimic the effect to be measured, and of possible systematic effects. They will also know the theory of the phenomena involved in the experiments. One might call this the recycling of expertise. The economic efficiency is obvious. Instrumental loyalty is less expensive. It is usually cheaper to use or modify an existing apparatus than to build a new one.[2] This is another aspect of the context of pursuit, the further investigation of a hypothesis or an experimental result (Franklin 1993b). In the episode discussed below the use of similar experimental apparatuses allowed the further investigation of a general subject area.[3]

In this paper I will examine the history of five experiments performed at the Princeton-Pennsylvania Accelerator (PPA) by the Mann-O'Neill collaboration[4]. The experiments were conducted over a period of four years and were designed to measure various aspects of K^+ meson decay. Each of these experiments was done with the same basic apparatus, with modifications for each of the specific measurements. We will see the increasing expertise of the experimenters as the experiments progressed. The later measurements were technically more difficult and built upon the acquired knowledge of how the experimental apparatus worked. In addition, we shall see that the data analysis in such experiments may last as long as, or even longer, than the data acquisition. The analysis of the earlier experiments did, in fact, benefit from some of the later work.[5] Conversely, some of the later work resulted from questions raised in the earlier experiments. The size of the group allowed a division of labor which allowed several experimenters to work on several of the experiments simultaneously. This is one way in which expertise is shared among experiments.[6]

The sequence of experiments also illustrates a style of doing physics, a style that is closely linked to the experimental apparatus. As discussed below, and illustrated by this episode, the experimental apparatus determines what quantities can be measured and influences both the analysis procedures and what type of results can be obtained.[7] Peter Galison (1987, p. 248; 1997) has noted the existence of two traditions in experimental high-energy physics; the visual and the electronic.[8] The visual tradition, illustrated recently by bubble chambers, usually has no selectivity. The bubble chamber accepts and detects all of the events produced. The electronic, logic tradition, illustrated by counters and spark chambers, is characterized by selectivity. By using various types of counters in combination to trigger the spark chambers one can select the type of event one wishes to observe.[9] Thus, in four of the five experiments discussed below, the experimenters used a Cerenkov counter to select only those K^+ meson decays that involved a positron, and these experiments are typical of the electronic tradition. The first data run used spark chambers in much the same manner as a bubble chamber, accepting all K^+ decays by triggering on any charged decay particle.[10]

The visual tradition, in which many details of an event are oberved and measured, often emphasizes singular, or "golden," events, those which clearly and unambiguously show the existence of a particular, usually rare, process.[11] The electronic tradition, particularly in counter experiments, emphasized the total number of certain types of event, with relatively little information about individual events. These general characterizations are neither rigid nor exclusive. During the 1970s, for example, bubble chamber experiments discovered the existence of new elementary particles by observing an excess in the number of events above that expected on phase-space considerations. Conversely, as discussed below, a spark chamber and counter experiment detected relatively rare K^+_{e2} events.

In chronological order of data taking the experiments performed by the Mann-O'Neill collaboration were: 1) measurement of the $K^+_{\mu 2}$, $K^+_{\mu 3}$, $K^+_{\pi 2}$, and K^+_{e3} branching ratios;[12] 2) measurement of the K^+_{e3} branching ratio and spectrum, using a slightly different apparatus; 3) measurement of the K^+_{e2} branching ratio; and 4) measurement of the form factor in K^+_{e3} decay; 5) measurement of the K^+_{e3} branching ratio and spectrum (A third measurement, using yet another variant of the experimental apparatus).

The motivation for the first experiment was the generally unsatisfactory situation, at that time, with respect to the measurements of the K^+ branching ratios.

> All present values of the K^+ branching ratios have been obtained from emulsion and bubble-chamber experiments. The emulsion experiments were pioneer ones, generally with poor statistics, and the separation of modes involving particles with momentum greater than 170 MeV/c was usually done with ionization and multiple-scattering measurements, which are sometimes ambiguous in their application to particles in that mass and momentum region. The bubble-chamber experiments are generally satisfactory statistically but the particle-detection efficiency is mode and momentum dependent and hence relatively elaborate event-weighting procedures have been required. In most instances, corrections due to the overlapping of products from the different decay modes and to the presence of undetected K^+ and π^+ decays in flight are not negligible and require a sophisticated analysis to obtain a meaningful result. A summary of the experimental situation indicates that certain of the

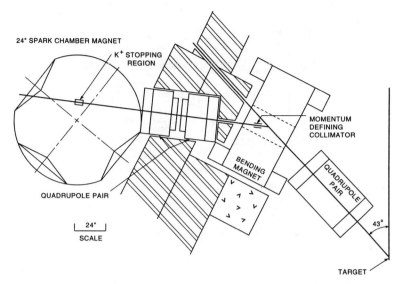

Figure 1. The beam transport system for all of the K^+ decay experiments. From Bowen et al. (1967).

branching-ratio values, particularly for $K^+_{\mu 3}$ and $K^+_{\pi 2}$, are in disagreement by two or more standard deviations, presumably indicating the presence of systematic effects which have not been accounted for. (Auerbach et al. 1967, p. 1506).

I. THE BASIC EXPERIMENT

A. THE BEAMLINE

An ample supply of stopped K^+ mesons was needed to perform these experiments. This was obtained using the beamline shown in Figure 1. A positive unseparated beam was secured from the Princeton-Pennsylvania Accelerator. The internal proton beam struck a platinum target and positive secondary particles emitted at a mean angle of approximately 46° were brought to a focus just beyond the bending magnet by the first quadrupole pair. The second quadrupole pair then focussed this image on the K^+ detection system shown in Figure 2. The beam was bent in the off-axis quadrupoles to discriminate against fast proton backgrounds and to provide control of the beam position.[13] The central momentum of the beam was 525 MeV/c with a 7% momentum spread (full width half maximum).

The positive beam consisted primarily of protons and pions, with a 2/1 ratio, and also contained small numbers of kaons, muons, and positrons. The pion to kaon ratio was about 300/1. The kaons needed were separated from the more numerous protons and pions by range in matter and by time of flight. The counter system used to identify stopped K^+ mesons consisted of scintillation counters C_1, C_2, C_3, and C_4. Sufficient

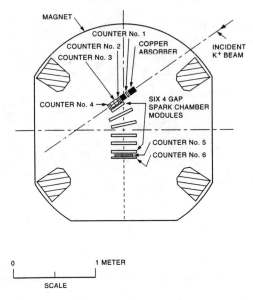

Figure 2. Details of the experimental apparatus for the K^+ decay experiments including the K^+ stopping telescope and the momentum chambers.

copper (7 cm) was placed in the beam telescope to stop the 525 MeV/c kaons in counter C_3, the stopping region. The copper placed in front of counter C_1 shielded it from beam protons and the total amount of copper in the beam removed virtually all of the protons in the beam.[14] Counter C_4 was used to veto pions, which have a longer range in matter than kaons of the same momentum, and therefore pass through the stopping region, C_3, and count in C_4. This veto eliminated approximately 75% of the pions. The signal ($C_1C_2C_3\overline{C}_4 = A_1$) ($C_1C_2C_3$ means counters C_1 and C_2 and C_3 fired in coincidence. $\overline{C}_4 =$ anticoincidence of C_4) was used to identify a stopped particle, which might be either a pion or a kaon.

Further identification of stopped kaons was provided by time of flight. The internal proton beam of the PPA had a bunched structure with bunches 1.5 ns wide separated by 34 ns. Thus, secondary particles were produced every 34 ns. A signal from the RF system of the accelerator was used to measure the time of flight of beam particles over the full length of the beam.[15] This was a momentum-selected beam so that pions and kaons, which have different masses, will have different speeds and therefore different times of flight. The time separation between pions and kaons travelling down the beam was 10 ns. A 4 ns wide coincidence between a beam counter and the RF oscillator of the synchrotron provided adequate discrimination between pions and kaons. A stopped kaon was indicated by the coincidence ($C_1C_2C_3\overline{C}_4$ + RF = A_2). The distribution of beam particles reaching the stopping region in shown in Figure 3. In the first two experiments an average of 200 stopped, identified kaons per second was obtained.

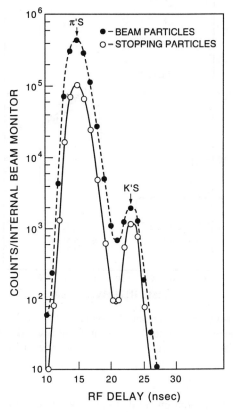

Figure 3. Time distribution of particles reaching the K^+ stopping region. From Bowen *et al.* (1967).

B. THE MOMENTUM CHAMBERS

Decay particles which left the stopping region at about 90° traversed a set of six thin-plate optical spark chambers located in a magnetic field. (The value of the magnetic field varied from 2–7.5 kilogauss in the different experiments). Decay particles were detected by a coincidence telescope C_5C_6 (Figure 2). To restrict the particles detected to kaon decays a 21 ns gate, triggered by a stopped kaon signal was used. (The mean life of the K^+ meson is 12.4 ns). This avoided any effects from other synchrotron bunches. If the decay particles were indeed due to kaon decay then the time between the stopping kaon signal and the decay signal should match the kaon lifetime. This is clearly seen in Figure 4. Pulses from counters C_3 and C_5 were displayed on an oscilloscope trace and photographed, so that the time between them, which measured the time interval between the K^+ stop and its decay, could be measured for each event. As seen in Figure 4 there is a small prompt peak at short decay times due to kaon decays in flight. These were

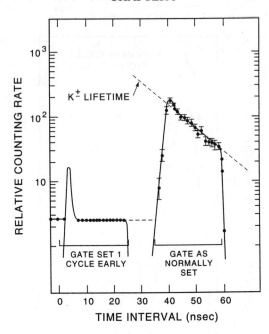

Figure 4. Time interval between the stopping K^+ signal and the decay particle signal, the K^+ decay time. A small prompt peak due to K^+ decays in flight is clearly seen. The signal for the gate set one cycle early shows accidental coincidences. From Auerbach et al. (1967).

completely eliminated by a requirement that the decay time of the event be greater than 2.5 ns after the stop of the K^+ meson.

This was the basic experimental apparatus. In each of the experiments the basic apparatus was modified. The momentum chambers were followed by either a range chamber to measure the range of the decay particles and to provide particle identification, or by a gas Cerenkov counter to identify decay positrons, or by both.[16]

II. THE K^+_{e3} BRANCHING RATIO AND MOMENTUM SPECTRUM

The first experimental result reported by the group used data obtained in the second data-taking run (Cester et al. 1966). This experiment was designed to measure both the K^+_{e3} branching ratio and the positron momentum spectrum for this decay. The latter measurement would set limits on the amount of scalar or tensor interaction present in the decay as compared to the dominant vector interaction.

The experimental apparatus for this experiment is shown in Figure 5. The major addition to the basic apparatus was a large aperture Cerenkov counter sensitive to positrons which was placed behind the momentum chambers. The spark chambers were pulsed by a stopped K^+ meson (A_2 signal) followed by a decay positron signal (C_5C_6 + Cerenkov counter, Co3 trigger). Other data were taken with the Cerenkov counter removed from the logic (Co2 trigger). The Co2 events were sensitive to all charged

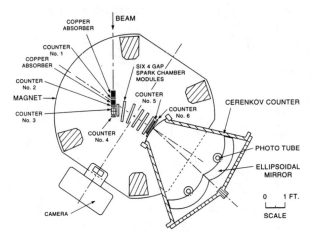

Figure 5. Experimental arrangement for the experiment to measure the K^+_{e3} branching ratio. From Cester et al. (1966).

decay modes of the K^+ meson, in particular the $K^+_{\mu 2}$ and $K^+_{\pi 2}$ decay modes. The charged particles from $K^+_{\mu 2}$ and $K^+_{\pi 2}$ decays each have a unique decay momentum, 236 MeV/c and 205 MeV/c, respectively. This was used to determine both the momentum calibration and the momentum resolution of the system. (See, for example, the $K^+_{\mu 2}$ peak in Figure 12). The ability of the apparatus to detect these known decays also gave good reason to believe that the apparatus was operating properly. As discussed below, this epistemological strategy was used in each of the experiments. After corrections for energy loss by the decay particles in the stopping region the centers of the momentum distribution for these decays agreed with the known momenta of these decays and the width of the distributions determined the momentum resolution of the chambers, which was found to vary from 2 to 4.5%, depending on the value of the magnetic field.

In order to cover the momentum spectrum of K^+_{e3} decay from 60 MeV/c to its endpoint, 228 MeV/c, and to achieve a momentum-independent efficiency, data were taken at three different magnetic fields, 2.1, 3.3, and 5.2 kilogauss, respectively.[17] To normalize the data taken at the different magnetic field settings and to determine the branching ratio, film from "Co2 trigger" runs was used obtain the relative rate of ($K^+_{\mu 2}$ + $K^+_{\pi 2}$) decays.[18] Film from the "Co3 trigger" runs provided the relative positron rate. For each run the number of K^+ decays (Co2 triggers) was recorded and all other rates were measured relative to the Co2 scaler reading. The same event acceptance criteria for both scanning and analysis were applied to $K^+_{\mu 2}$, $K^+_{\pi 2}$, and K^+_{e3} events.

A. POSITRON MOMENTUM DISTRIBUTION

The raw positron momentum spectrum was contaminated by two major backgrounds: (1) Events due to accidental coincidences between a charged particle decay and a noise

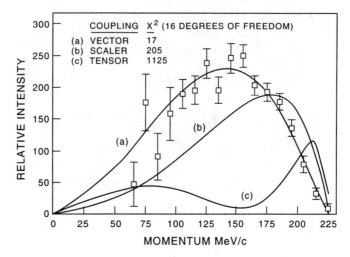

Figure 6. Corrected positron momentum spectrum above 80 MeV/c. The fitted distributions for pure vector (with radiative corrections), scalar, and tensor couplings are shown. From Cester *et al.* (1966).

pulse in the Cerenkov counter. (The efficiency of such detection was 0.7%). "This background was eliminated by subtracting the appropriate Co2 spectrum, normalized to the $K^+_{\mu 2}$ peak in the Co3 spectrum, from the Co3 spectrum" (Cester *et al.* 1966, p. 344).; (2) Positrons from Dalitz pairs and γ-ray conversion in the stopping region.[19] This background was estimated in a Monte Carlo simulation and subtracted from the observed momentum spectrum. The effect was less than 10% of the rate at any point. The data were also corrected for bremsstrahlung in the stopping region.[20]

The data taken at the three different settings of the magnetic field were combined into a single spectrum. The relative weights of the spectrum at each field setting were determined in two different ways: (1) Comparing the numbers of events in the momentum interval 150–228 MeV/c where the efficiency was both momentum and field independent; (2) Comparing the number of $(K^+_{\mu 2} + K^+_{\pi 2})$ events in the Co2 trigger spectrum obtained at each field setting. The two methods agreed. The data are summarized in Table I.

The final momentum spectrum is shown in Figure 6. The best fits to the spectra for vector, scalar, and tensor (V, S, T) interactions, assuming constant form factors in the decay (an assumption later tested in the fourth experiment in the series) are also shown.[21] The vector interaction is clearly dominant. At the 90% confidence level admixtures larger than 4% tensor and 18% scalar were rejected.

B. THE K^+_{e3} BRANCHING RATIO

The branching ratio R of K^+_{e3} to $(K^+_{\mu 2} + K^+_{\pi 2})$ is given by

Table I: Summary of data from Cester et al. (1966)

B field	p_{min} (MeV/c)	Co3 trigger # of K^+_{e3} events above p_{min}	Spectrum fraction A	Co3 counts	Co2 trigger ($K^+_{\mu 2}$ + $K^+_{\pi 2}$)	Co2 counts	Relative Ω_2/Ω_3	Weights (1)	Weights (2)
2.1	60	377	0.925	40760	289	1074	1.656	0.140	0.138
3.3	100	769	0.731	94393	1264	3852	1.644	0.361	0.375
5.2	150	533	0.357	115461	878	2679	1.655	0.498	0.487

Co3 trigger = Stopped K^+ meson + decay positron signal.
Co2 trigger = Stopped K^+ meson + charged decay particle (no positron signal required).
Ω_2/Ω_3 = Ratio of average solid angles for ($K^+_{\mu 2}$ + $K^+_{\pi 2}$) decays and for K^+_{e3} decay.
Relative Weights = Percentage of data taken at that magnetic field setting.

$$R = (K^+_{e3} \text{ events})/[(K^+_{\mu 2} + K^+_{\pi 2}) \text{ events}] \times \Omega_2/\Omega_3 \times 1/C_{EFF} \times 1/A,$$

where Ω_3 and Ω_2 are the average solid angles of acceptance for K^+_{e3} decays and for ($K^+_{\mu 2}$ + $K^+_{\pi 2}$) decays, respectively, C_{EFF} is the efficiency of the Cerenkov counter, and A is the fraction of the positron momentum spectrum actually measured.[22] The efficiency of the Cerenkov counter, C_{EFF}, was measured independently and found to be (97 ± 2)%. R was found to be 0.0589 ± 0.0016. Using the known $K^+_{\mu 2}$ and $K^+_{\pi 2}$ branching ratios of 0.632 ± 0.006 and 0.213 ± 0.006,[23] respectively, the group obtained a value of 0.0498 ± 0.002 for the branching ratio $R(K^+_{e3}/K^+)$, "in good agreement with the average of four previous experiments which yield $R(K^+_{e3}/K^+)$ = 0.049 ± 0.003."[24]

III. MEASUREMENT OF THE BRANCHING RATIOS $K^+_{\mu 2}$, $K^+_{\pi 2}$, $K^+_{\mu 3}$ AND K^+_{e3}

This experiment, actually the third result published by the group (Auerbach *et al.* 1967), was the first to take data. It shows several advantages of performing a sequence of experiments with essentially the same apparatus. The previous measurement of the K^+_{e3} branching ratio provided a check on the result for that same quantity measured in this experiment. In addition, results obtained during the subsequent experiment on the K^+_{e2} branching ratio, discussed in detail in the next section, provided important information used in the analysis of this experiment.

106 CHAPTER 3

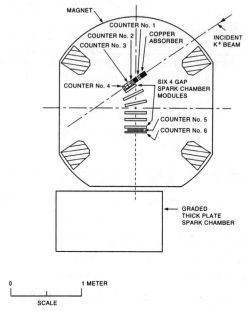

Figure 7. Experimental apparatus for the measurement of the $K^+_{\mu2}$, $K^+_{\pi2}$, K^+_{e3}, and $K^+_{\mu3}$ branching ratios. From Auerbach et al. (1967).

The experimental apparatus is shown in Fig. 7. In this experiment the momentum chambers were followed by a range chamber with plates of graded thickness. The plate thicknesses were chosen so that a pion and a muon with the same momentum would differ in range by about five gaps. Because this experiment was designed to measure the branching ratios of different decay modes with differing momenta the experimenters carefully checked that both the scanning efficiency for events as well as the selection criteria were independent of both decay mode and of momentum.[25]

A. SELECTION OF EVENTS FOR THE DIFFERENT DECAY MODES

Events associated with the various decay modes were selected by both momentum and by range in matter. A scatter plot of momentum versus range for accepted events is shown in Fig. 8. The four decay modes $K^+_{\mu2}$, $K^+_{\pi2}$, $K^+_{\mu3}$ and K^+_{e3} are clearly visible. Muons lose energy solely by ionization and for a given momentum have a well-defined range, with a small straggle. The muons from $K^+_{\mu2}$ decay have a unique momentum of 236 MeV/c and a range of 67 g/cm^2. The dense patch due to this decay mode is clearly visible at the upper right. The dense patch at about 200 MeV/c is due to pions from $K^+_{\pi2}$ decay, which have a unique momentum of 205 MeV/c and a range of 37 g/cm^2. Pions also interact strongly with nuclei and may have a shorter range. The vertical line below the dense patch is due to such interactions. Pions may also decay in flight into muons which, depending on their momentum, will have either a longer or shorter range than the pions, accounting for some of the events above and below the patch. The

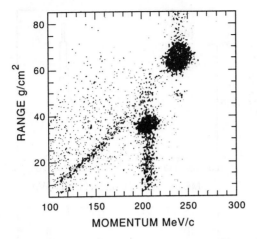

Figure 8. Scatter plot of momentum versus range for all measured events with decay times greater than 2.5 ns and which passed the spatial selection criteria. From Auerbach *et al.* (1967).

diagonal line from lower left to upper right is due to muons from $K^+_{\mu 3}$ decay. Such muons have a continuous momentum spectrum with an endpoint of 215 MeV/c and thus produce a line in the scatter plot. Positrons from K^+_{e3} decay (endpoint 228 MeV/c) account for the remaining events. Positrons do not have a well-defined range because, in addition to ionization loss, they lose energy by several different processes, including bremsstrahlung, which result in large energy losses (See range of positrons shown in Figure 14). The projection of the scatter plot for both momentum and range is shown in Fig. 9. The peaks expected for $K^+_{\mu 2}$ and $K^+_{\pi 2}$ decays are clearly visible.

Muons from $K^+_{\mu 2}$ decay have a momentum of 236 MeV/c.[26] The experimenters selected those events in the momentum range 220–260 MeV/c as their sample of this decay. Given their momentum resolution of 1.8%, obtained from an analysis of the $K^+_{\mu 2}$ peak, the number of $K^+_{\mu 2}$ events outside of this region, as well as the number of $K^+_{\pi 2}$, $K^+_{\mu 3}$ and K^+_{e3} events inside the region, was negligible in comparison to the total number of $K^+_{\mu 2}$ events. The final number of $K^+_{\mu 2}$ decays was 13,843 ± 132 (Table II). The number of $K^+_{\pi 2}$ events was obtained by selecting all of the events in the momentum interval 190–220 MeV/c (Recall that the momentum for this decay is 205 MeV/c). Contributions from $K^+_{\mu\nu\gamma}$, $K^+_{\mu 3}$, and K^+_{e3} were subtracted to obtain the final total of 4501 ± 80 events.

Table II: Summary of K^+ Branching Ratios from Auerbach *et al.* (1967)

Mode	Number of Events	Branching Ratio
$K^+_{\mu 2}$	13,843 ± 132	0.6344 ± 0.0044
$K^+_{\pi 2}$	4501 ± 80	0.2059 ± 0.0040
$K^+_{\mu 3}$	834 ± 64	0.0382 ± 0.0029
K^+_{e3}	1086 ± 35	0.0497 ± 0.0016

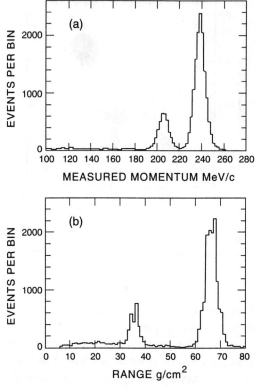

Figure 9. (a) Momentum distribution and (b) range distribution of all events in the data sample. From Auerbach *et al.* (1967).

The separation of the decay modes K^+_{e3} and $K^+_{\mu3}$ was more difficult. Unlike the previous experiment in which positrons from K^+_{e3} decay were identified by the Cerenkov counter, in this experiment the two decay modes had to be separated by range. As seen in Fig. 8, although muons from $K^+_{\mu3}$ form a line in the range-momentum scatter plot, positrons from K^+_{e3} decay cover an area which includes this line. Plots of the range distribution of particles in the momentum ranges 130–140 MeV/c and 140–150 MeV/c are shown in Fig. 10. Such graphs were obtained for each 10 MeV/c interval in the region from 130–190 MeV/c. This was the K^+_{e3} region. It was selected to avoid contamination from $K^+_{\pi2}$ decays at 205 MeV/c. A clear well-defined peak due to muons is visible. For each momentum interval the experimenters selected a range interval that safely included all the muons. Events outside this region were positrons from K^+_{e3} decay, with a small contamination from $K^+_{\pi2}$ decay, which was subtracted from the total.

To get the total number of positrons from K^+_{e3} decay one needed an estimate of the number of positrons in the muon region of the graphs. This was done using range measurements for positrons in the momentum regions of interest obtained in the K^+_{e2} experiment performed later, which included both the Cerenkov counter placed after the

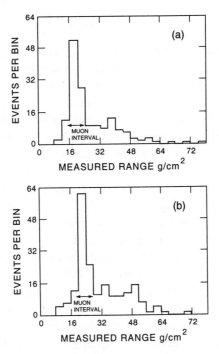

Figure 10. The range distribution of particles in the momentum interval (a) 130–140 MeV/c (b) 140–150 MeV/c. From Auerbach et al. (1967).

momentum chambers, followed by the range chamber. Because the Cerenkov counter positively identified the positrons this data provided an accurate range spectrum for positrons. Using this later measurement, the experimenters were able to calculate the number of positrons in the muon region of the graphs.[27] The length of time required for the analysis of this experiment worked to the experimenters' advantage. They were able to make use of the range measurements obtained later. Assuming that the K^+_{e3} spectrum was that given by the vector interaction with constant form factor, an assumption consistent with the results of their earlier measurement, the experimenters obtained a total of 1103 ± 74 K^+_{e3} events.

The number of $K^+_{\mu 3}$ decays in the momentum interval 130–190 MeV/c was obtained using the muon range spectrum for the 10 MeV/c momentum intervals discussed earlier. In this case one had to subtract the number of positrons in that region, along with small corrections for radiative decay and for pion decay in flight. The same procedure that calculated the number of positrons that should be added to the positron total from the region was also used to subtract the number of positrons from the muon total. There were several sources of background that affected all of the decay modes. These included K^+ decays in flight and backgrounds due to other beam particles. Several of these gave prompt events that were eliminated by a 2.5 ns decay time requirement. The others were calculated to be negligible.

B. THE BRANCHING RATIOS

Although the experimenters had obtained relatively clean samples of events from each of the decay modes several problems remained before the branching ratios could be calculated. Because the experimenters used only the momentum region 130–190 MeV/c for $K^+_{\mu 3}$ decay, as well as for K^+_{e3} decay, they needed to calculate the number of events they would have observed had they used the full momentum spectrum for each of the decays. Unlike the case of the positron spectrum from K^+_{e3} decay in which one could legitimately use the spectrum for the vector interaction with constant form factor to calculate the number of missing events (see earlier discussion), the spectrum for $K^+_{\mu 3}$ decay depended on two unknown form factors f_+ and f_-, even if one assumed the decay interaction was vector. Assuming $f_\pm(q^2) = f_\pm(0)$, where q^2 was the momentum transfer, the spectrum still depended on the parameter $\xi = f_-/f_+$. The data from the muon spectrum obtained in this experiment was insufficient to fix ξ, and data from muon polarization experiments was used to fix ξ, and to allow the calculation of the spectrum.

The final branching ratios were obtained as follows. Let $N_0 = N_{\mu 2} + N_{\mu 3} + N_{\pi 2} + N_{e3}$, where the N_j's are the corrected number of events for each decay mode obtained in the experiment.

$N_{TOTAL} = N_0 / [1 - (\tau + \tau')]$ where $\tau + \tau' =$ the branching ratios for $K^+ \to \pi^+\pi^+\pi^-$ and $K^+ \to \pi^+\pi^0\pi^0$, respectively. These two branching ratios had been measured in other experiments and were regarded as well-determined. The branching ratio for the jth decay mode is then given by $K^+_j = N_j/N_{TOTAL}$ The final results are shown in Table II. The previous experiment on the K^+_{e3} branching ratio was used as a check on the newly obtained value. The two results agreed, within statistics, and a weighted average of the two determinations was used to fix a final value for the K^+_{e3} branching ratio.

This experiment did not achieve its intent of resolving the questions concerning the branching ratios of the K^+ meson. There was still considerable uncertainty regarding the branching ratios. Recall the new and very different value obtained for the K^+_{e3} branching ratio obtained by Callahan and others. The experiment did, however, provide the most precise measurements of those quantities that had been made up to that time.

IV. MEASUREMENT OF THE K^+_{e2} BRANCHING RATIO[28]

The third experiment performed by the Mann-O'Neill collaboration was a measurement of the K^+_{e2} branching ratio, the fraction of all K^+ mesons that decayed into a positron and an electron neutrino ($K^+ \to e^+ + \nu_e$) (Bowen *et al.* 1967). This was the most technically demanding of the experiments and, as we shall see, made use of the expertise acquired in the first two experimental runs and the analysis of that data. In these earlier experiments the branching ratios measured had been of the order of several percent or higher. In this experiment the expected branching ratio was 1.6×10^{-5}, a factor of 1000 smaller.

The motivation for this measurement was that it would be a stringent test, in strangeness-changing decays, of the then generally accepted V-A theory of weak

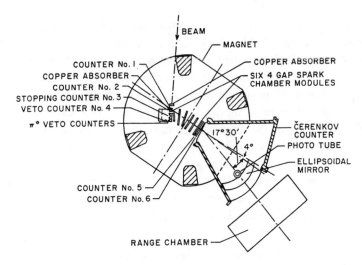

Figure 11. Experimental apparatus for the K^+_{e2} branching ratio experiment. From Bowen et al. (1967).

interactions.[29] Pure axial-vector coupling (A) predicted a ratio for $K^+_{e2}/K^+_{\mu 2}$ decays of 2.6×10^{-5}, corresponding to a branching ratio of 1.6×10^{-5}.[30] Pure pseudoscalar (P) coupling predicted a $K^+_{e2}/K^+_{\mu 2}$ ratio of 1.02. Even a coupling constant ratio, f_P/f_A, of 10^{-3} would increase the expected branching ratio by a factor of four. Thus, even a rough measurement of the K^+_{e2} branching ratio would be a stringent test for the presence of any pseudoscalar interaction in the decay, and of the V-A theory in general. The best previous measurement of the $K^+_{e2}/K^+_{\mu 2}$ ratio had set an upper limit of 2.6×10^{-3}, a factor of 100 larger than that predicted by V-A theory.

A. THE EXPERIMENTAL APPARATUS

The experiment used essentially the same beamline as had the previous two experiments, but with three major changes: (1) the momentum bite was increased from 7% to 13% (Full width half-maximum), about the 530 MeV/c central momentum; (2) the beam length was shortened from 8.8m to 7.2m; and (3) off-axis bending in the quadrupoles was eliminated. The new beam yielded an average of 500 stopped, identified K^+ mesons per second, averaged over several months, a considerable increase from the first two experiments in which the average rate was 200 per second.[31]

The apparatus for this experiment is shown in Figure 11. This apparatus incorporated both the Cerenkov counter, to identify positrons, as well as the range chamber to help eliminate background from other decay modes. Only half the Cerenkov counter was needed for this experiment and it had a measured efficiency of greater than 99% for positrons passing through the center of the front and rear windows of the counter. The efficiency fell to 95% for positrons passing through the extreme outer edge of the counter. The counter had a measured efficiency of 0.38% for other particles of comparable momentum.[32] The thick-plate range chamber was placed behind the Cerenkov counter which permitted measurement of the position of particles emerging from the counter as well as a measurement of their total range. The stopping K^+ mesons were identified, as they had been in the previous measurements, by range and by time of flight. The only additions to the beam telescope were two lead-scintillator counters placed above and below the stopping region, counter C_3. These counters, which were sensitive to γ-rays, were used to suppress events accompanied by a π^0 meson, e.g. $K^+ \to e^+ + \pi^0 + \nu_e$. (The π^0 decays into two γ-rays). The chambers were triggered by a coincidence between a stopped K meson ($C_1C_2C_3\overline{C}_4$+ RF) , a decay positron (C_5C_6 + Cerenkov counter), with the additional requirement of no γ-ray. The time between the K^+ stop and the decay was recorded for each event.

B. CALCULATION OF BACKGROUND

Because the expected K^+_{e2} branching ratio was so small, 1.6×10^{-5}, the experimenters needed an accurate calculation of expected background that might mask or mimic K^+_{e2} events to determine whether or not the experiment was feasible. If the background was too large the measurement could not be successfully carried out. The group's previous experience using the apparatus would be important for both the calculation of background and for performing the experiment.

The positron from K^+_{e2} decay has a momentum of 246.9 MeV/c in the kaon center of mass. This is higher than the momentum of any other direct product of K^+ decay. The closest competitor is the muon from $K^+_{\mu2}$ decay, which has a momentum of 235.6 MeV/c. The principal sources of high-momentum positrons that might mimic K^+_{e2} decay are

1. $K^+ \to e^+ + \pi^0 + \nu_e$, K^+_{e3} decay, with a maximum positron momentum of 228 MeV/c and a branching ratio of approximately 5%.

2. $K^+ \to \mu^+ + \nu_\mu$, followed by $\mu^+ \to e^+ + \nu_e + \overline{\nu}_\mu$, with a maximum momentum of 246.9 MeV/c , the same as that for K^+_{e2} decay, and a branching ratio of approximately 1.2×10^{-4} per foot of muon path. This decay rate per foot was considerably larger than the total expected K^+_{e2} decay rate. If this source of background could not be eliminated then the experiment could not be done.

Using the momentum resolution measured in this experiment of 1.9%,[33] and the known K^+_{e3} decay rate and momentum spectrum (both of which had been measured in previous group experiments) one could calculate that the number of K^+_{e3} events expected in the K^+_{e2} decay region, 242-252 MeV/c, was less than 5% of the expected K^+_{e2} decay rate. If the K^+ decayed in flight then the momentum of the positron from K^+_{e3} decay could be higher than 228 MeV/c. This possible source of background was

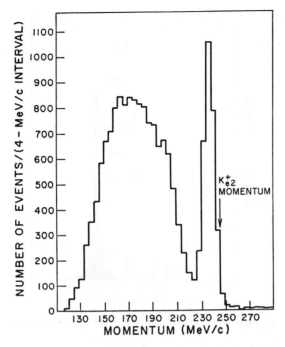

Figure 12. Momentum distribution for all K^+ decay events obtained with the Cerenkov counter in the triggering logic. From Bowen et al. (1967).

completely eliminated when "prompt" events were removed from the sample as discussed later.

The background due to K^+ decay into a muon, followed by muon decay into a positron, was also calculated. This involved a detailed calculation which included the decay rate, the momentum and angular distribution of the decay positrons relative to the muons, the extrapolation of the decay particle trajectory from the momentum chambers into the range chamber, and the momentum and angular resolution in the thin-plate momentum chambers. This last factor, which was quite important, was known from the previous experiments. The group calculated that the number of decay positrons with laboratory angles of less than 10° with respect to the muon flight path and with momentum greater than 225 MeV/c was less than 5% of the expected K^+_{e2} decay rate. The limits on momentum and angle were well within their previously determined experimental resolution. The total background rate was calculated to be approximately 15% of the expected K^+_{e2} decay rate in the K^+_{e2} momentum region (242-252 MeV/c). This strongly suggested that the experiment was feasible.[34]

C. REDUCTION OF DATA

Figure 12 shows the momentum distribution of 16,965 events obtained with the Cerenkov counter in the triggering logic. These events satisfied the following criteria:

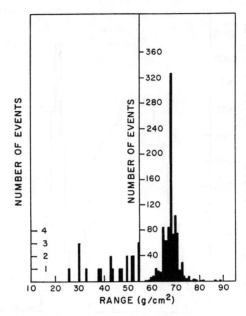

Figure 13. Range spectrum for muons from $K^+_{\mu 2}$ decay. From Bowen *et al.* (1967).

1) the track came from the stopping region, C_3; 2) the track passed through both the front and rear windows of the Cerenkov counter; and 3) a track traversing at least three plates was seen in the range chamber. This is the haystack from which the needle of a few K^+_{e2} events was to be found. The momentum for K^+_{e2} decay is shown. It is clear that if the K^+_{e2} events are present they are rather well hidden.

The large peak at 236 MeV/c and the smaller peak at 205 MeV/c are due to accidental coincidences between accelerator produced background in the Cerenkov counter and muons and pions from $K^+_{\mu 2}$ and $K^+_{\pi 2}$ decay, respectively. If the K^+_{e2} events were to be found then the background due to $K^+_{\mu 2}$ events had to be reduced. Because the momentum of these muons is known, this peak was used both to calibrate the momentum scale and to measure the momentum resolution.

The experimenters applied a set of selection criteria to eliminate unwanted background events while preserving a reasonable and known fraction of the K^+_{e2} events. The first criterion applied was that of range, the path length in the range chamber before the particle stopped or produced an interaction. Muons lose energy only by ionization loss and thus have a well-defined range in matter. The muons from $K^+_{\mu 2}$ had a mean measured range of 67 g/cm^2, with a straggle of about 4 g/cm^2 (Figure 13). The 1% of such events with a range less than 45 g/cm^2 is too large to be accounted for by range straggling, and was due to the occasional failure of the range chamber to operate properly. These apparatus failures gave rise to a background of about 15% of the expected K^+_{e2} rate. The experimenters measured the range distribution for positrons with momenta between 212 MeV/c and 227 MeV/c (Figure 14). These positrons differ

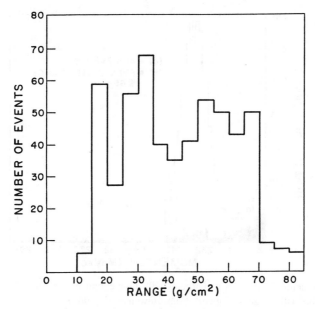

Figure 14. Range spectrum for positrons from K^+_{e3} decay with momentum between 212 and 227 MeV/c. From Bowen et al. (1967).

from K^+_{e2} positrons by only 10% in momentum, and were expected to behave quite similarly. Positrons do not have a well-defined range because they lose energy by several different processes, some of which involve large energy loss, and the distribution of ranges is approximately constant from about 15 g/cm^2 to 70 g/cm^2. The percentage of positrons with range less than a given value is shown in Table III. If events are required to have a range less than that of the muon from $K^+_{\mu2}$ decay, this will serve to minimize the background due to those events, while preserving a large, and known, fraction of the high energy positrons. A selection cut was made at 45 g/cm^2.[35] The effect of applying this criterion is shown in Figure 15. The haystack had gotten smaller.

As discussed earlier, a major source of background was decay of the kaon into a muon, followed by the decay of the muon into a positron. Most of these positrons are emitted at large angles to the muon path. If the decay occurred in the momentum chambers, it would have been detected by a kink in the track. Decays occurring between the end of the momentum chambers and the end of the Cerenkov counter, a very long

Table III: Range distribution of positrons from K^+_{e3} decay in the momentum region 212-227 MeV/c. From Bowen et al. (1967)

Range (g/cm^2)	% of events with a smaller range
40	45.6 ± 2.1
45	51.9 ± 2.1
50	59.2 ± 2.1
55	68.8 ± 1.9

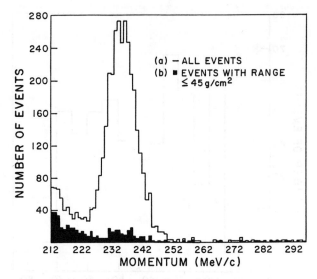

Figure 15. Momentum spectrum of particles with momentum greater than 212 MeV/c: (a) all events (b) events with range ≤ 45 g/cm^2. From Bowen et al. (1967).

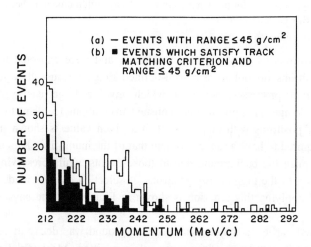

Figure 16. Momentum spectrum of particles with momentum greater than 212 MeV/c and range ≤ 45 g/cm^2: (a) all events (b) events satisfying track-matching criterion. From Bowen et al. (1967).

distance, could not be seen. Because of the large decay angle such decays could be detected by comparing the measured position of the particle when it entered the range chamber with the position predicted by extrapolating the momentum chamber track. If a decay had occurred, then the difference between the two positions would be large.

The accuracy of the comparison was limited by multiple scattering in the momentum chambers and the uncertainty in extrapolating the path through the fringe field of the magnet. Such a track-matching criterion was applied. In addition, because these decays

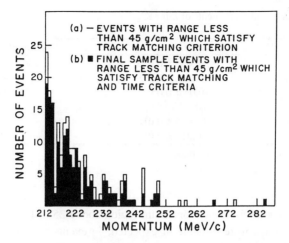

Figure 17. Momentum spectrum of particles with momentum greater than 212 MeV/c, range ≤45 g/cm², and satisfying the track matching criterion: (a) all events (b) events with K^+ decay time ≥ 2.75 ns. From Bowen et al. (1967).

occurred after the momentum chambers they would have a measured momentum equal to that of muons from $K^+_{\mu 2}$ decay, 236 MeV/c. These events would be further reduced by momentum cuts in the final analysis of the data. The effects of this track-matching cut are shown in Figure 16. Once again, the selection criteria served to preferentially reduce the events in the $K^+_{\mu 2}$ region relative to the events in the K^+_{e2} region. Both the range cut and the track-matching criteria were varied over reasonable limits and it was shown that the measured K^+_{e2} branching ratio was robust under these variations.[36]

There was one further major source of background. This was due to decays in flight of the K^+ meson. If the kaon decayed in flight then the momentum of the decay particle could be increased, leading to possible simulation of K^+_{e2} decays. Examination of the distribution of time intervals between the stopped kaon and the decay positron revealed the presence of a small peak due to such decays in flight. The peak had a base width of 2 ns. A cut was made removing all events with a time interval of less than 2.75 ns. This eliminated all of the decays in flight. The effect of this selection criterion is shown in Figure 17. The effect of this cut was to preferentially reduce the events in the K^+_{e2} region, indicating that decays in flight were indeed a source of simulated K^+_{e2} events.

A total of seven events remained in the K^+_{e2} region. Unfortunately, none of these had a label identifying them as a real K^+_{e2} decay. How one calculated the number of such decays and the branching ratio is discussed in the next section.

D. THE EXPERIMENTAL RESULT

The number of events in the K^+_{e2} region had to be corrected for various experimental effects to determine the final number of K^+_{e2} events. These included: 1) flat

Table IV: Summary of measurements of λ for K^+_{e3}

Experimenter	λ	Number of Useful Events
Brown et al. 1961	$+0.038 \pm 0.035$	
Jensen et al. 1964	-0.010 ± 0.029	407
Borreani et al. 1964	-0.04 ± 0.05	230
Kalmus[a]	$+0..028^{+0.013}_{-0.014}$	
Bellotti et al. 1966	$+0.025 \pm 0.018$	457
Imlay et al. 1967	$+0.016 \pm 0.016$	1393

[a] This value was from a preprint of a paper.

background, estimated from the momentum region above 252 MeV/c, where no events due to real K^+ decays were possible or expected; 2) events due to $K^+_{\mu 2}$ decay, which spilled over into the K^+_{e2} region because of the finite momentum resolution of the apparatus; and 3) K^+_{e2} events lost because of the finite momentum region selected for the decay and the finite momentum resolution. A final total of $6^{+5.2}_{-3.7}$ events were attributed to K^+_{e2} decay after these corrections.

The branching ratio, the rate compared to all K^+ decays, was calculated by normalizing the K^+_{e2} events to known K^+ decay rates by two different methods. The first used the upper end of the K^+_{e3} spectrum, the region from 212 MeV/c–228 MeV/c in Figure 17, which had been subjected to the same selection criteria as the K^+_{e2} events. To estimate the total number of K^+_{e3} events the experimenters needed to know the shape of the K^+_{e3} decay spectrum. This had, in fact, been measured in one of the earlier experiments in the sequence. The second method used the total sample of 16,965 K^+ decays given in Figure 12. The results for the branching ratio, using the two different methods, were $R = (2.0^{+1.8}_{-1.2}) \times 10^{-5}$ and $R = (2.2^{+1.9}_{-1.4}) \times 10^{-5}$, respectively. The two different methods, which had very different selection criteria, agreed and the final result given was their average, $R = (2.1^{+1.8}_{-1.3}) \times 10^{-5}$. This was in good agreement with the value predicted by the V-A theory, $R = 1.6 \times 10^{-5}$, (1.44×10^{-5} including radiative corrections). This also set an upper limit (at the two standard deviation level) for f_P/f_A, the pseudoscalar and axial vector coupling constants, of 3×10^{-3}.

V. THE ENERGY DEPENDENCE OF THE FORM FACTOR IN K^+_{e3} DECAY

The last of the experiments performed on K^+ decays by the collaboration was to measure the energy dependence of the form factor in K^+_{e3} decay (Imlay *et al.* 1967). The form factor took the form $f_+(q^2) = f_+(0)[1 + \lambda q^2/M_\pi^2]$, where λ was the energy-dependent parameter, q^2 the square of the four-momentum transferred to the leptons, and M_π the mass of the pion. Recall that in their earlier measurements of the branching ratio and the momentum spectrum in K^+_{e3} decay the group had assumed that the form factor was constant in their analysis. Although there was evidence supporting this assumption (see Table IV), all of the previous measurements had been done with

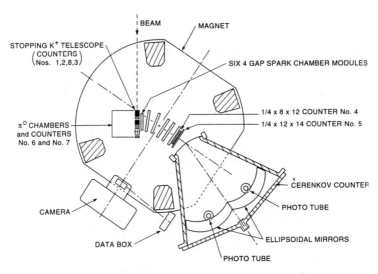

Figure 18. The experimental apparatus for measuring the form factor in K^+_{e3} decay. The π^0 spark chambers were located above and below the stopping K^+ telescope. From Imlay et al. (1967).

bubble chambers, and it was thought that a precision measurement using a different technique, namely spark chambers, would be useful. The new experiment would also be able to investigate the virtual strong interaction involved in the decay and muon-electron universality.

In this experiment we also see the increasing expertise of the group. By now the use of the momentum chambers and of the Cerenkov counter had become unproblematic and the group could concentrate on other additions to the apparatus which allowed the successful completion of this measurement.[37]

A. THE EXPERIMENTAL APPARATUS

There were several important changes made to the basic experimental apparatus for this experiment (Figure 18). The momentum spread of the beam was reduced from the 13% used in the K^+_{e2} branching ratio experiment to 7%, and a mercury plug was placed in the beamline downstream from the bending magnet to control the beam intensity. As discussed in detail below, background in the π^0 shower chambers was a significant problem and the beam intensity was reduced to approximately 100 stopped kaons per second.[38]

The stopping region was now a beryllium block rather than a scintillation counter. This was done to minimize energy loss by the decay positrons due to bremsstrahlung in the stopping region and to minimize conversion in the target of the γ rays produced by π^0 decay. A stopped K^+ meson was now identified by a coincidence ($C_1 C_2 \overline{C}_3$ + RF). The essential change in the experimental apparatus was the installation of shower spark

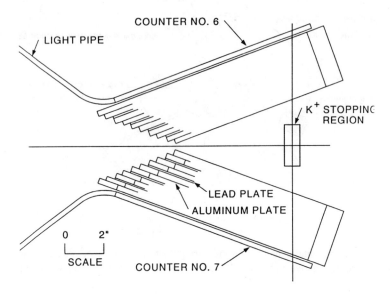

Figure 19. The π^0 spark chambers from the form factor experiment.

chambers above and below the beryllium stopping region. These were to detect the γ rays resulting from the decay of the π^0 produced in K^+_{e3} decay, $K^+ \to e^+ + \pi^0 + \nu_e$ (Figure 19).[39] Because of space limitations the chambers had only 10 gaps. The first two gaps of the chambers had thin plates, to minimize γ ray conversion, and acted as a veto for charged particles. If there were sparks in these gaps it indicated the presence of a charged particle coming from the stopping region and the event was rejected. These were followed by four lead plates to convert the γ rays into charged particles (showers), whose tracks could be observed. Counters C_6 and C_7 were placed behind the chambers to provide a signal for γ ray conversion. A K^+_{e3} decay event was identified, and the spark chambers triggered, by a coincidence between a K^+ stop ($C_1 C_2 \overline{C}_3$ + RF), a decay positron ($C_4 C_5$ + Cerenkov counter), and a count in either C_6 or C_7.[40]

It had been found in the earlier K^+_{e3} experiment that when the momentum chambers were placed in a 3 kilogauss magnetic field the efficiency for detecting positrons was uniform for momenta between 80 and 228 MeV/c, the maximum energy of a positron from K^+_{e3} decay. This was the magnetic field used. The momentum resolution with this field strength was 3.1%, determined by observing the muons and pions from $K^+_{\mu 2}$ and $K^+_{\pi 2}$, respectively.[41]

B. DATA ANALYSIS

1. Selection and Reconstruction of Events

Events were required to satisfy each of three selection criteria: (1) the positive track observed in the momentum chambers fit well to some momentum value, extrapolated back to the K^+ stopping region, and extrapolated into counter C_4 and the Cerenkov counter; (2) one and only one γ ray was identified in each of the two shower spark chambers; and (3) a signal was observed in the counter behind each shower chamber (counters C_6 or C_7) if the shower in that chamber contained sparks in one of the last two gaps. It was also required that one of the two counters, C_6 or C_7, gave a pulse at the time of the K^+ decay.

Each event was reconstructed using the charged particle momentum, the direction of the two γ rays determined by the conversion point of each γ rays and the stopping point of the K^+ meson, and the position of the stopped K^+ meson. The position of the stopped K^+ was known to lie along the positron trajectory, but was uncertain by ± 0.95 cm, one half the transverse thickness of the beryllium stopping region. The K^+ was assumed to stop at the intersection of the positron trajectory and the center line of the block. For the small fraction (10%) of events that failed to reconstruct properly as a K^+_{e3} decay, other positions within the block were tried. If the event failed to reconstruct with any of these positions it was rejected. An error of 0.95 cm in the position of the stopped K^+ typically introduced an error of 5 MeV in the calculated energy of the π^0. There was also a kinematic ambiguity in the reconstruction of the event which gave rise to two solutions for the π^0 energy. Both solutions were used in the subsequent analysis.

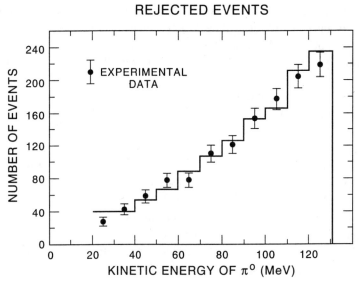

Figure 20. Distribution of π^0 kinetic energy for events rejected because they did not have a signal in the π^0 counters at the appropriate time, compared to the predicted distribution for spurious events. From Imlay *et al.* (1967).

2. Checks on Measurement Techniques and Background

Although not as severe a problem as was the case in the K^+_{e2} experiment, in which the signal had to be extracted from a much larger background, background was a problem in the K^+_{e3} form factor experiment. It was both reasonably large, approximately 23%, and the experimenters had to know its energy distribution so that it could be subtracted from the measured distribution, and a value of λ obtained. This background was primarily due to γ rays produced by beam particle interactions that were observed in the shower chambers and were not due to K^+_{e3} decay.

The experimenters first examined a sample of 738 $K^+_{\pi 2}$ events obtained with the Cerenkov counter removed from the triggering logic. These events were subjected to the same criteria as were the nominal K^+_{e3} events and were, in addition, required to have a momentum between 192–220 MeV/c. These events were extremely useful in obtaining a background estimate because they were overdetermined by two variables and could therefore be used to examine the consistency of the event reconstruction. In particular, the decays had to be coplanar. Of the total $K^+_{\pi 2}$ sample, 623 events were identified as real $K^+_{\pi 2}$ decays because they satisfied the coplanarity condition, had an opening angle between the γ rays of greater than $61.3°$,[42] and had a K^+ stopping position within the K^+ stopping region.

Of the 115 rejected events a total of 21 were expected, on the basis of a Monte Carlo calculation, to be due to backgrounds such as $K^+_{\mu 3}$ and K^+_{e3} decays. In the remaining 94 events rejected from the $K^+_{\pi 2}$ sample one of the two γ rays was not associated with K^+_{e3} decay, but arose from some interaction of a beam particle, such as π^+ charge exchange in the copper absorber. The sensitive time of the shower spark chambers was 500 ns so that any γ ray, even if it were not associated with K^+ decay, would be observed if it occurred during that time interval. Although such showers did not occur at the time of the K^+ decay and could be rejected by the pulse-time requirement in counters C_6 and C_7, occasionally a pulse would be present in the counters due to an accidental coincidence, or alternatively a pulse was legitimately absent because the shower did not penetrate to counters C_6 or C_7.

The experimenters hypothesized that the 94 rejected events were due to accidental coincidences between K^+ decay and a γ ray not associated with that decay. To check this hypothesis they examined a sample of $K^+_{\mu 2}$ events obtained with the Cerenkov counter removed from the triggering logic, with the additional criterion of a momentum between 231–252 MeV/c. Any γ ray associated with this decay, $K^+ \rightarrow \mu^+ + \nu_\mu$, must be spurious. A sample of 835 $K^+_{\mu 2}$ events was scanned for γ rays. Of these, 110 had a γ ray in one shower chamber, and 16 had γ rays in both chambers. Using the number of these γ rays that had an appropriate time pulse in counters C_6 or C_7 the group estimated that the background in the $K^+_{\pi 2}$ sample would be $(11.6 \pm 2.5)\%$, in comparison with the measured background of $(15.2 \pm 1.5)\%$. For K^+_{e3} decays the calculated background was $(20 \pm 3)\%$. More detailed estimates gave an estimated background of $(23 \pm 2)\%$ in the K^+_{e3} sample.[43]

To obtain the energy distribution of these background events the experimenters performed a Monte Carlo calculation generating spurious K^+_{e3} events. These simulated

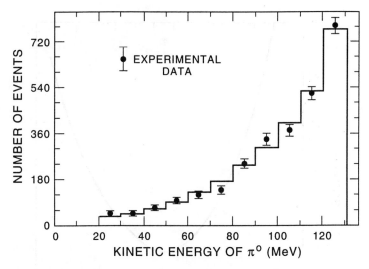

Figure 21. The corrected π^0 kinetic energy distribution for K^+_{e3} events compared to the predicted distribution for $\lambda = 0.016$. From Imlay et al. (1967).

events randomly placed one γ ray in the shower chambers. The validity of this calculation was checked by examining the distributions for γ-ray energy, positron momentum, and π^0 energy, for nominal K^+_{e3} events that were rejected by selection criterion 3, the requirement of an appropriately timed pulse in counters C_6 or C_7. These events were expected to be identical to the background events that were contained in the 1867 event nominal K^+_{e3} decay sample. The comparison between the calculated and measured π^0 energy distributions for rejected events is shown in Figure 20. The good agreement between the two distributions provided support for the Monte Carlo calculation.[44] The π^0 energy distribution, corrected for background was used to determine λ, the form factor energy-dependence parameter.[45]

Figure 21 shows the final π^0 energy distribution, the original measured spectrum with the appropriately normalized background spectrum subtracted, from which λ was obtained. The form factor, $f_+(q^2) = f_+(0)[\ 1 + \lambda q^2/M_\pi^2]$, is a function only of the π^0 energy.

3. Determination of λ

The final total of K^+_{e3} events was 1393 (1867 nominal K^+_{e3} events - background). The value of λ, the energy-dependence parameter, was found by comparing the experimental π^0 energy distribution (Figure 21) with distributions generated for various values of λ by an extensive and detailed Monte Carlo simulation of the experiment[46] and constructing a plot of χ^2 versus λ. For each value of λ approximately 6×10^4 Monte Carlo events were generated. This was a factor of 40 larger than the number of K^+_{e3} events in the final sample, so that the statistical accuracy of the Monte Carlo simulation would not be a problem. For $\lambda = 0.016$ the minimum value of χ^2, 12.2 for 10 degrees of

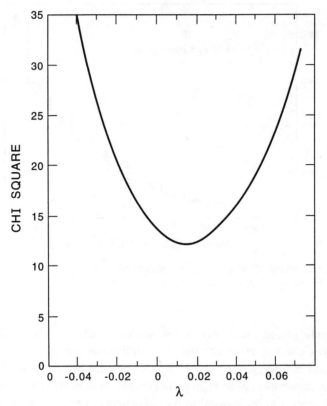

Figure 22. Goodness of fit for the vector interaction with a form factor $f_+(q^2) = f_+(0)[1 + \lambda q^2/M_\pi^2]$. A minimum χ^2 of 12.2 for 10 degrees of freedom occurred at $\lambda = 0.016$. From Imlay et al. (1967).

freedom, a reasonably good fit, was obtained (Figure 22). Uncertainties in the background gave rise to an uncertainty in λ. The 2% uncertainty in the magnitude of the background and the uncertainty in the background energy distribution led to an uncertainties of 0.005 and 0.006 in λ, respectively. The experimenters assigned a total uncertainty of 0.01 because of the uncertainties in the background.

The determination of the statistical uncertainty in λ was made more difficult because the two solutions for the π° energy were not statistically independent. This uncertainty was determined by examining 40 1500-event samples of Monte Carlo events. For each such sample the χ^2 versus λ curve was obtained in the same way as was the curve for the real data. The rms-deviation of the best-fit value of λ from the value of λ used to generate the 1500-event samples was then assigned to be the statistical uncertainty in λ. This was found to be 0.013. Combining the statistical uncertainty of 0.013 with the uncertainty due to background subtraction of 0.010 gave a final value $\lambda = 0.016 \pm 0.016$ in $f_+(q^2) = f_+(0)[1 + \lambda q^2/M_\pi^2]$

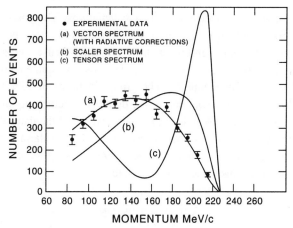

Figure 23. Corrected positron momentum spectrum above 80 MeV/c. The fitted istributions for pure vector (with radiative corrections), scalar, and tensor coupling are shown. From Eschstruth et al. (1968).

4. Conclusions

The group compared the Dalitz plot (a graph of π^0 energy versus positron energy for each event) experimentally obtained with that generated by an extensive and detailed Monte Carlo simulation of the experiment. The good agreement of the two supported the assumption of vector coupling for K^+_{e3} decays. This was in good agreement with previous experiments. The value of λ found, 0.016 ± 0.016, was in agreement with the weighted average of all measurements, including this one, $\lambda = 0.019 \pm 0.009$. If one attributed the energy dependence of the form factor to a single dominant intermediate state of mass M, then the value of M obtained from the value of λ was $M = 1180^{+\infty}_{-387}$ MeV. This was consistent with both the mass of the K^* resonance (890 MeV), but also consistent with no energy dependence.

VI. K^+_{e3} MOMENTUM SPECTRUM AND BRANCHING RATIO

As discussed earlier, the situation in 1967 with respect to the K^+_{e3} branching ratio was still somewhat uncertain (See Table IV). Although five experiments were consistent with one another, the two most precise measurements, that of Callahan et al. (1966) and that done previously by the Mann-O'Neill group (Cester et al. 1966), each claimed a precision of 4%, but differed by 25%. The Mann-O'Neill group decided to use part of the data taken in the K^+_{e3} form factor experiment to remeasure the K^+_{e3} branching ratio and momentum spectrum (Eschstruth et al. 1968).

The experimenters used only that portion of the data taken in the form factor experiment that did not require one of the π^0 counters (Counter C_6 or C_7) in the triggering logic. This avoided the additional uncertainty that would result from including the π^0 detection efficiency in their calculations. The analysis of the 4638 events obtained was identical to that used in the group's previous measurement (See Section II). In this analysis the group was truly recycling their expertise. Their results agreed with those they had obtained previously. The positron momentum spectrum from this experiment is shown in Figure 23. The group set limits of 15% and 4% for the admixture of scalar and tensor coupling to the dominant vector coupling, respectively. Their previous values were 18% and 4%. Further analysis of the data taken with the π^0 counters in the triggering logic set limits of 5% and 7% for scalar and tensor coupling respectively.[47] The group also obtained a K^+_{e3} branching ratio of (5.20 ± 0.19)%, consistent with their previous value of (4.98 ± 0.20)%.

VII. DISCUSSION

This history of this sequence of experiments done by the Mann-O'Neill collaboration illustrates both the normal practice of science and the practice of normal science. It shows both instrumental loyalty and the recycling of expertise.[48] The group used the same basic experimental apparatus in each of the experiments; an apparatus that was modified so that different measurements could be performed. As we have seen, the later measurements benefitted from the expertise acquired earlier. By the time of the K^+_{e2} branching ratio measurement, the use of the momentum chambers, the range chamber, and the Cerenkov counter was well-enough understood so that they could be used straightforwardly in that more difficult measurement. Similarly the measurement of the form factor in K^+_{e3} decay, which involved the addition of the π^0 chambers and counters and the attendant analysis problems, made use of knowledge and expertise acquired in earlier experiments. The last measurement of the K^+_{e3} momentum spectrum and branching ratio was the application of previously acquired expertise (the analysis procedures were identical) to analyze data acquired for another purpose.

Another interesting aspect of a sequence of experiments using essentially the same experimental apparatus to investigate very similar phenomena is that one can use the same effect to calibrate the experimental apparatus and to argue that it is working properly. In this sequence it was the observation of the momentum peaks due to $K^+_{\mu 2}$ and $K^+_{\pi 2}$ decays. In each of these two-body decays the decay particles have a fixed and unique momentum, 205 MeV/c and 236 MeV/c, for $K^+_{\pi 2}$ and $K^+_{\mu 2}$ decays, respectively. The fact that the apparatus could detect both of these peaks argued that it was working properly and also provided a calibration of the momentum scale. In addition, the measured width of the peaks also determined the momentum resolution of the apparatus, a quantity needed for the calculation of the experimental results. In the first experiment, which accepted all charged decay modes of the K^+ meson, the events due to these decays were automatically present. In the later experiments, in which a decay positron was selected for by using a Cerenkov counter, some provision had to be made to detect them. In the K^+_{e3} experiments the Cerenkov counter was removed from the triggering logic to allow a sample of those decays to be collected. In the K^+_{e2}

experiment, accidental coincidences between those decays and noise in the Cerenkov counter provided sufficient events.

Another virtue of such a sequence was that because the analysis of some of the data acquired earlier took so long that that analysis could benefit from knowledge acquired later. Thus the early measurements of K^+ decays using only the range chamber benefitted from the information acquired in the K^+_{e2} experiment in which positrons were identified by the Cerenkov counter and their range measured. All of the experiments were on the same general topic, K^+ decays, so that the group could use the same theoretical framework for the entire sequence.

Although the accepted V-A theory of weak interactions was further tested in these experiments, the experiments were essentially independent of theory.[49] The sequence of experiments seemed to proceed by a logic of its own. The first experiment, which measured the branching ratios of the major K^+ decay modes, used only the basic experimental apparatus plus the range chamber. The second experiment, which measured the K^+_{e3} branching ratio, used the Cerenkov counter to select and identify positrons from this specific decay mode. As we have seen, this was the decay mode that was the most difficult to isolate using only the range chamber. This was because the positrons from the decay, unlike the muons from $K^+_{\mu 2}$ and $K^+_{\mu 3}$ decays and the pions from $K^+_{\pi 2}$, do not have a well-defined range. Both the Cerenkov counter and the range chamber were needed in the experiment to measure the K^+_{e2} branching ratio to isolate the very small number of events due to that decay mode from a large background. Even though the Cerenkov counter had only a small efficiency (0.38%) for detecting particles other than positron, the very small branching ratio made this background substantial, to say the least. The final two experiments on aspects of K^+_{e3} decay added π^0 shower chambers to the basic apparatus used in the second experiment and made use of the expertise acquired earlier.

This episode also makes clear the distinction between experimental data and an experimental result. As we have seen, the production of a result from the data is a long, careful process, which includes calculation of backgrounds, efficiencies, and experimental resolution. The result is not immediately given by inspection of the data. I note that in several of these experiments, the measurement of the branching ratios of the major decay modes, measurement of the K^+_{e2} branching ratio, and measurement of the form factor in K^+_{e3} decay, the analysis of the data took longer than its acquisition.

This sequence of experiments had their origin in the generally unsatisfactory situation in the early 1960s with respect to the K^+ branching ratios. At the end of the sequence the situation was improved, but not resolved. The group had made useful contributions to the measurement of these quantities having added new measurements that lowered the statistical error of the branching ratios. In addition, the group had made the first actual measurement, as opposed to setting an upper limit, of the K^+_{e2} branching ratio, and had measured the form factor in K^+_{e3} decay. These experiments had a life of their own.

NOTES

[1] Every sequence of experiments must start with a first experiment, and thus involve the construction of a new apparatus. Sometimes experimenters may construct apparatus to perform a single experiment.

[2] This is not to say that only experimentalists recycle expertise. Theorists also recycle. They may very well work in a given area for some time or even use a general computer program, with specific modifications for each problem.

[3] For a related discussion see Nebeker (1994).

[4] The group was known as the Mann-O'Neill collaboration after the senior members of the group, Alfred Mann and Gerard O'Neill.

[5] The first published result was actually obtained in the second data run. The analysis of the data obtained in the first run was not completed until after the results of both the second and third experiments were obtained. The results of the first data run were actually the third to be published. This illustrates the distinction between experimental data and an experimental result.

[6] The number of experimenters working on the different experiments ranged from seven to nine. This was a reasonably sized group in the late 1960s. In contrast, experimental groups now number hundreds of experimenters. The group members who worked on more than one of the experiments, along with the number of experiments they worked on, follows: A.K. Mann (5), W.K. McFarlane (4), P.T. Eschstruth (4), G.K. O'Neill (3), A.D. Franklin (3), E.B. Hughes (3), R.L. Imlay (3), D.H. Reading (3), D.R. Bowen (3), J.M. Dobbs (2), D. Yount (2), R. Cester (2), and D.H. White (2).

[7] The apparatus also determines what cannot be measured. In the experiments discussed below requiring a Cerenkov counter signal eliminated those decay modes that did not include a positron. Similarly, in an experiment that discovered the weak neutral current, the original trigger signal required the presence of a muon and would not have been able to detect such neutral current events. Only when the trigger was changed could that measurement have been done; see Galison 1987, Chapter 4.

[8] In recent times these traditions have merged in electronic imaging apparatuses such as the Time Projection Chamber. See Galison (1997).

[9] Bubble chamber experiments can select events to measure during the scanning process. Events may be selected by topology, for example, a four-prong event in which four outgoing particles are observed. This can, of course, only be done after the bubble chamber photographs have been taken. There were attempts to trigger bubble chambers using various counters, but they were unsuccessful.

[10] This is not strictly true. Two decay modes, $K^+ \to \pi^+\pi^+\pi^-$ and $K^+ \to \pi^+\pi^0\pi^0$, were not detected because of their low decay momenta.

[11] For an illustration of this see Galison's discussion of the observation of a neutrino-electron scattering event (1987, pp. 180-85).

[12] These decay modes are

1) $K^+_{\mu 2}$, $K^+ \to \mu^+ + \nu_\mu$

2) $K^+_{\mu 3}$, $K^+ \to \mu^+ + \pi^0 + \nu_\mu$

3) $K^+_{\pi 2}$, $K^+ \to \pi^+ + \pi^0$

4) K^+_{e3}, $K^+ \to e^+ + \pi^0 + \nu_e$

[13] This was changed in the later experiments.

[14] Protons have a shorter range in matter than either kaons or pions.

[15] There was a 34 ns ambiguity, but this was inconsequential.

[16] Other counters and spark chambers were added to the apparatus in the experiment to measure the form factor in K^+_{e3} decay.

[17] The value of p_{min} given in Table I shows the minimum value of the momentum for which the efficiency was constant.

[18] These branching ratios were both larger than that of K^+_{e3} and better known. Hence they were used for the normalization procedure.

[19] Dalitz pairs are electron-positron pairs produced in the decay $\pi^0 \to \gamma + e^+ + e^-$. The dominant decay mode of the π^0 is into 2 γ rays. Interaction of γ rays with matter may also produce electron-positron pairs.

[20] Bremsstrahlung are high-energy γ rays produced by the interaction of charged particles with matter.

[21] This assumption of constant form factors was reasonably well supported at this time (Table IV).

[22] This was calculated assuming a pure vector interaction and a constant form factor for the decay.

[23] The acccepted values for these branching ratios was used.

[24] A footnote to the paper noted that they had recently received a preprint from A.C. Callahan which gave a K^+_{e3} branching ratio of 0.0394 ± 0.0017. Thus, there were two measurements of the K^+_{e3} branching ratio each of which claimed 4% uncertainty, but which differed by 25%. This led to another measurement of the K^+_{e3} branching ratio by the Mann-O'Neill collaboration later in the experimental sequence.

[25] Events accepted as satisfactory after computation were required to satisfy the following six criteria: (1) the decay time of the event was greater than 2.5 ns; (2) the projected track obtained from the analysis of the momentum chamber sparks came from the stopping region; (3) the projected track passed through counters 5 and 6; (4) the projected track traverse a certain area at the front of the range chamber; (5) the fitted momentum chamber track had a good fit; and (6) the projected momentum-chamber track matched the observed range-chamber track at the first plate of the range chamber within certain limits.

[26] All of the events discussed satisfy the selection criteria discussed in note 25. I will not discuss any of the technical details of corrections unless they relate specifically to the sequence of experiments.

[27] They could have obtained a somewhat less accurate estimate by fitting the positron spectrum outside the muon interval and then extrapolating through the muon interval.

[28] For a more detailed discussion of this experiment see Franklin (1990, pp. 118-131).

[29] The V-A theory of weak interactions had strong experimental support at this time, although it had not been severely tested in strangeness-changing decays. See Franklin (1990, Chapter 5).

[30] This does not include radiative corrections.

[31] Part of the improvement in the beam was due to an inadvertent error in the initial set-up of the beam. When the program to set the beam parameters had been run initially an incorrect field

length was used for the quadrupole magnets. The use of the wrong value for the field length resulted in a reduced beam rate. Before the error was found and corrected the experimenters empirically investigated the properties of the beamline, in particular the location of the collimator in the bending magnet. They were able to improve the beam rate. When the error was found and corrected, the empirical improvements were retained, helping to raise the rate of stopping K^+ mesons. This improvement in the beam intensity allowed the measurement of the K^+_{e2} branching ratio, as opposed to merely setting an upper limit for the decay.

[32] The alert reader will have noticed that this rate is approximately half the 0.7% rate found earlier. This was due to the fact that only half the Cerenkov counter, which had only half the phototubes and thus half the noise rate, was used in this experiment.

[33] The momentum resolution was measured using the muon peak from $K^+_{\mu2}$ decays in coincidence with noise pulses in the Cerenkov counter shown in Figure 12.

[34] There were other sources of background due to experimental effects which were not *a priori* calculable. These are discussed below.

[35] The limits on this, and on each of the other selection cuts, was varied and it was found that the final result was robust against reasonable changes in the limits set on these cuts.

[36] The robustness of the result under the reasonable variation of selection cuts or criteria is important. One might worry that the result was due only to particular cuts.

[37] I note that because of the difficulties introduced by the need to detect γ rays and the subsequent analysis, this experiment was technically more difficult than the measurement of the branching ratio and the momentum spectrum in K^+_{e3} decay.

[38] Recall that in the first two experiments the intensity had been 200/second and in the K^+_{e2} experiment the intensity was 500/second.

[39] In the K^+_{e2} experiment the experimenters wished to eliminate events which had γ rays. In this experiment they wanted to guarantee that the events contained γ rays.

[40] A count in C_6 or C_7 was required in the trigger signal for only 75% of the data.

[41] The larger momentum resolution was due to the lower magnetic field used in this experiment.

[42] This opening angle was required by the π^0 decay kinematics.

[43] See Imlay *et al.* (1967, pp. 1208-9) for details.

[44] Similar graphs were obtained for the positron momentum spectrum and for the γ-ray energy spectrum. The calculated and measured distributions also agreed very well.

[45] The experimenters also checked that the requirement of a pulse in counter C_6 or C_7 did not introduce any significant bias that was dependent on γ-ray energy.

[46] For details see Imlay *et al.*(1967, p. 1210).

[47] This analysis was completed after the group's paper on the form factor in K^+_{e3} decay was published.

[48] I note that experimenters do not always use the same type of apparatus or work in the same general subject area. I was a member of the Mann-O'Neill collaboration and worked on the last three of these experiments, which were all on K^+ meson decays and weak interactions. My previous graduate work had been on the investigation of the photoproduction of η^0 and ρ^0 mesons. Although these experiments used spark chambers and scintillation counters, the same experimental technique used in the K meson sequence, they were in the general area of electromagnetic, rather than weak, interactions. After I left Princeton and went to the University of Colorado I worked on yet another subject, namely strong interactions, using bubble chambers,

a very different experimental technique. Although physicists may be loyal to both their experimental apparatus and to a particular subject area, that loyalty is not absolute.

[49] This is not to say that there were no theoretical implications of this work or that the group was unaware of them. As we have seen, the K^+_{e2} did set a stringent limit on the presence of a pseudoscalar interaction in the weak interactions. The other experiments also supported the V-A theory.

CHAPTER 4

THE RISE OF THE "FIFTH FORCE"

1. INTRODUCTION

One of the interesting questions in the history and philosophy of science is how a hypothesis is proposed and how it acquires sufficient plausibility to be considered worthy of further theoretical and experimental investigation by the scientific community.[1] I am not denying that the suggestion of hypotheses or theories is the free creation of an individual scientist, but I doubt that such creative events occur in a vacuum. Thus, although Newton's thoughts on the universality of gravitation may have been triggered by the apple falling on his head, it seems unlikely that it would have had that effect had he not already been thinking about gravitation and the motion of the moon.[2] I suggest that when a scientist offers a hypothesis they, or the rest of the scientific community, may have been considering the problem for a time. In addition to solving the problem, the hypothesis is also likely to be supported by other empirical evidence, or has some theoretical plausibility because it resembles previous successful solutions of other problems.[3] It may also fit in with an existing research program or look like a fruitful or interesting line of research. Another factor may be that the theory has desirable mathematical properties. For example, the Weinberg-Salam unified theory of electroweak interactions did not receive much attention until 1971 when 't Hooft showed it was renormalizable.[4] These may also be reasons why a theorist may pursue an hypothesis.

An experimentalist planning to investigate an hypothesis may have similar reasons for their work. In addition, there may be what one might call experimental reasons for such pursuit. These may include the fact that the proposed measurement can be done with existing apparatus or with small modifications of it. The measurement may fit in with an existing series of measurements in which the experimenter(s) have expertise or the experimenter may think of a clever way to perform the measurement. If the hypothesis is sufficiently important the experimenter may even construct an entirely new apparatus. At this point the cost of the experiment, the availability of research funds, as well as the perceived interest and importance of the experiment and hypothesis will certainly enter into the decision to do the experiment, but that is left for future discussion.

As we shall see below, the suggestion of a "Fifth Force" in gravitation occurred after the authors had been worrying about the problem for some time, did have some empirical support, and also resembled, at least in mathematical form, Yukawa's previous successful suggestion of the pion to explain the nuclear force. It also fit in with the previous work on modifications of gravitational theory by Fujii and others.

In this paper I will look at how and why Ephraim Fischbach, Sam Aronson, Carrick Talmadge, and their collaborators came to suggest modifying gravitational theory by adding such a force.[5] I will also examine the evidential context at the time that led at least a segment of the physics community to investigate this hypothesis. At the present time, I believe it is fair to say that the majority of the physics community does not believe that such a force exists. The current experimental limits on the strength of such a force are approximately 10^{-4} that of the normal gravitational force (This depends on the choice of coupling, i.e. baryon, isospin, etc., and on the assumed range of the force). Although some experimental anomalies do remain they are not presently regarded as serious.[6] There are, in addition, other experimental results which contradict the anomalous results and the overwhelming preponderance of evidence is against the existence of the Fifth Force.

2. K MESONS AND CP VIOLATION

The story of the Fifth Force begins with a seeming digression because it involves not a modification of gravitational theory, but rather new predictions and tests of that theory. In 1975 Colella, Overhauser, and Werner had measured the quantum mechanical phase difference between two neutron beams caused by a gravitational field.[7] Although these experiments showed the effects of gravity at the quantum level, they did not, in fact, distinguish between General Relativity and its competitors, as Fischbach pointed out (1980; Fischbach and Freeman 1979). This was because these experiments were conducted at low speeds, and in the non-relativistic limit all existing gravitational theories, such as General Relativity and the Brans-Dicke theory, reduced to Newtonian gravitation. Fischbach went on to discuss how one might test general relativity at the quantum level by considering gravitational effects in hydrogen.

In this work, partly as a result of conversations with Overhauser, Fischbach went on to consider whether or not gravitational effects might explain the previously observed violation of CP symmetry (combined charge-conjugation or particle-antiparticle symmetry and parity or space reflection symmetry) in K°_L decays. He had shown that an external gravitational field resulted in an admixture of atomic states of opposite parity. For a two-fermion system, such as positronium or charmonium, this also leads to a change in the eigenvalue of CP. This made "it natural to attempt to connect $V\sigma$ [the gravitational effect] with the known CP-violating K_L decays" (Fischbach 1980, p. 371. Although, as Fischbach noted, there were both experimental and theoretical reasons against gravity as the source of CP violation, the relevance of the arguments to his case were not clear.

The arguments Fischbach was referring to concerned attempts to explain CP violation and will be relevant to the later history as well. Bell and Perring (1964) and Bernstein, Cabibbo and Lee (1964) had speculated that a long-range external field that coupled differently to the K° and $\overline{K}_0$, a hyperphoton, could explain the observation. Such a field predicted that the effect would be proportional to the square of the energy of the K mesons. Weinberg (1964) had pointed out that because neither strangeness nor isotopic spin, the supposed origins of the field, were absolutely conserved, the hyperphoton must have a finite mass, related to the range of the interaction. Assuming that the range of the interaction was the size of our galaxy, he calculated the ratio ($K^o_2 \to 2\pi$ + hyperphoton)/($K^o_2 \to 2\pi$) as 10^{19}. This implied that the K meson and all strange particles would be totally unstable, in obvious disagreement with experiment. He could explain the observations if he assumed that the range of the interaction was the size of the earth, which he regarded as implausible. The issue became moot when the experiments of Galbraith et al. (1965) and of De Bouard et al. (1965) at very different energies from both each other and from the original experiment of Christenson et al. (1964) failed to show the predicted energy-squared dependence. In fact, the experiments indicated that the CP violation was constant as a function of energy for the energy range 1-10 GeV.

Fischbach was also motivated by what he took to be a "remarkable numerical relation." Using his calculated energy scale for the gravitational effect, δm, the known K_L-K_S mass difference, and an enhancement factor of $m_K/\delta m$, for which no justification was given, he found that his calculation of the gravitational effect for K mesons = 0.844×10^{-3}, whereas the CP violating parameter 1/2 Re ε was equal to $(0.83 \pm 0.03) \times 10^{-3}$ or $(0.80 \pm 0.03) \times 10^{-3}$, for K_L semileptonic decay or decay into two pions, respectively. This seems indeed to be a remarkable coincidence because there is no known connection between gravity and CP violation, or any accepted explanation of CP violation itself.

Fischbach continued to work on the question of how to observe gravitational effects at the quantum level. A relativistic version of the Colella, Overhauser, and Werner experiment using neutrons did not seem feasible, so he turned his attention to K mesons, where such experiments did seem possible. He began a collaboration with Sam Aronson, an experimental physicist with considerable experience in K meson experiments. At this time Aronson and his collaborators had been investigating the regeneration of K_S mesons (Bock et al. 1979; Roehrig et al. 1977). Although the published papers stated that "the data are consistent with a constant phase [of the regeneration amplitude]" (Bock 1979, p. 351-2),[8] Aronson and Bock, two members of the group were troubled by what seemed to be an energy dependence of the phase. In fact, Bock had investigated whether changes in the acceptance could account for the effect. They couldn't. The data are shown in Figure 1, along with the constant phase prediction. Although the data are consistent with a constant phase, there is at least a suggestion of an energy dependence. The low energy points have a larger phase than the high energy points.

Aronson and Bock then asked Fischbach if there was a theoretical explanation of the effect. Fischbach had none to offer. This suggested energy dependence led them to examine the possible energy dependence of the parameters of the K^o-$\bar{K}_0$ system in some detail. They found suggestive evidence for such a dependence (Aronson et al. 1982). They examined Δm, the K_L-K_S mass difference, τ_S, the lifetime of the short-lived K meson, η_{+-}, the magnitude of the CP-violating amplitude, and tan ϕ_{+-}, the tangent of the phase of the CP-violating amplitude. They fitted these parameters to an energy dependence of the form $x = x_0[1 + b_x \gamma^N]$, $N = 1,2$ and $= E_K/M_K$. They found that the

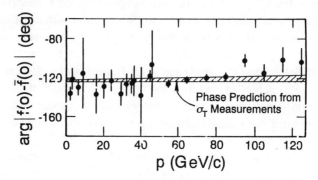

Figure 1. The phase of the regeneration amplitude as a function of momentum. From Bock *et al.* (1979)

coefficients differed from zero by 3, 2, 2, and 3 standard deviations, for the quantities noted above, respectively. One of their fits is shown in Figure 2.

The fit to ϕ_{+-} depended on the value one assumed for ϕ_{21}, the phase of the regeneration amplitude. (Recall that it was the suggestion of an energy dependence in this quantity that led to this investigation). The measured quantity, in fact, depends on $\phi_{21} - \phi_{+-}$. One could attribute the energy dependence to either one of them separately, or to both of them. All theoretical models at the time (See Aronson 1983b for details) predicted that over this energy range, 30-110 GeV, the change in ϕ_{21} would be less than 2°. The observed change of approximately 20° was then attributed to an energy dependence in ϕ_{+-}.

The group continued their study and presented a more detailed analysis that included data from other experiments. The most significant energy dependence, that in ϕ_{+-}, is shown in Figure 3.[9] They concluded, "The experimental results quoted in this paper are of limited statistical significance. The evidence of a positive effect in the energy dependences of Δm, τ_S, η_+, and ϕ_{+-} is extremely tantalizing, but not conclusive. The evidence consists of $b_x^{(N)}$'s which are different from zero by at most 3 standard deviations"(Aronson *et al.* 1983a, p. 488).

A second paper (Aronson *et al.* 1983b) examined possible theoretical explanations of the effects.[10] They found, "Using this formalism we demonstrate that effects of the type suggested by the data [energy dependences] cannot be ascribed to an interaction with kaons with an electromagnetic, hypercharge, or gravitational field, or to the scattering of kaons from stray charges or cosmological neutrinos" (Aronson 1983b, p. 495). They suggested that a tensor field mediated by a finite mass quantum might explain the effects and concluded, "It is clear, however, that if the data...are correct, then the source of these effects will represent a new and hitherto unexplored realm of physics" (Aronson *et al.* 1983b, p. 516).[11]

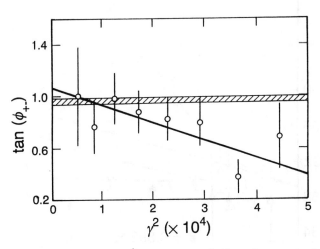

Figure 2. Tan ϕ_{+-} as a function of γ^2 (energy squared). From Aronson et al. (1982).

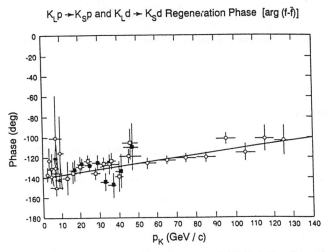

Figure 3. ϕ_{+-} as a function of mementum From Aronson et al. (1983a). See discussion in note 9.

The subsequent history of measurements of these quantities seems to argue against any energy dependence.[12] Coupal et al. (1985) measured η_{+-} at 65 GeV/c and found a value $\eta_{+-} = (2.28 \pm 0.06) \times 10^{-3}$ in good agreement with the low energy (5 GeV) of $(2.274 \pm 0.022) \times 10^{-3}$, and in disagreement with $(2.09 \pm 0.02) \times 10^{-3}$ obtained by Aronson et al. (1982).[13] Grossman et al. (1987) measured τ_S, the K_S lifetime, over a range 100-350 GeV/c. Their results, along with those of Aronson et al. (1982), are shown in Figure 4. The fits obtained by Aronson for possible energy dependence are also shown. They concluded, "No evidence was found for the momentum dependence suggested by

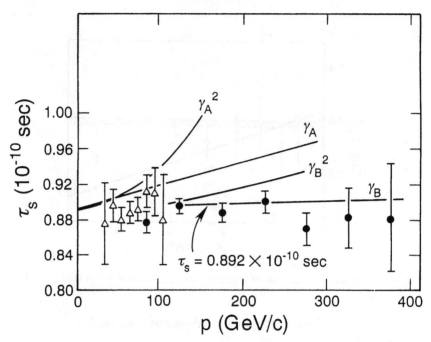

Figure 4. τ_S, the K_S lifetime as a function of momentum. From Grossman *et al.* (1987).

the intermediate-range 'fifth-force' hypothesis" (Grossman *et al.* 1987, p. 18). Still, at the time of our story, the suggested energy dependence remained a "tantalizing" effect.

3. MODIFICATIONS OF NEWTONIAN GRAVITY

A second strand of our story concerns the recent history of alternatives to, or modifications of, standard gravitational theory.[14] For a time, at least, this strand was independent of the K meson story. The Fifth Force story, per se, began when the two strands were joined. Newtonian gravitational theory and its successor, Einstein's general theory of relativity, although strongly supported by existing experimental evidence,[15] have not been without competitors.[16] Thus, Brans and Dicke (1961) offered a scalar-tensor alternative to General Relativity. This theory contained a parameter ω, whose value determines the relative importance of the scalar field compared to the curvature of spacetime. For small values of ω the scalar field dominates, while for large values the Brans-Dicke theory is indistinguishable from General Relativity. By the end of the 1970s experiment favored a large value of ω, and thus, favored General Relativity. For example, the lunar laser-ranging experiments required $\omega > 29$, while the Viking time-delay results set a lower limit of $\omega > 500$ (Will 1981).

In the early 1970s, Fujii (1971, 1972, 1974) suggested a modification of the Brans-Dicke theory that included a massive scalar exchange particle in addition to the usual massless scalar and tensor particles. He found that including such a particle gave rise to a force that had a short range (of the order 10 m – 30 km) depending on details of the model. In Fujii's theory, the gravitational potential had the form V = -GmM/r[1 + $\alpha e^{(-r/\lambda)}$], where α was the strength of the interaction and λ its range. The second term was Fujii's modification. O'Hanlon (1972) suggested the same potential. This model also predicted a gravitational constant G that varied with distance.[17] Fujii calculated that the gravitational constant at large distances G_∞ would be equal to 3/4 G_{LAB}, the value at short distances. Fujii also looked for possible experimental tests of this theory. Most interestingly for our story, he discussed the famous experimental test of Einstein's equivalence principle that had been performed by Eötvös, Pekar and Fekete (1922). (This experiment, which will be very important later in our history will be discussed below). He noted that if his new field coupled equally to baryons and leptons there would be no effect, while if the field did not couple to leptons such an effect would be observed. He calculated that for an Eötvös-type experiment on gold and aluminum there would be a change in angle η = 0.07×10^{-11}, for an assumed range of 40 km. The best experimental limit at the time was that of Roll, Krotkov, and Dicke (1964) of 3×10^{-11}, although that experiment which measured the equality of fall toward the sun cast no light on his short range force. A note added in proof remarked that he had learned that the best estimate of the range of such a force was considerably smaller. For an assumed 1km range he found that the change in angle was 0.5×10^{-9}. This was still smaller than the limit set by Eötvös, whose experiment did apply to his theory. He suggested redoing the Eötvös experiment and other possible geophysics experiments, although he noted that mass inhomogeneities would present difficulties. Other modifications of gravitational theory were suggested by Wagoner (1970), Zee (1979, 1980), Scherk (1979), and others. Zee's modification had a much shorter range than that of Fujii, whereas Scherk suggested a repulsive force with a range of about 1 km.

Long (1974) considered the question of whether or not Newtonian gravity was valid at laboratory dimensions. He "found that past G [the gravitational constant] measurements in the laboratory set only very loose limits on a possible variation in G and that present technology would allow a considerable improvement" [p. 850]. He also made reference to the suggestions of Wagoner, of Fujii, and of O'Hanlon. Long (1976) proceeded to test his hypothesis experimentally, and found a small variation in G of the form $G(R) = G_0[1 + 0.002 \ln R]$, where R is measured in centimeters.

Long's work led Mikkelsen and Newman (1977) to investigate the status of G.[18] They used data from laboratory measurements, orbital precession, planetary mass determinations, geophysical experiments, and solar models. They concluded, "Constraints on G(r) in the intermediate distance range from 10m < r < 1km are so poor that one cannot rule out the possibility that $G_c[G_\infty]$ differs greatly from G_0 [G_{LAB}]" (p. 919). They pointed out that their analysis "does not even rule out Fujii's suggested value $G_c/G_0 = 0.75$" (p. 924).

The experimental study of possible violations of Newtonian gravity continued. Panov and Frontov (1979) found G(0.3m)/G(0.4m) = 1.003 ± 0.006 and G(10m)/G(0.4m) = 0.998 ± 0.013. They concluded that, despite the fact that their experimental uncertainty was larger than Long's measured effect, "These results do not confirm the data of D.R.

Long, according to which spatial variations of G do exist" (p. 852). Spero *et al.* (1980) agreed. Their measurements at distances of 2cm-5cm had the required sensitivity to check Long's result and "The results support an inverse-square law. Assuming a force deviating from inverse square by a factor [1 + ε ln r] [Long's suggested form] it is found that $\varepsilon = (1 \pm 7) \times 10^{-5}$" (p. 1645).

Long (1981) surveyed the literature and concluded that within the quoted uncertainties all the results, including that of Panov and Frontov, were consistent with his observation. He argued that Spero's result did not, in fact, contradict his. This was because his suggested cause of the deviation, a quantum gravity vacuum polarization effect, would be significant only for a non-zero gravitational field, whereas Spero's experiment was conducted in a zero gravitational field.

The most important summary of this work, from the point of view of the subsequent history of the Fifth Force, was that given by Gibbons and Whiting (1981). Their results are shown in Figure 5 for both attractive and repulsive forces. In both cases α is restricted to lie below each curve, except for curve b, which is Long's result, and in which α must lie between the two curves. They stated, "However, conventional vacuum polarization effects in quantum gravity do not lead to the behavior required by Long: such effects are insignificant" (p.636). Curve a was Spero's data, calculated at 1 standard deviation (s.d.); c was from Panov and Frontov; d from Mikkelsen and Newman using lunar surface gravity and Mercury and Venus flybys; e a comparison of satellite and geodesy data by Rapp (1974, 1977), the upper curve assumed agreement to 0.1 ppm (parts per million) and the lower curve 1 ppm. Rapp had reported an agreement to 2 ppm.

A different type of experiment, that of measuring gravity in either a mine (Stacey *et al.* (1981) or in submarines (Stacey 1978) was discussed in curve f. The curves were calculated, not measured, assuming a mine experiment with an accuracy of 1% (upper curve) and a submarine experiment with an accuracy of 0.1% (lower curve), both for a depth of 1 km. Stacey *et al.* (1981) had measured G and found it to be G = (6.71 ± 0.13)×10^{-11} $m^3 kg^{-1} s^{-2}$, in agreement with the laboratory value of (6.672 ± 0.004)×10^{-11}. Their quoted uncertainty included an estimate of possible systematic effects, which increased the uncertainty by about a factor of three. They also surveyed other mine and borehole measurements of G and found them to be, in general, systematically slightly high but "tantalizingly uncertain" because of possible mass anomalies. A somewhat later paper (Stacey and Tuck 1981) gave numerical details of that survey and reported values of G, calculated in two different ways, based on a comparison of sea floor and sea surface measurements of G = (6.730 ± 0.010)×10^{-11} and (6.797 ± 0.016)×10^{-11} $m^3 kg^{-1} s^{-2}$, where the uncertainty is purely statistical and does not include possible systematic effects. Once again the results were higher than the laboratory value, but because of the uncertainty about possible systematic effects no firm conclusion could be drawn.

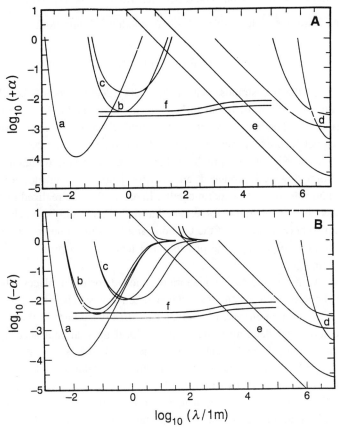

Figure 5. $\log_{10}\alpha$ versus $\log_{10}(\lambda/1m)$. α is constrained to lie below the curves. From Gibbons and Whiting (1981).

Gibbons and Whiting summarized the situation as follows.

> It has been argued that our experimental knowledge of gravitational forces between 1m and 10km is so poor it allows a considerable difference between the laboratory measured gravitational constant and its value on astronomical scales -- an effect predicted in theories of the type alluded to above [These included Fujii and O'Hanlon, whose work was also cited in the experimental papers] ....it can be seen that for 3m < λ < 10km α is very poorly constrained (p. 636).
>
> We conclude that there is very little scope for a theory which allows deviations > 1% from Newton's law of gravitational attraction on

laboratory or larger length scales.... . Further large scale experiments are essential to improve bounds on α between 1m and 10km (p. 638).

There was, however, a clear, although small, window of opportunity.

4. THE FIFTH FORCE

Until early 1983 the two strands, that of the energy dependence of the K^0-$\bar{K}_0$ system parameters and that of modifications of Newtonian gravity and their experimental tests, had proceeded independently. At about this time Fischbach became aware of the discrepancies between experiment and gravitational theory.[19] He made no connection, at this time, between the two problems because he was still thinking in terms of long-range forces, which produced an energy-squared dependence of the K^0-$\bar{K}_0$ parameters, and was ruled out experimentally. In early 1984, he realized that this would not be the case for a short-range force, and that the effect could be much smaller.[20] At this time he also became aware of the Gibbons and Whiting summary and realized that such a short-range force might be possible and that the two problems might have a common solution.

Fischbach, Aronson, and their collaborators looked for other places where such an effect might be seen with existing experimental sensitivity. They found only three: 1) the K^0-$\bar{K}_0$ system at high energy, which they had already studied; 2) the comparison of satellite and terrestrial determinations of g, the local gravitational acceleration;[21] and 3) the original Eötvös experiment, which measured the difference between the gravitational and inertial masses of different substances. If a short-range composition-dependent force existed it might show up in this experiment. They noted that the very precise modern experiments of Roll, Krotkov, and Dicke (1964) and of Braginskii and Panov (1972) would not have been sensitive to such a force because they had compared the gravitational accelerations of pairs of materials toward the sun, and thus looked at much larger distances.

The apparent energy dependence of the K^0-$\bar{K}_0$ parameters along with the discrepancy between gravitational theory and experiment led Fischbach, Aronson and their colleagues (Fischbach *et al.* 1986) to reexamine the original data of Eötvös *et al.* (1922) to see if there was any evidence for a short-range, composition-dependent force.[22] By this time they knew of Holding and Tuck's (1984) result which gave G measured in a mine as $G = (6.730 \pm 0.003) \times 10^{-11}$ m^{3}kg^{-1}s^{-2} in disagreement with the best laboratory value of $(6.6726 \pm 0.0005) \times 10^{-11}$. This result was still uncertain because of possible regional gravity anomalies. Fischbach used the modified gravitational potential $V(r) = -Gm_1m_2/r[1 + \alpha e^{(-r/\lambda)}]$. They remarked that such a potential could explain the geophysical data quantitatively if $\alpha = (-7.2 \pm 3.6) \times 10^{-3}$, with $\lambda = 200 \pm 50$ m. (This was from a private communication from Stacey. Details appeared later in Holding, Stacey and Tuck (1986)). This potential had the same mathematical form as that suggested much earlier by Fujii. Recall also that Fujii had suggested redoing the Eötvös

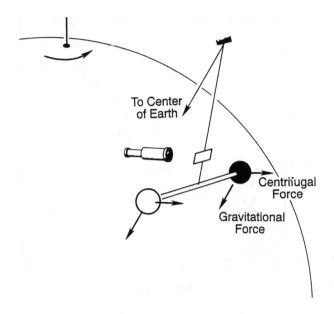

Figure 6. A schematic view of the Eötvös experiment. From Will (1984).

experiment. Fujii's work does not seem to have exerted any direct influence on Fischbach. No citations of it are given in this paper.[23]

The apparatus for the Eötvös experiment is shown schematically in Figure 6.[24] The gravitational force is not parallel to the fiber due to the rotation of the earth. If the gravitational force on one mass differs from that on the other the fiber will rotate. Reversing the masses should give a rotation in the opposite direction.

Fischbach attempted to combine the gravitational discrepancy with the energy dependence of the K^0-$\overline{K}_0$ parameters. They found that if they considered a hypercharge field with a small, finite mass hyperphoton (the K^0 and $\overline{K}_0$ have different hypercharges) they obtained a potential of the same form as shown above.[25] They also found that $\Delta k = \Delta a/g$, the fractional change in gravitational acceleration for two substances, would be proportional to $\Delta(B/\mu)$ for the two substances, where B was the baryon number of the substance (equal, in this case, to the hypercharge) and μ was the mass of the substance in units of the mass of atomic hydrogen.

They plotted the data reported by Eötvös as a function of $\Delta(B/\mu)$, a quantity unknown to Eötvös, and found the results shown in Figure 7. The linear dependence visible is supported by a least-squares fit to the equation $\Delta k = a\, \Delta(B/\mu) + b$. They found a = $(5.65 \pm 0.71) \times 10^{-6}$ and b = $(4.83 \pm 6.44) \times 10^{-10}$. They concluded, "We find that the Eötvös-Pekar-Fekete data are sensitive to the composition of the materials used, and that their results support the existence of an intermediate-range coupling to baryon number or hypercharge" (Fischbach *et al.* 1986, p. 3).[26] They calculated the coupling

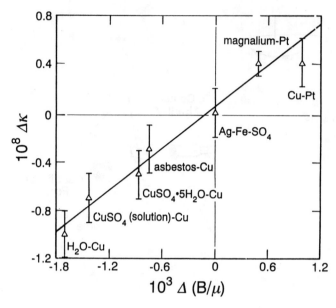

Figure 7. Δk as a function of Δ(B/μ). From Fischbach et al. (1986).

constant for their new interaction for both the Eötvös data and for the geophysical data and found that they disagreed by a factor of 15, which they found "surprisingly good" in view of the simple model of the earth they had assumed.

Not everyone was so sanguine about this.[27]

It seems fair to summarize the Fischbach paper as follows. A reanalysis of the

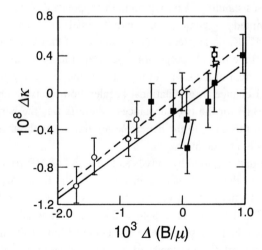

Figure 8. (See note 27.) Final summary reported by Eötvös and Fischbach's reanalysis.

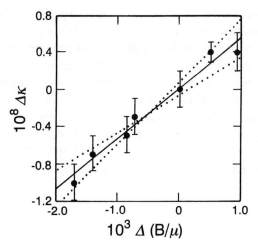

Figure 9. (See note 27). 95% confidence level in Fischbach's data.

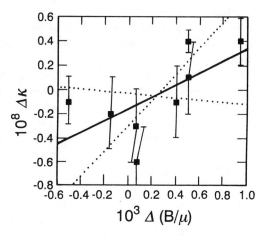

Figure 10. (See note 27). 95% confidence level in Eötvös' data.

original Eötvös paper presented suggestive evidence for an intermediate-range, composition-dependent force, which was proportional to baryon number or hypercharge. With a suitable choice of parameters, one could relate this force to anomalies in mine measurements of gravity and to a suggested energy dependence of the parameters of the K^0-$\bar{K}_0$ system. It was these three pieces of suggestive, although certainly not conclusive, evidence that led others to extend the investigation of the Fifth Force, both experimentally and theoretically.[28] The fact that it also fit in with the tradition of modifying gravitational theory of Fujii and others no doubt strengthened that decision. The context of pursuit is not devoid of evidence.

The fact that the subsequent history has argued against the existence of the Fifth Force should not cause us to overlook the fact that the suggestion of that force was the result of a sequence of reasonable and plausible steps. This started with the Colella, Overhauser, and Werner measurement, Fischbach's attempt to connect CP violation and gravity, and the subsequent observation of suggestive energy dependence of the K^0-$\bar{K}_0$ parameters. At the same time the work on the modifications of Newtonian gravity and the tantalizing results on the measurement of gravity in mines were proceeding. When these two strands were joined together it led the collaborators to reanalyze the original Eötvös experiment, where again a suggestive effect appeared. The suggestion of the Fifth Force then followed. There may not be a logic of discovery, but, at least in this case, it is not a totally mysterious process.

NOTES

[1] For a discussion of similar issues see "Discovery, Pursuit, and Justification."

[2] Professor Sam Westfall, the noted Newton scholar and biographer, believes the story of Newton and the apple is true. He reports that Newton repeated it on at least four occasions during his lifetime. (Private communication).

[3] I do not suggest that these are necessary or sufficient conditions for such hypothesis creation, after all an individual might come up with a solution of the first try, but my speculation is that these are the usual circumstances.

[4] Pickering (1984b, p. 106) noted that the citation history of Weinberg's paper clearly shows this. 't Hooft showed the theory was renormalizable in 1971. The citations were 1967, 0; 1968, 0; 1969, 0; 1970, 1; 1971, 4; 1972, 64; 1973, 162.

[5] These three physicists played the leading roles in the formulation of the Fifth Force hypothesis. I will refer to papers written by them and their collaborators by the first author listed. This may give the impression that only a single author was involved. This is definitely not the case. Virtually all of these papers had multiple authors.

[6] More informally, during a discussion held at the 1990 Moriond Workshop, Orrin Fackler, a member of the Livermore Tower group, remarked, "The Fifth Force is dead." No dissenting voice was heard. The group included many of those who have been active in working on the Fifth Force including Ephraim Fischbach and Sam Aronson, who originally suggested it. (I was present at the discussion). Perhaps a bit more formally, the Fifth Force did not appear on the program for the 1991 Moriond Workshop. There are some who might question whether or not the force was ever alive.

[7] Overhauser and Colella (1974) contains a discussion of the theoretical aspects of this experiment

[8] A stronger statement had appeared earlier. "The results are clearly consistent with constant phase..." (Roehrig *et al.* 1977, p. 1118).

[9] The graph actually shows the energy dependence of ϕ_{21}, assuming ϕ_{+-} was constant. If ϕ_{21} is considered to be a constant, the graph shows the energy dependence of ϕ_{+-}.

[10] An earlier paper (Fischbach 1982) had examined the same question, although in less detail, and reached the same conclusion

[11] These results were not greeted with enthusiasm or regarded as reliable by everyone within the physics community. Commenting on the need for new interactions to explain the effects, an anonymous referee remarked, "This latter statement also applies to spoon bending." (A copy of the referee's report was given to me by Fischbach). The paper was, however, published.

[12] This possible energy dependence played an important role in the genesis of the Fifth Force hypothesis, as discussed in detail below. It may very well have been a statistical fluctuation

[13] There is a further oddity in this history. The six measurements of η_{+-} made prior to 1973 had a mean value of $(1.95 \pm 0.03) \times 10^{-3}$. The value cited by Coupal et al. was the mean of post-1973 measurements. For details of this episode see Franklin (1986, ch. 8).

[14] I will be discussing here modifications of the Newtonian inverse square law, and not the well-established relativistic post-Newtonian corrections, which are of order $GM/c^2 r$.

[15] For an excellent and accessible discussion of this see Will (1984). For more technical details see Will (1981).

[16] The history of gravitational theory is not a string of unbroken successes. Newton himself could not explain the motion of the moon in the *Principia* and his later work on the problem, in 1694-5, also ended in failure. (Westfall 1980, pp. 442-3, 540-8). The law was also questioned during the 19th century when irregularities were observed in the motion of Uranus. The suggestion of a new planet by Adams and LeVerrier and the subsequent discovery of the planet Neptune turned the problem into a triumph. During the 19th century it was also found that the observed advance of the perihelion of Mercury did not match the predictions of Newtonian theory. This remained an anomaly for 59 years until the advent of Einstein's General Theory of Relativity, the successor to Newtonian gravitation

[17] Some readers might worry that a variable constant is an oxymoron, but it does seem to be a useful shorthand

[18] The influence of Long's work is apparent in the first sentence of the abstract. "D.R. Long and others have speculated that the gravitational force between point masses in the Newtonian regime might not be exactly proportional to $1/r^2$" (Mikkelsen and Newman, 1977, p. 919).

[19] Fischbach [private communication] attributes this to a conversation with Wick Haxton, who told him about it.

[20] Fischbach's first calculation was for a δ-function force.

[21] Rapp (1974, 1977) had already found $\Delta g/g \sim (6 \pm 10) \times 10^{-7}$. For the proposed Fifth Force parameters (see below) the predicted effect would be approximately 2×10^{-7}.

[22] Because the energy dependence of the K^0-$\overline{K}_0$ parameters might have indicated a violation of Lorentz invariance, Fischbach (1985) had looked at the consequences of such a violation for the Eötvös experiment.

[23] Fischbach keeps detailed chronological notes of papers read and calculations done. He reports that he has notes on Fujii's work at this time, but does not recall it having any influence on his work

[24] Eötvös was originally interested in measuring gravity gradients so the weights were suspended at different heights. This introduced a source of error into his tests of the equivalence principle, the equality of gravitational and inertial mass.

[25] Fischbach noted that in the limit of infinite range their suggested force agreed with that proposed by earlier by Lee and Yang on the basis of guage invariance.

[26] In a later paper (Aronson 1986) the group suggested other experiments, particularly on K meson decay, that might show the existence of such a hyperphoton

[27] An interesting sidelight to this reanalysis is reported in a footnote to the Fischbach paper. Instead of reporting the observed values of Δk for the different substances directly, Eötvös and his colleagues presented their results relative to platinum as a standard. "The effect of this combining say $\Delta k(H_2O\text{-}Cu)$ and $\Delta k(Cu\text{-}Pt)$ to infer $\Delta k(H_2O\text{-}Pt)$ is to reduce the magnitude of the observed nonzero effect [for water and platinum] from 5σ to 2σ" (Fischbach *et al.* 1986, p.6). Δk $(H_2O\text{-}Cu) = (-10 \pm 2) \times 10^{-9}$ and $\Delta k(Cu\text{-}Pt) = (+4 \pm 2) \times 10^{-9}$, respectively. Adding them to obtain $\Delta k(H_2O\text{-}Pt)$ gives $(-6 \pm 3) \times 10^{-9}$. Figure 8 shows both the final summary reported by Eötvös as well as Fischbach's reanalysis, along with best-fit straight lines for both sets of data separately. (This is my own analysis). Although several of the experimental uncertainties have increased, due to the calculation process, the lines have similar slopes. The major difference is in the uncertainty of the slopes. If one looks at the 95% confidence level, as shown separately for the Fischbach and Eötvös data, respectively, in Figures 9 and 10, one finds that at this level the original Eötvös data is, in fact, consistent with no effect, or a horizontal straight line. This is certainly not true for the Fischbach reanalysis. A skeptic might remark that the effect is seen only when the data are plotted as a function of $\Delta(B/\mu)$, a theoretically suggested parameter. As De Rujula remarked, "In that case, Eötvös and collaborators would have carried their secret to their graves: how to gather ponderous evidence for something like baryon number decades before the neutron was discovered" (1986, p. 761). It is true that theory may suggest where one might look for an effect, but it cannot guarantee that the effect will be seen. Although one may be somewhat surprised, along with De Rujula, that data taken for one purpose takes on new significance in the light of later experimental and theoretical work, it is not unheard of. There is a possibility that Eötvös and his collaborators might actually have seen something of this effect, but discounted it. They report, "The probability of a value different from zero for the quantity$\times[\Delta k]$ even in these cases is vanishingly little, as a review of the according observational data *shows quite long sequences with uniform departure from the average* [emphasis added], the influence of which on the average could only be annulled by much longer series of observations" (Eötvös, Pekar, and Fekete 1922, p. 164. The original summary gives an average value for $\times = (-0.002 \pm 0.001) \times 10^{-6}$, which seems to justify Eötvös's original conclusion, "We believe we have the right to state that$\times$relating to the Earth's attraction does not reach the value of 0.005×10^{-6} for any of these bodies" (Eötvös, Pekar, and Fekete 1922, p. 164).

[28] Not everyone in the physics community was impressed by the evidence presented. Glashow, a Nobel Prize-winning particle theorist, was quite negative. "Unconvincing and unconfirmed kaon data, a reanalysis of the Eötvös experiment depending on the contents of the Baron's wine cellar [a humorous allusion to the importance of local mass inhomogeneities], and a two-standard-deviation geophysical anomaly. Fischbach and his friends offer a silk purse made out of three sows' ears, and I'll not buy it" (quoted in Schwarzschild 1986, p. 20). For further details see Franklin (1993a).

CHAPTER 5

THERE ARE NO ANTIREALISTS IN THE LABORATORY

One of the continuing polemics in the philosophy of science has been the battle between the realists and the antirealists. Although there are probably as many variants of these positions as there are adherents of them, I shall adopt, and defend, Bas van Fraassen's characterization of realism, "Science aims to give us, in its theories, a literally true story of what the world is like; and acceptance of a scientific theory involves the belief that it is true" (1980, p. 8). I shall argue that we have good reasons to believe in both the truth of scientific laws and theories and in the reality of the entities involved in those theories.

The battle has flared up since van Fraassen's seductive account of an antirealist position, that of constructive empiricism.[1] In his own words, "Science aims to give us theories which are empirically adequate; and acceptance of a theory involves as belief only that it is empirically adequate," (1980, p. 12), where empirical adequacy means that what the theory "says about the observable things and events in the world, is true -- exactly if it 'saves the phenomena'" (1980, p. 12).[2] I shall discuss this view in some detail later.

My own position, which one might reasonably call "conjectural" realism, includes both Sellars's view that "to have good reason for holding a theory is *ipso facto* to have good reason for holding that the entities postulated by the theory exist," (Sellars, 1962, p. 97), and the "entity realism" proposed by Cartwright (1983) and by Hacking (1983). Both Hacking and Cartwright emphasize the manipulability of an entity as a criterion for belief in its existence. "We are completely convinced of the reality of electrons when we regularly set out to build -- and often enough succeed in building -- new kinds of device that use various well-understood causal properties of electrons to interfere in other more hypothetical parts of nature" (Hacking, 1983, p.265).[3] Cartwright also stresses causal reasoning as part of her belief in entities. In her discussion of the operation of a cloud chamber she states, "...if there are no electrons in the cloud chamber, I do not know why the tracks are there." (Cartwright, 1983, p.99). In other words, if such entities don't exist then we have no plausible causal story to tell. Both Hacking and Cartwright grant existence to entities such as electrons, but do not grant "real" status to either laws or theories, which may postulate or apply to such entities.

In contrast to both Cartwright and Hacking, I suggest that we can have good reasons for belief in the laws and theories governing the behavior of the entities, and that several of their illustrations implicitly involve such laws.[4] I shall present an illustration of my own later. I agree with them, however, that we can go beyond Sellars and have good reasons for belief in entities even without such laws.

1. ARE THERE REALLY K MESONS?

Before discussing some of the philosophical issues, I would like to present an example from contemporary physics which illustrates my view. I will argue that this experiment provides good reasons to believe in both the existence of K mesons, as well as in the truth of several laws involved in the operation of the experimental apparatus.

I do not, however, wish to imply by this that experiment always provides good reasons for belief in entities and laws.[5] I do wish to argue that it can, and does, do something I believe Van Fraassen would deny. One must also be careful about which entities the experiment argues for and which are assumed on the basis of prior knowledge. Thus, in the example that follows, the experiment provides reasons for belief in K mesons, but the existence of electrons, protons, and pions and their properties is regarded as unproblematic. One should also note that the status of an entity may change during the course of an experiment. In the case of the experiments designed to demonstrate the existence of weak neutral currents, the existence of the currents was shown only after all the data was taken and an argument constructed that the events seen could not be due to neutron background or to the failure of the apparatus to detect muons.[6] (See Galison (1987) for details). As David Cline, one of the experimenters, remarked toward the end of one of the experiments, "At present I don't see how to make these effects [the neutral currents] go away (Galison, 1987, p. 235)." I note also that in this experiment the existence of neutrinos, neutrons, pions, muons, and electrons was taken as given.

The evidence for the reality of entities may also change over time. Thus, when Gell-Mann gave a seminar on the quark model in its early stages, before any experiments were conducted, he suggested that one might regard them either as useful mathematical devices or as real entities. After experiments on deep inelastic electron scattering were completed there was more reason to believe that quarks were real. Subsequent experiments strengthened that belief.

The experiment I wish to discuss was designed to measure the K^+_{e2} branching ratio, the fraction of all K^+ mesons that decay into a positron and a neutrino. (Bowen *et al.* 1967). The first order of business was to obtain a supply of stopped K^+ mesons. The experimental apparatus is shown in Figure 1. The group obtained an unseparated positive beam of momentum 530 MeV/c from the Princeton-Pennsylvania Accelerator (PPA). This beam included pions and protons, in addition to the K^+ mesons (kaons) needed. The kaons were identified by their range in matter and by time of flight. The beam telescope consisted of four scintillation counters, C_1, C_2, C_3, and C_4, with 6.7 cm of copper placed before the stopping region, which was counter C_3. A stopped particle was indicated by a coincidence between C_1, C_2, and C_3 with no pulse in C_4 ($C_1 C_2 C_3 \overline{C_4}$). The copper eliminated virtually all of the protons, which have a shorter range in matter than kaons, before they reached C_3. In addition, protons do not decay into positrons.[7] Pions were a more serious problem. There were about 1000 times as many pions as kaons in the beam. Most of the pions, which have a longer range than kaons, passed through the stopping region and counted in C_4, and were eliminated, reducing the ratio

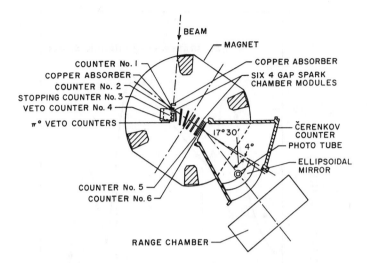

Figure 1. Plan view of the experimental apparatus used to measure the K^+_{e2} branching ratio. From Bowen et al (1967)

of pions to kaons to about 100 to 1. Time of flight provided additional discrimination. Particles of the same momentum (this was a momentum selected beam) but different masses have different velocities and therefore different times of flight. The internal proton beam at the PPA consisted of bunches of protons separated in time by 34 ns. Thus, particles were produced every 34 ns. A signal from the RF (radio-frequency) system of the accelerator signalled the production of particles and could therefore be used to time the beam particles. For the beam transport system used in this experiment the difference in time of flight between pions and kaons was 8 ns, so a narrow coincidence of 3 ns ($C_1C_2C_3\overline{C}_4$ + RF) was used to separate kaons from pions. The background of unwanted pions was reduced to approximately 5%.[8]

Particles from decays at approximately 90^0 to the incident beam were detected by two scintillation counters, C_5 and C_6, and then passed through a gas Cerenkov counter, which was set to detect high energy positrons. The time between the K^+ stop and the decay particle was recorded for each event.

A K^+ decay was identified by a coincidence between a stopping K^+ signal ($C_1C_2C_3\overline{C}_4$ + RF) and a decay particle pulse (C_5C_6). If the events were really due to K^+ decays the time distribution between the K^+ stop and the decay pulse should match the known K^+ lifetime. This experimental check was performed, the lifetime measured, and the positive results found are shown in Figure 2. An electronic gate was used to eliminate pion background, as shown in the figure.

Thus, the experimenters determined that the particles had a definite charge, mass, and lifetime[9] which, in addition, agreed with the known properties of the kaon.[10] If there are no K mesons, then we have no plausible explanation of what was being observed. It would seem odd, in such circumstances, to refer to the kaons as merely useful fictions,

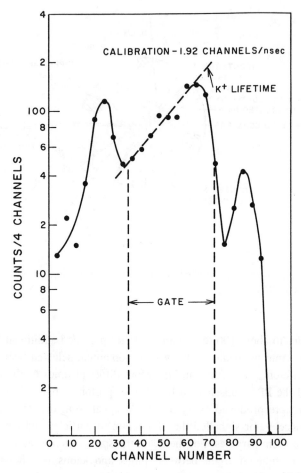

Figure 2. The decay time spectrum obtained by Bowen et al (1967), The K^+ lifetime is shown.

even if the particles could not be directly observed in Van Fraassen's sense of unaided human sense perception. I note that the three properties were sufficient to identify the particle as a kaon.[11]

This procedure also seems to me to be quite similar to identifying a person by noting that measurements on them of all the properties listed on a driver's license give definite values. Suppose we had an entity who had a definite height, weight, gender, hair color, eye color, date of birth, and home address, which were, in addition, exactly those listed on the driver's license of Bas van Fraassen. Would we not be justified in concluding not only that there is a real Bas van Fraassen, but that the entity we are observing is one and the same person. Yes, the skeptical reader, or an antirealist, might reply, but we can observe the entity directly. Suppose, however, that we had determined all of these quantities without such direct viewing, as well we might have.[12] Would we not still be justified in believing in the real existence of the philosopher? This seems to me to be by far the best explanation of the observations. If there is no Bas van Fraassen then we are faced with a remarkable and bizarre set of coincidences -- similarly for K mesons.[13]

It might be argued that there is a difference between the the arguments for the existence of K mesons and the existence of Bas van Fraassen. Van Fraassen is an individual whereas K mesons are a type.[14] I do not believe that this is a valid objection. I could, for example, name my K mesons. Lest this be regarded as totally facetious, let me point out that scientists are, in fact, already referring to individual elementary particles by name.

> 'Here...in the center of our Penning trap resides positron (or anti-electron) Priscilla, who has been giving spontaneous and command performances of her quantum jump ballets for the last 3 months.' There can be little doubt about the identity of Priscilla during this period, since in ultrahigh vacuum she never had the chance to trade places with a passing antimatter twin. The well-defined identity of this elementary particle is something fundamentally new, and deserves to be recognized by being given a name, just as pets are given names of persons (Dehmelt 1990, p. 539).

A further example of the existence of, at least, a small number of entities, namely xenon atoms is shown in Figure 3. A group of scientists used a scanning, tunneling electron microscope to manipulate a group of xenon atoms to spell out "IBM." (The company the scientists worked for is clear). This is an example in which the manipulability of the objects themselves argues for their existence. As Hacking might say, "If you can spell IBM with them, they are real."

In establishing the observed properties of the kaons we have also made use of, and a commitment to, several physical laws. Thus, the momentum of the particles was fixed by requiring them to travel in a circle of fixed radius in a known magnetic field and then determining the momentum by using $F = ma$, $a = v^2/r$, and $F = qvB$ where F is the force on the particle, q, m, v, and a are its charge, relativistic mass, velocity, and acceleration, r is the radius of curvature, and B is the known magnetic field. From this we find that momentum $= mv = qBr$. We have also made use of the dependence of the range of charged particles on their charge, mass, and velocity. Our use of these laws to establish the properties of kaons seems to me to give them the same epistemic status as the particles and their properties. The successful performance of the measurements gives us reason to believe not only in the kaons, but also in the laws. If the laws weren't valid it is hard to imagine that the measurements would be possible.[15] In contrast to Sellars's view, these laws do not involve K mesons *per se*, but are laws obeyed by all charged particles. Belief in these particular laws does not give us good reasons for belief in K mesons.

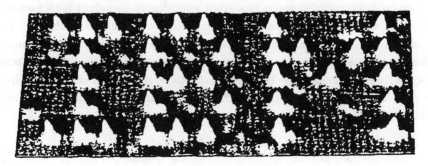

Figure 3. A magnified picture of xenon atoms spelling out IBM.

An example where the support for a law does give us good reason to believe in an entity is the following. The discovery of the Ω^- particle supported the eightfold way, a particle classification scheme, which both predicted and accounted for the existence of certain elementary particles, including the Ω^-. This discovery also gave us good reason for belief in kaons, which are part of the scheme. This will be discussed further below. Similar support would be provided for the existence of elements by the discovery of new elements predicted by Mendeleev's Periodic Table.[16]

2. IS CONSTRUCTIVE EMPIRICISM EMPIRICALLY ADEQUATE?

A) IS IT EMPIRICAL?

How might van Fraassen react to the argument given above for the existence of K mesons?[17] His discussion in *The Scientific Image* (1980, pp. 75-77) states that measurement of a particle's properties, such as the charge on the electron, does not imply that the particle exists. In his view a theory leaves blanks for experiment to fill in and that this type of experiment, "*shows how that blank is to be filled in if the theory is to be empirically adequate*" (1980, p. 75). I shall discuss in detail later why I believe that this analysis, based on Millikan's experiments, is historically inadequate. I also do not believe that this is a sufficient answer to the argument in favor of the existence of kaons. First, there is no theory of kaons for which the blanks have to be filled in, unless it is the statement that every particle has a mass, a charge, and a lifetime.[18] Second, the experimenters measured three properties of the particle simultaneously, rather than just one, providing more evidence for its existence. It is hard to imagine filling in three blanks simultaneously without believing that there is actually something that has the properties.

Van Fraassen would also regard the K meson or its properties as unobservable because the story I have recounted involves instrumental detection and not unaided human perception. "A calculation of the mass of a particle from the deflection of its trajectory in a known force field is not an observation of that mass" (1980, p.15). There is a real question as to why van Fraassen privileges unaided sense perception in arguing for the existence of entities and for the validity of measurements.[19] Human perception is notoriously unreliable. It can be influenced by weather conditions (mirages), the state of the body (alcohol, drugs, etc.), stress, and so on. Eyewitness identification in trials has been shown to be far from infallible, and optical illusions do occur. It seems to me that the arguments one gives for the validity of human observations are, in fact, the same arguments one gives for the validity of instrumental observations and measurements and that neither is privileged over the other. (For details of the strategies used to establish the validity of experimental results see Franklin (1986, Chapter 6) and Franklin and Howson (1988). A similar point on instrumental detection has been made by Bogen and Woodward (1988)).

Despite this argument, van Fraassen might very well still deny the existence of the kaon on the grounds that we have only measured or calculated its properties and not observed it directly. It is not clear to me, however, how he would explain the measurements if not by the existence of a particle. As Cartwright might say, "If there are no kaons then we have no explanation of our measurements."[20]

Van Fraassen might respond that, although there are indeed K mesons, constructive empiricism deals only with the attitude one should take toward theories that involve kaons. He might further state that the evidence in support of such theories only gives us reason to believe the theory is empirically adequate and does not give us reason to believe in the existence of kaons, whereas the measurements I have cited do support such existence. I don't believe that such an evidential distinction can be maintained, that between evidence in support of the existence of a particle such as the kaon and evidence that supports a theory of kaons, and therefore their existence.

Consider the eightfold way mentioned above. One found evidence in favor of the existence of the Ω^- particle, i.e. its mass and charge. The existence of such a particle was predicted by the eightfold way and therefore its observation supports, at the very least, the empirical adequacy of the theory. But, the content of the theory is precisely the existence of certain particles and their properties. Therefore, I believe that van Fraassen would have to accept that detection of the Ω^- provided support for the existence of K mesons. I don't believe that having accepted this argument that van Fraassen is then justified in excluding evidence that supports a theory, but does not bear directly on the entities involved in the theory. For example, observation of atomic parity violation supports the Weinberg-Salam unified theory of electroweak interactions. It also supports the existence of the the intermediate vector bosons that the theory uses to explain this effect.[21] It would seem then that he would have to agree with Sellars and myself that any evidence that supports a theory gives ipso facto good reasons for believing in the entities postulated by that theory. Once van Fraassen has accepted the arguments given in favor of the existence of K mesons, as I believe he should, then he has lost the battle.

A second point concerns the question of what van Fraassen means when he says that acceptance of a theory involves only the claim that what it says about "observable phenomena" (1980, p. 57) is true. He states earlier that "the term 'observable' classifies putative entities (entities that may or may not exist)" (1980, p. 15). It seems clear that he cannot regard only entities as observables, for then theories would make very few statements concerning observables. Theories also make statements about measurable quantities and certainly acceptance of a theory must involve, at the very least, the belief that what it says about such measurable quantities is true. If a constructive empiricist does not subscribe to at least this much then the science they believe in is so impoverished as to scarcely merit the name.

To make this clearer, let us consider a kinetic theory which consists of a hard sphere model of gas molecules and Newton's laws. One consequence of this theory is that PV = constant, at constant temperature, where P is the pressure of the gas, and V is its volume. We would certainly regard V as measurable, if not observable. For a reasonably sized container, which has the shape of a rectangular solid, we can determine the volume using only a ruler and the naked eye. What then of pressure? We can certainly detect pressure with our unaided senses. All one needs to do is to dive beneath the surface in a swimming pool. This would seem to make pressure an observable. In order to see whether or not what the theory says about these observables is true we must not only be able to detect pressure, we must be able to measure it. To do this we will need measuring devices or instruments which not only will go beyond the unaided senses[22] but will also, no doubt, involve a theory of the apparatus.[23] I don't see how a constructive empiricist can avoid allowing quantities like pressure and volume to be observables. Otherwise theories will have virtually no empirical content. Once, however, it is granted that pressure is an observable then other quantities detected only with instruments also become observable, such as the mass, lifetime, and charge etc. of particles. That being granted, we can then say that we have good reasons, based on observation, for the existence of entities.

B) IS IT ADEQUATE?

One of Van Fraassen's arguments in favor of constructive empiricism is that it provides an accurate description of scientific practice. "However, there is also a positive argument for constructive empiricism – it makes better sense of science, and scientific activity, than realism does and does so without inflationary metaphysics" (1980, p. 73). I wish to dispute this claim.

If scientists are interested only in empirical adequacy then we cannot explain why so much effort is devoted to resolving a contradiction between two theories, each of which is empirically adequate within a certain range of phenomena. Why should scientists have worried about the contradiction between Maxwell's electromagnetic theory and Bohr's atomic theory? Both were, after all, empirically adequate within their own ranges of phenomena. It was the fact that Maxwell's theory predicted that the Bohr atom would be unstable (the Bohr atom should decay in 10^{-9} seconds). Why not accept the empirical adequacy of the theories for their separate phenomena, and let it go at that? This is not what scientists do, or what they did in this case. They work to resolve or eliminate the contradiction. The scientific realist has an explanation of this. The reason scientists act as they do is because when they accept a theory they believe in its truth, and truth cannot be contradictory.

Van Fraassen might respond that Bohr's theory shows that Maxwell's is empirically inadequate, or vice versa. But this will not work. It is true that the two theories are incompatible, but there is nothing in the available data on atomic spectra that contradicts Maxwell's theory. The theory didn't say anything about atomic spectra. It is true that there was no classical electromagnetic explanation of the phenomena, but that might just provide a reason for looking for one. Still, there already existed an empirically adequate theory of atomic spectra, namely Bohr's theory. Why search for another one if your aim is only saving the phenomena?

An antirealist might respond that the removal of the contradiction is necessary in order to construct a unified theory of the phenomena, and that unified theories have, in the past, turned out to be more empirically adequate, i.e. they fit more phenomena than the separate theories they replace. I agree that the search for unified theories is part of scientific practice, and that it has sometimes, although certainly not always, been successful in the past. Recall Einstein's or Heisenberg's futile searches for a unified field theory late in their careers.[24] I do not believe the antirealist position offers any advantage here over the realist view in the explanation of scientific practice. At best, it seems equally good. This judgment of equality depends, of course, on believing that the search for a unified theory is usually, if not always, successful, and that such unification results in theories that are more empirically adequate. I am not convinced that this is always the case. The unifications we remember in the history of science do satisfy this criterion, but we tend to forget those that don't.

My suggestion of equality does not hold for Van Fraassen's explanation of the episode of Millikan's measurement of the charge on the electron. As mentioned earlier, he views Millikan's measurement of the charge on the electron as filling in a blank in atomic theory. Although he admits that the experiment was "a test of the theory that there exists this elementary electrical charge, it was not surprising at this time that such tests should bear out theory" (1980, p. 75). He concludes that, "In a case such as this one, *experimentation is the continuation of theory by other means*" (1980, p. 77).

While I agree with him that at the time most, but certainly not all, physicists believed that there was an elementary electrical charge, the question was an open one. We can look at the introductions to Millikan's own papers during this period to see what he regarded as the important questions and answers. (I note that at this time the question of whether or not there was a fundamental unit of electrical charge and what the value of that unit was were closely related).

In 1911 Millikan reported that it had been found possible

> "To present direct and tangible demonstration through the study of the behavior in electrical and gravitational fields of this oil drop, carrying its captured ions, of the correctness of the view advanced many years ago and supported by evidence from many sources that all electrical charges, however produced, are exact multiples of one definite elementary, electrical charge, or in other words, that an electrical charge instead of being spread uniformly over the charged surface has a definite granular structure, consisting, in fact, of an exact number of specks, or atoms of electricity, all precisely alike, peppered over the surface of the charged body.
>
> To make an exact determination of the value of the elementary electrical charge which is free from all questionable theoretical assumptions and is limited inaccuracy only by that attainable in the measurement of the coefficient of viscosity of air." (Millikan 1911, p. 350).

The ordering of these statements seems to indicate Millikan's view of their relative importance.

By 1913 Millikan regarded the question of charge quantization as settled on the basis of his own previously published (1911) work. "The total number of changes which we have observed would be between one and two thousand, and *in not one single instance has there been any change which did not represent the advent upon the drop of one definite invariable quantity of electricity, or a very small multiple of that quantity*" (Millikan, 1911, p. 360). His 1913 paper begins, "The experiments herewith reported were undertaken with the view of introducing certain improvements into the oil drop method of determining e and N and thus obtaining a higher accuracy than had been in the evaluation of these most fundamental constants." (Millikan 1913, p. 109). He is, in fact, filling in the blank, but only because he regarded the more important question of charge quantization as having already been answered.

During the course of these measurements Millikan was engaged in a controversy with Felix Ehrenhaft concerning the quantization of charge. (For details see Holton (1978) and Franklin (1986, Chapter 5)). During 1912 and 1913 a lull occurred in the controversy and opinion, as well as other experimental results, was generally favorable to Millikan. Ehrenhaft and two of his students, Zerner and Konstantinowsky, returned to the attack in 1914 and 1915. (Ehrenhaft 1914; Konstantinowsky 1915; Zerner 1915). Millikan answered the criticism, I believe successfully, in a paper entitled "The Existence of a Subelectron?" (Millikan 1916).

The vast preponderance of evidence and the opinion of the physics community strongly favored Millikan and he was awarded the Nobel Prize in Physics in 1924. The presentation speech by Gullstrand further emphasizes the importance of experimentally establishing charge quantization.

> Millikan's aim was to prove that electricity really has the atomic structure, which, on the basis of theoretical evidence, it was supposed to have... By a brilliant method of investigation and by extraordinarily exact experimental technique Millikan reached his goal... Even leaving out of consideration the fact that Millikan has proved by these researches that electricity consists of equal units, his exact evaluation of the unit has done physics an inestimable service, as it enables us to calculate with a higher degree of exactitude a large number of the most important physical constants. (Gullstrand 1965).

The filling in of the blank of the value of e, as van Fraassen has it, while of great importance, is secondary to the issue of charge quantization. Even three years after Millikan had received the Nobel Prize, O.D. Chwolson, a respected physicist, wrote "It [the Millikan-Ehrenhaft dispute] has already lasted 17 years and up to now it cannot be claimed that it has finally been decided in favor of one side or the other, i.e. that all researchers have adopted one or the other of the two possible solutions of this problem. The state of affairs is rather strange" (Chwolson 1927). Although by that time, most physicists would have disagreed with this assessment, it indicates that the issue had not been closed by 1911 as Van Fraassen states.

I believe that the realist account that Millikan "discovered" or gave strong evidence for the existence of a fundamental unit of electricity, and then made a precise measurement of that unit, is a far better description of this episode than van Fraassen's view that Millikan was merely filling in a blank left by theory.

3. CONCLUSION

I believe that the discussion of kaons has shown that we can have good reasons to believe in both the existence of entities and in laws, and that our belief in the laws is a belief that it is true. If my arguments are correct then constructive empiricism is not philosophically justified. I also believe that I have cast doubt on the empirical adequacy of constructive empiricism. It has a vague notion of "observable," which I believe is far too strict to apply to the practice of science. If one extends the notion as I have suggested then it supports a realist position. Constructive empiricism is, at best, no better than the realist account in discussing the elimination of contradictions between theories or the unification of theories. It also fails to give an adequate account of Millikan's experiment, which is better explained by a realist position.

Nevertheless, showing that constructive empiricism is not satisfactory provides only a small amount of support for the realist view. The elimination of one of many alternative explanations (and there are many of them) only slightly strengthens our belief in the remaining alternatives.[25] I have, however, also given positive arguments for the particular version of realism that I support. This includes the discussion of the existence of kaons, the elimination of contradictions, and the historical account of Millikan's experiments.

Supporting a realist position does not, however, mean that I believe in either the absolute truth of the laws or in the "real" existence of the entities. It means only that I think we have good reasons for believing in the truth of the laws and in the existence of the entities.

NOTES

[1] See, for example, the recent book on realism edited by Leplin (1984) and the volume devoted to discussion and criticism of van Fraassen's view edited by Churchland and Hooker (1985).

[2] Van Fraassen admits that "observable" is a vague predicate but argues that one can still make use of it. I believe, as will be discussed later, that his characterization of observable is, in fact, too strict.

[3] Morrison (1990) has argued that manipulability is not sufficient to establish belief in an entity. She discusses particle physics experiments in which particle beams were viewed not only as particles, but also as beams of quarks, the particle constituents, even though the physicists involved had no belief in the existence of quarks. Although I believe that Morrison's argument is correct in this particular case, I do think that manipulability can, and often does, give us good reason to believe in an entity. See, for example, the discussion of the microscope in Hacking (1983). More recently scientists have used the scanning tunneling microscope to spell out IBM with xenon atoms. This seems to me to be very good evidence for such atoms. To paraphrase Hacking, "If you can spell IBM with them, they are real.

[4] In Cartwright's discussion of the electron track in the cloud chamber, for example, she can identify the track as an electron track rather than as a proton track only because she has made an implicit commitment to the law of ionization for charged particles, and its dependence on the mass and momentum of the particles. The momentum is measured by curvature in a known magnetic field in a way similar to that discussed below for K mesons. A similar point has been made by Morrison (1990). See also my review of Cartwright in *Foundations of Physics*, 1984.

[5] I am grateful to Bob Ackermann for raising this point

[6] I note that it is the argument that is constructed, not the weak neutral currents, as some have stated (see Pickering 1984).

[7] Although modern theory does predict proton decay into a positron, both the measured proton lifetime and the predicted rate give a decay rate that is too small to be of any significance in this experiment.

⁸ The reader may note that the pion also decays into an electron and a neutrino, with a branching ratio approximately seven times that of the kaon. These pion decays could easily be separated from kaon decays by their momentum. The momentum of an electron from pion decay is 69.8 MeV/c, while that for an electron from kaon decay is 246 MeV/c. The experimental apparatus had a momentum resolution of 1.9%, which could easily distinguish between the two decays.

⁹ In fact, the experimenters had assumed that the charge on the particle was e. They did, however, measure the range, the radius of curvature in a known magnetic field, and the time of flight of the particles. From these three measurements, the three unknown quantities, the charge, mass, and velocity of the beam particles could have been calculated.

10 The reader may object that using agreement with known properties of the kaon already assumes that the kaon exists. I believe, however, that establishing that a particle has a definite mass, charge, and lifetime is sufficient to establish its existence. Recall J.J. Thomson's "discovery" of the electron by showing that cathode rays had a definte charge to mass ratio.

¹¹ There are of the order of 100 elementary particles, each with different properties (in particular each particle has a unique mass) so that specifying three properties serves to identify the particle

¹² One could, for example, measure the weight of the entity by using a scale which was located behind an opaque screen, but which had a remote readout visible to the observer, and so on.

¹³ I note that this story agrees with Cartwright's causal view, but differs slightly from Hacking's because it is the K mesons themselves that are under investigation.

¹⁴ I am grateful to Zeno Switinjk for pointing this out and also for providing the quotation from Dehmelt, which argues against his own point.

¹⁵ I have used the term valid to describe the laws. I don't think the laws have to be true in order to have a successful measurement, and I don't know how to make sense of the term "approximately true." What I mean by valid is that to within the required experimental accuracy the laws give correct results. This does not mean that I do not feel that we are justified in our belief that the laws are true. Observation of evidence entailed by a theory should, and indeed does, strengthen our belief that the theory is true. It also strengthens our belief that the theory is empirically adequate, but that is beside the point.

¹⁶ For a discussion of this historical episode and a discussion of the evidential value of prediction and accommodation see Howson (1991).

¹⁷ For the sake of economy I am attributing the arguments that might hypothetically be offered by a defender of constructive empiricism to its foremost proponent, Bas van Fraassen. He might not, of course, agree that he would offer any of these arguments

¹⁸ This is not strictly true. The current theory of strongly interacting particles, quantum chromodynamics (QCD), does, in principle, predict the mass of the kaon and other elementary particles, given the masses of the up, down, and strange quarks. Because there are more particles than quarks, one can use the observed masses of some of the particles to calculate the quark masses, and then proceed to calculate the masses of the other particles. These are very difficult and complex calculations and, at present, are accurate to approximately 20%. This theory was not available in 1967, when the experiment was performed.

¹⁹ Van Fraassen actually privileges vision over the other senses. No one would regard hearing something as evidence for the existence of an entity. Recall the old television commercial "Is it live or is it Memorex?" in which a listener cannot distinguish between a recording and a live person. The sense of touch also does not provide reliable evidence for the existence of an entity.

Remember the story of the five wise men each touching a different part of an elephant and reaching different conclusions as to its nature.

[20] Cartwright (1983, pp. 87-99) argues that when we have a causal explanation of a measurement or an observation then we are justified in making an inference to the most probable cause, i.e. an entity. I agree. Van Fraassen might ask why we need an explanation at all.

[21] I note that more direct evidence for the existence of these particles was found later.

[22] It may very well be true that any detection device, such as a mercury barometer, will have an output detectable by unaided human senses, but that is not the point here.

[23] For a discussion of how one comes to believe in an experimental result see Franklin (1986, Chapter 6) and Franklin and Howson (1988). For further discussion of the theory-ladenness of measurement see Franklin (1989).

[24] The search for unification also applies to noncontradictory and quantum electrodynamics. These were not contradictory. Each was empirically adequate for its own range of phenomena. The search for unification was successful, leading to the Weinberg-Salam unified theory of electroweak interactions. This unified theory was widely hailed as a major achievement and led to the prediction and obervation of new phenomena, weak neutral currents, atomic parity violation, and the observation of intermediate vector bosons, the W and Z_o particles. Similarly, Maxwell's theory unified the separate theories of electricity and magnetism and led to the observation of electromagnetic radiation.

[25] For a discussion of this see Franklin and Howson (1988).

CHAPTER 6

DISCOVERY, PURSUIT, AND JUSTIFICATION

In earlier discussions of the philosophy of science, philosophers such as Reichenbach (1938) distinguished between the "context of discovery" and the "context of justification." The context of justification was concerned with the evidential basis for belief in scientific hypotheses and with the logical structure of science, whereas the context of discovery was left to psychology. More recently, philosophers of science such as Laudan (1980) and Nickles (1980b) have suggested a third classification, that of pursuit, the further investigation of a theory or an experimental result. In addition, recent studies of science, which emphasize scientific practice, have emphasized that science is a complex activity demanding a richer description than just discovery and justification.[1]

In this essay I will add my own support for a tripartite classification scheme of scientific activity: discovery, pursuit, and justification. By discovery I mean the process by which a theory or hypothesis is generated and proposed. Pursuit is the further investigation of a theory or of an experimental result.[2] Justification is the decision process by which the scientific community comes to accept or reject a theory or an experimental result as part of the corpus of scientific knowledge.[3] By supporting this scheme I am not claiming that it is an adequate account of scientific activity, but only that using it is a start on such an account. Scientific practice is far too complex an activity for such a limited description.

There are some obvious problems associated with such a scheme. These practices or activities cannot always be clearly or easily separated from one another (see note 2). Thus, the evidence that might lead a scientist to propose a hypothesis may also provide support for it, or may provide grounds for further experimental or theoretical investigation. It may also be difficult in practice to distinguish between using a theory as the basis for further work and acceptance of that theory as scientific knowledge. One may not have access to the scientist's attitude towards the theory. Later I will argue on the basis of his own statement that Dydak was using certain experimental results and theory only as the basis for further calculations, without being seriously committed to their truth, or to their status as knowledge.[4] In pursuing an investigation, as discussed later, scientists may have very different attitudes toward the truth of a hypothesis. These may range from believing that the hypothesis is false to thinking that the issue has already been resolved and the hypothesis is true, or encompass any attitude in between. It may also not be clear which activity is occurring. Thus, one might regard Fischbach's work on the fifth-force hypothesis as either discovery, or pursuit of a solution to the problem of the cause of CP violation.

Despite the difficulties involved, I support the threefold classification of scientific activities because it will provide us with a more adequate description of the practice of science, it can help to illuminate the varied roles that experimental evidence plays in science, and it may help to partially resolve the differences between those investigators such as myself, who believe that experimental evidence plays a decisive role in justification, and those who deny that the process of justification exists or who minimize the role that experimental evidence plays in that process. This view has been associated with social constructivism, and for my purposes is best illustrated by the strong programme of the Edinburgh school, particularly the work of Andrew Pickering.

To be fair, the social constructivists claim that they do not deny that experimental evidence plays a role in theory choice, confirmation, or refutation, but only that it cannot compel such decisions. "...it is untrue that I deny that science is a reasonable enterprise, or that evidence has a constitutive role toplay in the production of scientific knowledge.[5] The problem...is rather that there are *too many reasons*..."(Pickering 1991, p. 459). "Social constructivists do *not* say that experimental evidence is irrelevant to theory choice,confirmation, or refutation. Nor do they argue that there are no good reasons for belief in the validity of evidence. Instead they argue that experimental evidence does not compel acceptance of a theory or in Pickering's (1984a, p.5) terms, 'experiment cannot oblige scientists to make a particular choice of theories.' "(Lynch 1991, pp. 476-7). As discussed below, much depends on how one interprets "compel." Nevertheless, in the studies presented thus far (see, e.g., Pickering 1984a, 1984b, 1991), evidence does not seem to play a major role. The reasons offered for such decisions include the opportunity for future work, consistency with existing community commitments, career interests, and the recycling of expertise. Pickering says, "Quite simply, particle physicists accepted the existence of the neutral current because they could see how to ply their trade more profitably in a world in which the neutral current was real" (1984b, p. 87). The emphasis on future opportunities for research, or pursuit, is clear. Here Pickering is identifying a reason for pursuit with grounds for belief, or justification.

It also seems to be the case, in these episodes, that it is always agreement with existing, accepted theory that provides more opportunity for future work.[6] "Scientific communities tend to reject data that conflict with group commitments and, obversely, to adjust their experimental techniques and methods to 'tune in' on phenomena consistent with these commitments" (Pickering, 1981, p. 236). Whether this is, in fact, the case will be discussed later.

In this essay I would like to suggest that the differences between these competing views on the nature of science may be clarified and partially resolved by closer attention to the actual practice of science and to scientists' activities; to discovery, pursuit, and justification. I wish to suggest that many of the considerations given by the social constructivists do play an important role in pursuit, but not in justification. This does not seem likely to completely solve the problem because Pickering and others, although not using the term, do apply their analysis to justification. Nor are all of their arguments suspect because they do not make this distinction.[7]

I will discuss this issue using two episodes; one has already been discussed in some detail by the proponents of both views, that of atomic parity-violation experiments and their relation to the acceptance of the Weinberg-Salam (W-S) unified theory of electroweak interactions (Pickering 1984a, 1991; Ackermann 1991; Lynch 1991; and

Franklin 1990, Chapter 8), and the second, that of the proposal, pursuit, and ultimate rejection of the hypothesis of a Fifth Force in gravity (Franklin 1993a). This essay will also provide an opportunity to respond to comments and criticisms of my account of the atomic parity-violation episode (see references above).

1. ATOMIC PARITY VIOLATION EXPERIMENTS

I begin with the episode of atomic parity-violation experiments. In 1957, it had been experimentally demonstrated that parity, or left-right symmetry, was violated in the weak interactions. This feature of the weak interactions had been incorporated into the W-S unified theory of electroweak interactions. The theory predicted that one would see weak neutral-current effects in the interactions of electrons with hadrons, the strongly interacting particles. The effect would be quite small when compared to the dominant electromagnetic interaction, but could be distinguished from it by the fact that it violated parity conservation. A demonstration of such a parity-violating effect and a measurement of its magnitude would test the W-S theory. One such predicted effect was the rotation of the plane of polarization of polarized light when it passed through bismuth vapor. Such a rotation is possible only if parity is violated. This was the experiment performed by the Oxford and Washington groups.

Pickering and I agree about the early part of the story.[8] In 1976 and 1977, experimental groups at Oxford University and at the University of Washington reported results from such atomic parity-violation experiments that disagreed with the predictions of the Weinberg-Salam theory. At the time the theory had other experimental support, but was not universally accepted. In 1978 and 1979 a group at the Stanford Linear Accelerator (the SLAC E122 experiment) reported results which confirmed the W-S theory, and on the basis of those results, combined with the previous support, the scientific community accepted the W-S theory. (For more detailed history see Pickering 1984a, pp. 294-302; Franklin 1990, Chapter 8).

At this point the accounts diverge. Pickering notes that the W-S theory was regarded as established despite the fact that "there had been no *intrinsic* change in the status of the Washington-Oxford experiments" (Pickering 1984a, p. 301). In his view, "particle physicists *chose* to accept the results of the SLAC experiment, *chose* to interpret them in terms of the standard model (rather than some alternative which might reconcile them with the atomic physics results), and therefore *chose* to regard the Washington-Oxford experiments as somehow defective in performance or interpretation" (1984a, p. 301). Pickering regards the Washington-Oxford results and those of SLAC E122 as having the same evidential weight,[9] and that the reason the physics community chose to accept the SLAC results and the W-S theory it supported was because they provided more opportunity for future work and were also consistent with existing commitments.[10]

My view is different. I regard the two experimental results as having quite different evidential weights. The initial Washington-Oxford results (there were later ones) used new and untested experimental apparatus and had large systematic uncertainties (as large as the predicted effects). In addition the initially reported results were internally inconsistent and by 1979 there were other atomic parity-violation results that confirmed the W-S theory (These were done at Berkeley (Conti 1979) and Novosibirsk (Barkov and Zolotorev 1978a, 1978b, 1979)). The overall situation with respect to the atomic parity results was quite uncertain. The SLAC experiment, on the other hand, although also using new techniques, had been very carefully checked, and had far more evidential weight. As Bouchiat, an atomic physicist,[11] remarked, "I would like to say that I have been very much impressed by the care with which systematic errors have been treated in the experiment [SLAC E122]. It is certainly an example to be followed by all people working in this very difficult field" (Bouchiat 1980, pp. 359-60). Bouchiat's comments are particularly important. He was giving a summary talk on the evidential situation concerning the W-S theory at a conference attended by representatives of virtually all the experimental groups working on atomic parity violation, including Washington, Oxford, Novosibirsk, as well as the SLAC E122 group. Such talks do, in fact, convey the consensus of those at the conference.

Faced with this situation the physics community chose to accept the SLAC results, which supported the Weinberg-Salam theory, and to await further developments on the uncertain atomic parity-violation results. I note here that the SLAC experiment also tested a hybrid model and found it to be experimentally refuted.

I should emphasize here that Pickering and I agree that the W-S theory was accepted on the basis of the results of the SLAC E122 experiment. We do, however, disagree about the reasons for that decision. We also differ in our interpretation of how the physics community dealt with the discordant atomic parity results of Washington-Oxford and Novosibirsk-Berkley. As discussed earlier, Pickering thinks that the Washington-Oxford results were simply regarded as wrong, whereas I believe that the community suspended judgment on the discordant results pending further work.

The reader may well ask whether or not Pickering and I are both discussing justification and merely offering different explanations. This is partially correct, but a more detailed examination of Pickering's views indicates that he has also conflated pursuit and justification. In his most recent comments on this episode (1991), he argues against my view (that the decision of the physics community to accept the W-S theory on the basis of reasonable evidence) fails because there were too many reasons. He presents four alternative scenarios, which he regards as equally reasonable. None of these alternatives actually occurred, leading Pickering to argue that because reason was unable to decide the issue, his account, based primarily on future research opportunities, is correct. Three of the alternatives involve questions about the evaluation of experimental evidence; 1) the physics community might have decided that the atomic parity-violation results of Washington-Oxford were wrong and were therefore excluded,[12] 2) the physics community might have lumped the atomic parity-violation results together with those of SLAC E122 and concluded that they neutralized each other,[13] and 3) the community might have waited until E122 had been replicated before making a decision.[14] I refer the reader to the detailed accounts (Pickering 1984a, 1991 and Franklin 1990, Chapter 8) to decide whether or not these alternatives are equally reasonable. I do not agree that they are, and the physics community seems to have agreed with my evaluation. Pickering might, however, deny that they engaged in such an evaluation of the experimental evidence (see, however the quotations given below by Dydak and Bouchiat, which do evaluate the evidence). In addition, the atomic parity-violation experiments, including repetitions of the original Wahington-Oxford experiments on bismuth, as well as others, have continued through the 1980s and into the 1990s, reaching agreement with the predictions of the W-S theory.[15] If the original Oxford-Washington results were simply regarded as wrong, there seems little reason for this to have happened. If, however, judgment was suspended concerning which of the discordant results was correct, then the subsequent experimental work certainly makes sense, and even seems to be required.

The fourth alternative does, however, clearly involve pursuit. Pickering asks why a theorist might not have attempted to find a variant of electroweak gauge theory that might have reconciled the Washington-Oxford atomic parity results with the positive E122 result. (What such a theorist was supposed to do with the supportive atomic parity results of Berkeley and of Novosibirsk is never mentioned). "But though it is true that E122 analysed their data in a way that displayed the improbability [the probability of the fit to the hybrid model was 6×10^{-4}] of a particular class of variant gauge theories, the so-called 'hybrid models', I do not believe that it would have been impossible to devise yet more variants" (Pickering 1991, p. 462). Pickering notes that open-ended recipes for constructing such variants had been written down as early as 1972 (Pickering 1991, p. 467). I agree that it would certainly have been possible to do so, but one may ask whether or not a scientist might have wished to do so. If the scientist agreed with my view that one had reliable evidence (E122 and others) that supported the W-S theory and a set of conflicting and uncertain results from atomic parity-violation experiments that gave an equivocal answer on support of the W-S theory, what reason would they have had to invent an alternative?[16]

This is not to suggest that scientists do not, or should not, engage in speculation, but rather that there was no necessity to do so in this case. Theorists often do propose alternatives to existing, well-confirmed theories. As we shall see in the fifth-force episode, the hypothesis arose, in part, out of the tradition of modifications to the Brans-Dicke theory of gravity, which was itself an alternative to the strongly believed and strongly supported general theory of relativity, the currently accepted theory of gravitation.

Some of the evidence that Pickering cites to support his interpretation of the atomic parity episode also indicates a failure to adequately distinguish between pursuit and justification. He cites Dydak's (1979) conference summary talk "It is difficult to choose between the conflicting results in order to determine the *eq* [electron-quark] coupling constants. *Tentatively*, we go along with the positive results from Novosibirsk and Berkeley groups and hope that future developments will justify this step (*it cannot be justified at present, on clear cut experimental grounds*)" (Dydak 1979, p. 35, emphasis added). Pickering concludes, "Having decided not to take into account the Washington-Oxford results, Dydak concluded that parity violation in atomic physics was as predicted in the standard model" (1984a, p. 300).[17] I find no justification for this conclusion. To calculate the coupling constants Dydak had to assume that one of the set of conflicting results was correct. There was no reasonable way in which he could have used all of the results. He *tentatively* accepted those that agreed with the standard model, and admitted he had no good justification for this. He was not saying that the predictions of the standard model were correct. He was clearly accepting those results as the basis for further calculations, or for pursuit. It would certainly have seemed odd if he had accepted results that disagreed with the standard model, when he was using that model as the basis of his calculations. He was not discussing justification. I also note that his summary talk does indicate an uncertain attitude toward the experimental results on atomic parity violation.

Pickering and others are relying here on two philosophical points. (See Nelson 1994 for discussion of these issues). The first is the underdetermination of theory by evidence, the fact that one can always find an alternative theoretical explanation for a given experimental result. Pickering alludes to this when he states, "I do not believe it would have been impossible to devise yet more variants" (1991, p. 462). The second is the Duhem-Quine thesis, that if an experiment seems to refute a theory it, in fact, refutes the conjunction of both that theory and background knowledge and one doesn't know where to place the blame (see Franklin 1990, pp. 144-61 for details). One may save a hypothesis or theory from refutation by suitable changes in one's background knowledge. MacKenzie goes even further and suggests not only that experiment cannot resolve points of theoretical controversy, but that experimental results themselves may be questioned indefinitely.

> Recent sociology of science, following sympathetic tendencies in the history and philosophy of science, has shown that no experiment, or set of experiments however large, can on its own compel resolution of a point of controversy, or more generally, acceptance of a particular fact. A sufficiently determined critic can always find a reason to dispute any alleged "result." If the point at issue is, say, the validity of a particular theoretical claim, those who wish to contest an experimental proof or disproof of the claim can always point to the multitude of auxiliary hypotheses (for example about the operation of instruments) involved in drawing deductions from the given theoretical statement to a particular experimental situation or situations. One of these auxiliary hypotheses may be faulty, critics can argue, rather than the theoretical claim apparently being tested. Further, the validity of the experimental procedures can also be attacked in many ways (MacKenzie 1989, p. 412).

Pickering agrees. In discussing a Monte Carlo calculation needed for the analysis of an experiment on weak neutral currents he states, "My object here is simply to demonstrate that assumptions were made which could legitimately be questioned: one can easily imagine a determined critic taking issue with some or all of the assumptions"

(1984b, p. 96). Despite their disclaimers, both Pickering and MacKenzie write as if they were such determined critics.

Much depends here on the word "compel." No one would dispute the logical point that any hypothesis can be saved from refutation by suitable changes in one's background knowledge, that several theories may explain the same evidence, or that a sufficiently determined critic might be willing to question an experimental result *ad infinitum*. The question is what price one has to pay in terms of one's background knowledge in order to maintain such skepticism. If one reads "compel," as MacKenzie seems to, as "entail," then I agree that no finite set of confirming instances can entail a universal statement. It seems to me, however, that a more plausible meaning for "compel" is having good reasons for belief, and this is what I have argued was true in the episode of atomic parity violation. The physics community made a reasoned judgment concerning the available evidence and then based its evaluation of that theory on that judgment.

In the case of experimental results, I have previously suggested an epistemology of experiment, a set of strategies that provides grounds for belief in the validity of results, that can be philosophically justified, and which is also used in the practice of science. These include: 1) experimental checks and calibration, in which the experimental apparatus reproduces known phenomena; 2) reproducing artifacts that are known in advance to be present; 3) intervention, in which the experimenter manipulates the object under observation; 4) independent confirmation using different experiments; 5) elimination of plausible sources of error and alternative explanations of the result (the Sherlock Holmes strategy); 6) using the results themselves to argue for their validity; 7) using an independently well-corroborated theory of the phenomena to explain the results; 8) using an apparatus based on a well-corroborated theory; and 9) using statistical arguments (see Franklin 1986, Chapter 6; 1990, Chapter 6 for details).[18]

In the case of the Duhem-Quine problem, I argue that scientists never confront all the logically possible explanations of a given result. There is usually only a reasonable number of plausible or physically interesting alternatives on offer. In this case scientists evaluate the cost of accepting one of these alternatives in the light of all the existing evidence. One of these alternatives may be better supported by the evidence than any of the others. The alternatives themselves may also be tested, subject to the usual difficulties, and we may be left with only one explanation.

Bob Ackermann (1989) has asked whether or not one can deal with this question in the absence of a theory of plausibility, and others have questioned whether the limiting of the hypothesis space of plausible or interesting alternatives precludes the reasonableness of science. Ideally one would like to have a theory of plausibility to justify the limits placed on the hypothesis space of alternative explanations. In the actual practice of science, however, the limits placed on the hypothesis space do not seem very stringent. Thus, when the experiment of Christenson *et al.* (1964) detected K^0_2 decay into two pions, which seemed to show that CP symmetry (combined particle-antiparticle and space inversion symmetry) was violated, no fewer than 10 alternatives were offered.[19] These included 1) the cosmological model resulting from the local dysymmetry of matter and antimatter, 2) external fields, 3) the decay of the K^o_2 into a K^o_1 with the subsequent decay of the K^o_1 into two pions, which was allowed by the symmetry, 4) the emission of another neutral particle, "the paritino," in the K^o_2 decay, similar to the emission of the neutrino in beta decay, 5) that one of the pions emitted in the decay was in fact a "spion," a pion with spin one rather than zero, 6) that the decay was due to another neutral particle, the L, produced coherently with the K^o, 7) the existence of a "shadow" universe, which interacted with our universe only through the weak interactions, and that the decay seen was the decay of the "shadow K^o_2," 8) the failure of the exponential decay law, 9) the failure of the principle of superposition in quantum mechanics, and 10) that the decay pions were not bosons. As one can see, the limits placed on alternatives were not very stringent. By the end of 1967, all of the alternatives had been tested and found wanting, leaving CP symmetry unprotected. Here the differing judgments of the scientific community about what was worth proposing and pursuing led to a wide variety of alternatives being tested.[20]

One question that does not seem to be asked is why Pickering's, or anyone else's, constructivist account is any more compelling than one based on evidence. Surely, other reasonable, constructivist alternatives can be constructed. The underdetermination of theory by evidence would seem to guarantee it, even for historical data. Thus, in the case of atomic parity violation, would it not have been reasonable for a scientist to think that all the important results of the W-S theory had already been both calculated and measured, and that the best opportunity for future research would be if a new theory was required? Therefore they would have accepted the Washington-Oxford results and rejected both E122 and the atomic parity violation results that supported the W-S theory. This might well be the case for any well-accepted and well-supported theory. While constructivist accounts do seem to favor accepted theories, the history of science shows that the scientific community often does accept the refutation of such theories, and that these refutations seem to lead to large amounts of work for the community.[21] In the case of the fifth force, as we shall see, even the suggestion of a plausible alternative to an accepted theory seemed to generate opportunities for further research.

Ackermann (1991) has also raised an interesting and important question as to whether or not my normative view[22] that experiments can be weighted as good and bad can be done in "sufficient time to effect valid discriminations between rival scientific theories" (1991, p. 452). "But his [Franklin's] weight of evidence may always come too late to catch the constructionist account opening a window on scientific irrationality" (Ackermann 1991, p. 455). My view is that this sort of decision not only can be done in real time, but also that in the case of atomic parity violation, it was done in real time. Thus, Bouchiat after hearing a report on the Soviet Novosibirsk experiment and the ensuing extensive questioning concerning systematic error, concluded, "*As a conclusion on this Bismuth session,*[23] *one can say that parity violation has been observed roughly with the magnitude predicted by the Weinberg-Salam theory*" (1980, p. 365). He was evaluating all of the existing experiments and stating that, in his view, the positive Novosibirsk result was the most believable. Note, however, that he was only arguing for an effect of roughly the right size, and by implication suggesting further work on the subject.[24] Recall also his evaluation of the SLAC E122 experiment as "an example to be followed by all people working in this difficult field" (1980, pp. 359-60).

Sometimes the evaluation cannot be completed immediately. In the case of the fifth-force hypothesis, to be discussed in detail later,[25] the two initial experimental reports differed dramatically. Thieberger (1987a, 1987b) found a result in agreement with the suggested force, while Stubbs et al. (1987) found no evidence of such a force. In this case, however, no immediate decision was made. Careful contemporary analysis of the two experiments found no error in either one,[26] leading to numerous repetitions of the experiments, as well as to new ones, over the next three years. The final conclusion, based on an overwhelming preponderance of evidence, was that the fifth force does not exist. Rationality sometimes takes awhile. Discordant results are grounds for agnosticism and future work, barring an obvious error in one of the experiments. This was the case in both the atomic parity-violation and the fifth force episodes.[27] In both cases one had further experimental investigation.

Sometimes further analysis can uncover errors. Thus, the positive fifth force results of Eckhardt (1988) and of Stacey and Tuck and their collaborators (Stacey and Tuck 1981; Stacey, Tuck, and Moore 1987a; Stacey et al. 1987b) were shown to be wrong because the theoretical model used for comparison with the measurements did not take the local terrain into account adequately.

Ackermann (1989, 1991) has also pointed out that my accounts of experiment need to be more closely tied to the experimental apparatus itself. In a similar vein, Lynch criticizes my inattention to the distinction between epistemological argument and situated practical reasoning (1991, p. 472). To support his argument he presents a transcript of what experimenters were saying during a flight to record the intensity of light from a star as the star was eclipsed by the planet Uranus. He contrasts this with the published account. "In their published article Elliot, Dunham, and Millis (1977, p. 14) give a 'Franklinian' interpretation of the transcript. Alongside the transcript they present commentaries on what they later determined they were seeing at the time. For instance, just before Dunham exclaims 'What was that?' the article tells us 'First secondary occultation appears on the chart record, but is not noticed for almost a minute'" (Lynch 1991, p. 482). Although Lynch regards the published account as a plausible conceptualization, he claims that it does not capture what the experimenters thought at the time. "The 'epistemological' strategies are deeply imbedded in the practical situation" (p. 482).

I beg to differ. The experimenters were watching a chart recorder, and their immediate reactions, however interesting, are not epistemological arguments. What they observed, or perhaps more properly, what their experimental result was, could only be determined after considerable thought and analysis. A chart recording's meaning is not obviously given. I agree with Lynch that some epistemological strategies are indeed embedded in the experimental situation, but this includes the apparatus itself, as well as the experimental practices such as experimental checks and calibration. Without such checks and calibration the chart recording by itself has no meaning. What would have happened if the experimenters in Lynch's example had performed such checks and the apparatus had failed the tests? They would have concluded, sadly, that their apparatus was not functioning properly, and that their data was unusable and could not yield a valid experimental result.[28] Ackermann's suggestion that we need to pay more attention to the experimental apparatus is correct.

Let me illustrate this with the example of an experiment to measure the K^+_{e2} branching ratio, the fraction of all K^+ mesons that decay into a positron and a neutrino. The positron resulting from such a decay was to be identified by a count in a Cerenkov counter, by its momentum (247 MeV/c), and by its range in matter. This would separate the particle from the far more prevalent $K^+_{\mu2}$ decay (for details see Franklin 1990, pp. 115-31).[29] Thus, when I, as a member of the group, found the first event that had a Cerenkov count, a momentum of 247 MeV/c, and a range consistent with that of an electron, I was elated. I remember thinking that we had found one K^+_{e2} decay, and I said so at the time. My enthusiasm was, of course, premature. What we had in fact found was a candidate for such a decay. It was only after an analysis process that lasted for a year after data taking was completed, that we could say that we had an excess of such events above background, and had measured the branching ratio, my comments notwithstanding.

I believe Lynch has misunderstood his own example. Without further analysis the experimenters, as in the case above, did not have a result. They had data. These are not the same.[30] It is, in fact, very rare that an experimental result is immediately available.[31]

Even a videotape of data taking would still not address the question of justification. When the scientific community makes a judgment about the validity of an experimental result, all they usually have available is the published paper, which, as even Lynch and Pickering admit, contain the epistemological arguments that I think are crucial, and this is the basis on which the community makes its judgment about the validity of experimental results.[32] These valid results are then used as the basis for reasonable theory choice, confirmation, and refutation, or for justification.

2. THE FIFTH FORCE[33]

A) DISCOVERY

The story of the fifth force, a proposed modification of the law of gravity, began not as Kuhn ([1962], 1970) might have it, with an anomaly in gravitational theory, but with a successful experimental confirmation of that theory. Colella, Overhauser, and Werner (1975) had shown the existence of gravitational effects at the quantum level by observing interference between two neutron beams which had passed through different gravitational potentials. This test, performed at low velocity, did not, however, distinguish between general relativity (the accepted gravitational theory) and any of its competitors, notably the Brans-Dicke theory (Brans and Dicke 1961).

After this experiment, Ephraim Fischbach,[34] partially as a result of conversations with his colleague Overhauser, began thinking about how one might perform such a test at high velocity, and thus distinguish between the competing theories. He also thought about whether or not gravitational effects might explain CP violation. A high velocity neutron beam was not feasible and he thought of K mesons, another type of neutral particle,[35] for which high velocity beams could be produced. This was also the system in which CP violation, the violation of the combined symmetry operation of left-right and particle-antiparticle interchange, had been observed.[36]

Fischbach was encouraged by what he referred to as a "remarkable numerical relation." His calculation of the gravitational effect was equal in size to the CP violating effect. This was truly remarkable because there is no known connection between gravity and CP violation and also because his calculation of the gravitational effect included an enhancement factor, for which no justification was given, of $m_K/\Delta m$, a factor of 1.4 x 10^{14} (Δm is the mass difference between the short-lived and long-lived K meson). He began a collaboration with Sam Aronson, an experimentalist with considerable experience with K mesons. Aronson suggested that there might be an energy dependence of the parameters of the K^0-$\bar{K}_0$ system (an alternative description of the K^0 mesons), and subsequent analysis (Aronson et al. 1982) found a "tantalizing" two or three standard-deviation effect in those parameters.[37]

A second strand of the story involves the history of alternatives to, or modifications of, standard gravitational theory. In 1961, Brans and Dicke had proposed a scalar-tensor alternative to the tensor theory of general relativity. In the early 1970s, Fujii (1971, 1972, 1974) and others had suggested a modification of Brans-Dicke theory that included a massive scalar particle, in addition to the massless scalar and tensor particles of that theory. This gave rise to a modified gravitational potential V = -GmM/r [1 + αe-r/λ], where α is the strength of the additional term, and, λ its range. The second term is Fujii's modification. It is also identical, in mathematical form, to the fifth-force hypothesis. Fujii also predicted a gravitational constant that varied with distance,[38] where the gravitational constant at large distances G_∞ = 3/4 G_0, the value at short distances. He suggested various experimental tests of his theory, particularly redoing the Eötvös experiment, the classic experiment that had shown that the ratio of gravitational to inertial mass was the same for different substances (Eötvös, Pekar, and Fekete 1922). Subsequent surveys (Mikkelsen and Newman 1977; Gibbins and Whiting 1981) of the status of G, the gravitational constant, found that a force with a range 10 m to 1 km, and with a strength approximately one percent that of gravity, was not ruled out by the measurements. Mikkelsen and Newman noted that their analysis "does not even rule out Fujii's suggested value G_∞/G_0 = 0.75" (p. 924). At approximately the same time a series of gravity measurements in mines and boreholes was giving values of G that were systematically high, but "tantalizingly uncertain" because of possible local mass anomalies (Stacey and Tuck 1981). Stacey also showed that these anomalies were consistent with a small short-range addition to the gravitational force. The two strands came together in 1983-4 when Fischbach became aware of the mineshaft gravity anomalies and began thinking about a possible short-range force, rather than his previous work on a long-range force.[39] Fischbach, Aronson, and their collaborators looked for other places where the effects of such a force might be observed and found

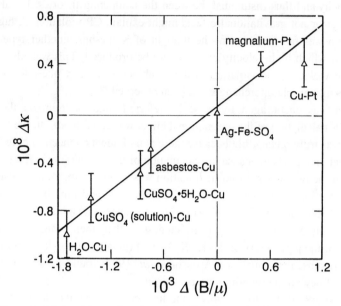

Figure 1. Δk ($\Delta a/g$) as a function of $\Delta (B/\mu)$. From Fischbach et al. (1986).

only three: 1) the K^0-$\bar{K}^0$ system at high energy, which they had already studied,; 2) a comparison between satellite and terrestrial determinations of g, the local gravitational acceleration; and 3) the Eötvös experiment. They reanalyzed Eötvös's original data and found a surprising linear dependence between $\Delta a/g$, the fractional change in gravitational acceleration and $\Delta(B/\mu)$, where B is the baryon number and μ is the mass of the substance in units of the mass of atomic hydrogen, for two different substances (Figure 1) (Fischbach et al. 1986).

Thus, there were three pieces of evidence that could each be explained by a substance dependent Fifth Force with a strength approximately one percent that of gravity and with a range of about 100 m; the reanalyzed Eötvös experiment, the energy dependence of the K^0-$\bar{K}^0$ parameters, and the mineshaft gravity anomalies.

B) PURSUIT

The publication of the results of the reanalysis of the Eötvös experiment made an impact, even in the popular press, with an article in the *New York Times* and an editorial in the *Los Angeles Times*. It also generated immediate comments and criticism within the physics community. These emphasized the importance of local mass asymmetries and suggested improved sensitivity of proposed experiments by placing them in an environment where such asymmetries existed, such as on a cliff or a hillside (Thodberg 1986; DeRujula 1986a, 1986b; Newfeld 1986; Thieberger 1986; Bizzeti 1986; and

Milgrom 1986).[40] Others suggested that there was an error in the reanalysis and that there was therefore nothing to be explained (Keyser, Niebauer, and Faller 1986; Elizalde 1986; and Kim 1986). Chu and Dicke (1986) suggested that the effect could be explained by conventional physics, a thermal gradient across the apparatus. Fischbach and his collaborators offered answers to these criticisms that satisfied both themselves and a segment of the physics community (see Paik's comments below). Theorists were also busy attempting to explain the origin of the proposed force and looking for possible observable effects in other areas.

At the end of 1986 the attitude within the physics community toward the fifth force ranged from outright rejection to enthusiasm. Glashow, a particle theorist, was quite negative. "Unconvincing and unconfirmed kaon data, a reanalysis of the Eötvös experiment depending on the contents of the Baron's wine cellar [an allusion to the importance of local mass inhomogeneities], and a two-standard-deviation geophysical anomaly! Fischbach and his friends offer a silk purse made out of three sows' ears, and I'll not buy it" (quoted in Schwarzschild 1986, p. 20).

Others cited the fruitfulness of the suggestion. Bars and Visser cited Fischbach's analysis as the original motivation for their work. "In the course of our work we became convinced that forces similar to the reported one are likely to exist as remnants of higher dimensions" (1986, p. 25). Although they had some reservations, they regarded Fischbach's suggestion as having had a positive effect. Maddox noted that, "Fischbach et al. have provided an incentive for the design of better measurements by showing what kind of irregularity it will be sensible to look for" (1986, p. 173). An important feature of experimental design is knowing how large the observed effect is supposed to be.

A much more positive view was, "considerable, and justified, excitement has been provoked by the recent announcement [by Fischbach] -- that a reanalysis of the celebrated Eötvös experiment together with recent geophysical gravitation measurements supports the existence of a new fundamental interaction" (Lusignoli and Pugliese 1986, p. 468).

Paik's summary of the situation seems reasonable. "It is clear that the recent announcement of the possible discovery of a 'Fifth Force' (Fischbach et al.) stimulated great interest on the part of experimentalists to resume, improve and accelerate old experiments, as well as to plan new experiments. After the storm of criticisms, the essential claim of Fischbach et al. that the original Eötvös data show a strong correlation with chemical composition seems to be intact. Whether this represents a new physics or is an artifact of statistical fluctuation, only time will tell" (Paik, 1987, p. 394).

It seems clear, judging by the substantial amount of work published in 1986, that a significant segment of the physics community thought the fifth-force hypothesis was worth further investigation. This was about to become even more apparent. Although almost invisible in the published literature, experiments were being designed, performed, and analyzed. The results would start to appear in early 1987.

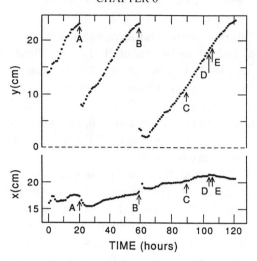

Figure 2. Position of the center of the sphere as a function of time. The y axis points away from the cliff. The motion of the sphere in the y direction indicates the presence of the fifth force. From Thieberger (1987a).

C) JUSTIFICATION

As mentioned earlier, the first two experimental results on the fifth force appeared in early 1987, first at the Moriond Workshop and then in the published literature (Raab 1987, Stubbs et al. 1987, and Thieberger 1987a, 1987b).[41] They disagreed. The Washington group (they referred to themselves as the Eöt-Wash group) used a torsion balance on a hillside and found no evidence for a fifth force. Thieberger, using a float experiment located on a cliff, found positive results consistent with such a force. (See Figures 2 and 3). No obvious error was found in either experiment.

Although the most obvious conclusion was that one of the two experiments was wrong, some theorists attempted to reconcile the two, either by increasing the mathematical complexity of the model, or by proposing a more complex source for the force, other than baryon number. This second suggestion was found wanting by an Eöt-Wash experiment (Adelberger et al. 1987).

The evidential situation became even more confused when Niebauer, McHugh, and Faller (1987) performed a Galilean experiment, dropping objects made of two different materials but finding no evidence for a fifth force, while Boynton (1987), using a torsion pendulum, found evidence for the force, at about the 3-SD level.[42] In addition, geophysics continued to provide support for a fifth force (Stacey et al. 1987a, 1987b).

The 1988 Moriond Workshop added to the confusion, with both positive and negative results being reported. Eckhardt et al.(1988a, 1988b) presented results from measurements of gravity at various heights on a 600 m tower, that gave evidence of a significant deviation from accepted gravitational theory. The Eöt-Wash group (Adelberger et al.1988) set even more stringent limits on the presence of the fifth force,

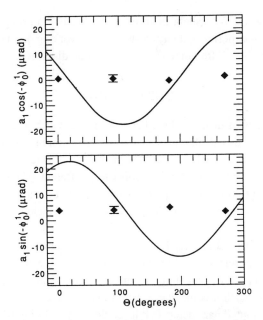

Figure 3. Deflection signal as a function of θ, the variable angle of the stage. The theoretical curves correspond to the signal expected for a fifth force with strength, $\alpha = 0.01$ and range, $\lambda = 100$ m. From Raab (1987).

which were also inconsistent with both Boynton's and Thieberger's results. Bizzeti *et al.* (1988) used a float experiment similar to that of Thieberger and found no evidence of a fifth force. As Fischbach (1988) pointed out, all the experimental results could be reconciled, but only at the cost of increasing complexity and a very restrictive set of theoretical parameters, for which no independent justification existed.

The 1988 Moriond Workshop marked the point of maximum evidential confusion. Although a few later experimental results would be compatible with the existence of such a force, they would also be compatible with alternative explanations that did not include it. Most of the further measurements would confirm Newtonian gravity, and set more stringent limits on the presence of the fifth force.[43]

In addition, doubts would be cast on some, but not all, of the earlier positive results. Bartlett and Tew (1989a, 1989b, 1990) argued that the failure to take local terrain into account properly might explain the positive results of both the Eckhardt tower experiment and the Australian mineshaft measurements of Stacey and collaborators. Further analysis by the original experimenters themselves showed that this was correct (Jekeli, Eckhardt, and Romaides 1990; Eckhardt 1990; and Tuck 1989). By the end of 1989 only two positive Fifth Force results remained, Thieberger's result, which by then even he doubted (Thieberger 1989) and Boynton's original 1987 result, which seemed to be superseded by his latest work (1990). These were opposed by a very large number of negative results (see note 43).

At the 1990 Moriond Workshop, in an informal discussion group attended by many of the theoretical and experimental physicists actively working on the Fifth Force, Orrin Fackler announced, "The Fifth Force is dead."[44] No one disagreed.[45]

3. DISCUSSION

One of the advantages of distinguishing between the activities of discovery, of pursuit, and of justification is that it illuminates some of the different roles that experimental evidence plays in science. We have seen that the weight of evidence used in these three contexts may be, and usually is, quite appropriately different. The evidence that might encourage a scientist to propose a hypothesis may be less convincing than that required to further pursue it, which will be, in general, far less convincing than that required for belief or justification.

Thus, Fischbach was encouraged to continue his work on the fifth force by a "remarkable numerical relation" between his calculation of the gravitational effect for K mesons and the observed CP violating parameter, even though it needed an unjustified factor of 1.4×10^{14}. Such a coincidence would hardly have justified belief. Similarly, Dydak tentatively adopted parameters that he could not justify in order to continue with his theoretical calculations.

It was three pieces of evidence; the energy dependence of the K^0-$\bar{K}^0$ parameters, the geophysical gravity anomalies, and most importantly the reanalysis of the Eötvös experiment, that provided grounds for pursuit. Recall, however, the differing judgements, made at the time, concerning the fifth force. These differing judgments about the pursuit worthiness of an hypothesis led to a larger hypothesis space of alternatives being explored. Although I suspect that most of the physics community was skeptical of the existence of the fifth force, a considerable amount of effort was devoted to its investigation. Even an alternative to a well supported community commitment may offer opportunity for future work.

Belief in a hypothesis may not even be required for pursuit. In the case of the fifth force, interviews with several of those who performed experiments (Adelberger, Bartlett, Boynton, Eckhardt, Faller, and Newman) indicated that their belief varied from thinking it was definitely false (Eckhardt) to the most positive view that it had a twenty or thirty percent chance of being correct (Newman). Interestingly, Eckhardt, at first, reported a positive fifth force result, while Newman's experiments set some of the most stringent limits on the presence of such a force. Contrary to Pickering's "tuning in to community commitments" view, scientists do not always find what they expect to find.

In both episodes, the fifth force and atomic parity violation, discordant results led to further experimental work. Without an obvious error, even the admitted uncertainty of the original Washington-Oxford results was not seen as sufficient to reject them.[46] We have also seen that discordant results may also lead to theoretical attempts to reconcile them. This was true in both the case of the fifth force and in the case of Weinberg-Salam theory and atomic parity violation.

These episodes indicate that the reasons offered by social constructivists for justification are important factors in pursuit. Certainly the opportunity for future work enters into pursuit decisions. Pursuit is, after all, future work. So does the recycling of expertise. In the case of the fifth force the previous experience with gravity experiments, as well as the availability of experimental apparatus, was an important part of the decisions of Eckhardt, Faller, and Newman to pursue that investigation.

The fact that the subsequent history has argued against the existence of the Fifth Force should not cause us to overlook the fact that the suggestion of that force was the result of a sequence of reasonable and plausible steps. This started with the Colella, Overhauser, and Werner measurement, Fischbach's attempt to connect CP violation and gravity, and the subsequent observation of suggestive energy dependence of the K^0- $\overline{K}^0$ parameters. At the same time the work on the modifications of Newtonian gravity and the tantalizing results on the measurement of gravity in mines were proceeding. When these two strands were joined together it led the collaborators to reanalyze the original Eötvös experiment, where, again, a suggestive effect appeared. The suggestion of the fifth force then followed. There may not be a logic of discovery, but, at least in this case, discovery is not a totally mysterious process.

I believe that the histories of these two episodes have demonstrated that evidence is decisive in justification as well as the usefulness of distinguishing between discovery, pursuit, and justification. Such discussion can illuminate the various roles that evidence plays in science. The history presented has also shown that the reasons offered by constructivists, that is opportunity for future work, consistency with community commitments, recycling of expertise, and career interests, do play a role in pursuit. I also believe that the history has shown that these reasons do not influence what the experimental results are, the acceptance of those experimental results, or their use in justification.[47] Experimental evidence may not compel a decision, but it does provide good reasons for it.

NOTES

[1] For references to some of that work see Franklin (1993c).
[2] Pursuit may occur either before or after justification, or may indeed overlap with it. It is the evidence provided by pursuit that leads, in large part, to justification. It is often the case that scientists investigate an accepted theory in order to articulate it or to calculate its further implications. I will, however, more typically mean pursuit before a theory has been accepted. Pursuit after a theory has been accepted may also lead to anomalies that will start the next round of discovery, pursuit, and justification
[3] This is the meaning of acceptance I shall use in this essay.
[4] Scientists may also propose hypotheses as possible explanations of a phenomenon without being seriously committed to their truth. One might call these "interesting speculations" as opposed to serious suggestions. For examples see Franklin (1986, Chapter 3).
[5] Although this quotation seems to imply that evidence has an essential role in the production of scientific knowledge, it is not clear to me exactly what Pickering believes that role is.

[6] This has led Peter Galison to refer to such scholars as "theory firsters" (quoted in Pickering 1991, p. 463).

[7] I believe that there are other grounds for thinking these arguments are wrong, and some of these will be discussed later.

[8] Pickering and I have presented the two different historical accounts of the episode, while Ackermann and Lynch have commented on those accounts.

[9] Pickering might object to my use of the term "evidential weight," something he seems to deny exists, but he does put both the Washington-Oxford and SAC results on an equal footing.

[10] Pickering regards the original Oxford-Washington results as mutants slain by the SLAC E122 experiment. As discussed below, I believe the mutants died of natural causes. They were uncertain to begin with, and later results that confirmed the W-S theory were more reliable. Valid experimental evidence decided the issue.

[11] In his 1984 account Pickering makes much of the institutional power of the high energy physics community (those doing work like E122) as compared to that of the community of atomic physicists. "Matched against the mighty traditions of HEP [high-energy physics], the handful of atomic physicists at Washington and Oxford stood little chance" (Pickering 1984a, p. 302). Hence the importance of Bouchiat's group affiliation.

[12] Although I suspect that a majority of the physics community were skeptical of the Washington-Oxford results, there were no obvious reasons for believing they were wrong, although the systematic uncertainties that they themselves cited did make the results uncertain. As discussed below, there was a reasoned evaluation of the discordant results.

[13] Although both the atomic parity-violation experiments and SLAC E122 used new techniques, they were, in fact, quite different apparatuses, subject to different backgrounds and sources of error and uncertainty. There were no good reasons to lump them together

[14] As noted by Bouchiat, and discussed in detail in Franklin (1990, Chapter 8), the SLAC E122 experiment was very carefully checked. There were good reasons to believe the result was valid without waiting for an expensive and time consuming replication. Contrary to Jacqueline Susann, once may be enough

[15] The calculations of the effect predicted by the W-S theory have also changed. Recent calculations have reduced the size of the expected effect.

[16] Pickering's view seems to depend on accepting the Washington-Oxford results and rejecting those of Novosibirsk and Berkeley. He offers no reasons to support this judgment

[17] Pickering persists in his interpretation. "Thus, in what I think was the first major review talk to follow the announcement of E122's results, the reviewer discounted the negative findings (from groups working at the Universities of Oxford and Washington) while including the positive findings (from groups at Novosibirsk and Berkeley) in his calculations of the phenomenological parameters describing the electroweak interaction" (1991, p. 461). How this is consistent with a tentative adoption, that was admittedly unjustified, for the purposes of calculation, escapes me.

[18] Lynch has objected to my use of the word valid. What I mean by valid is that we have good reasons for belief in the correctness of the results. I prefer it to "correct" because we may have good reasons to believe a result that is later shown to be incorrect. Thus, in my view, the early atomic parity violation results were invalid.

[19] For details of this episode see Franklin (1986, Chapter 3).

[20] I note here that the physics community was accepting a result that seemed to refute a strongly supported and well-established symmetry law, and we can see that it led to large amounts of both theoretical and experimental work.

[21] See, e.g., the histories of parity and CP violation in Franklin (1986, Chapters 1, 3).

[22] I believe it to be both normative and descriptive.

[23] The Washington, Oxford, and Novosibirsk experiments were all performed on atomic bismuth.

[24] The reader will notice that Dydak and Bouchiat have made slightly different evaluations of the evidence on atomic parity violation. Dydak regards the evidence as uncertain, whereas Bouchiat has made a judgment as to which result he thinks is most reliable, i.e., the Novosibirsk result that reported a positive result on parity violation. Neither judgment, however, states that the evidence agrees with W-S theory. In Bouchiat's more positive evaluation the result is only "roughly the magnitude predicted by the Weinberg-Salam theory." The situation had changed slightly between the two evaluations. Bouchiat's later evaluation took place several months after Dydak's and was at a conference in which the Novosibirsk group had not only presented a paper, but had also been subjected to severe questioning concerning possible systematic errors.

[25] For a detailed history of this episode, along with a comparison between possible constructivist accounts and my own evidence model account, see Franklin (1993a).

[26] That is still the case today, although the vast preponderance of evidence has convinced both Thieberger and the physics community that something must have been wrong with his experiment, even if no mistake has been found.

[27] The atomic parity-violation case was even more uncertain because of the admitted, large systematic uncertainties in the results.

[28] They might, e.g., have measured their output signal as a test light source was gradually covered. Suppose that their signal had not changed, or that it had changed when the source was held constant. They would have rejected their data as unreliable. A case from the actual practice of science occurred in an experiment to measure the spin of He^6 (Commins and Kusch 1958). A beam of He^6 atoms passed through a long inhomogeneous magnetic field. If the spin were 1 then the counting rate with the field on would be a factor of three lower than with the field off. This was observed in the first experimental run. (I was an assistant on the experiment). There was, fortunately, an experimental check on the apparatus. The beam was diverted so that it reached the detector without passing through the magnetic field. In this case the counting rates with the field on and off should have been equal. The factor of three difference persisted, indicating that the magnetic field lowered the efficiency of the detector by an unfortunate factor of three. When the detector was better shielded against magnetic fields the test succeeded. The result of the experiment was that the spin of He^6 was zero.

[29] The muon decay occurred 63 percent of the time, while the expected positron decay ratio was 1.6×10^{-5}. The currently accepted value for the branching ratio is $(1.55 \pm 0.07) \times 10^{-5}$.

[30] Jim Bogen and Jim Woodward (1988) have made this clear in their distinction between data and phenomena. Their point is that theory predicts phenomena, which is the same as my term "experimental result."

[31] A survey of my colleagues who work in the areas of experimental condensed matter physics, high energy physics, and atomic physics supports this view.

[32] For examples of epistemological strategies used in several famous experiments see Franklin (1986, Chapter 7). Social constructivists might regard these as arguments used to persuade and not as epistemological.

[33] For a detailed history of this episode see Franklin (1993a). A detailed technical discussion of the fifth force is in Fischbach (1988).

[34] Fischbach, Sam Aronson and Carrick Talmadge were the originators of the fifth-force hypothesis.

[35] The neutrality of the particle eliminated electromagnetic interactions, which were much larger than the gravitational effects of interest.

[36] See Franklin (1986, Chapter 3) for a detailed discussion of both CP violation and its discovery.

[37] The parameters were Δm, the K_L-K_S mass difference, τ_S, the lifetime of the short-lived K^o meson, η_{+-}, the magnitude of the CP violating amplitude, and $\tan \phi_{+-}$, the tangent of the phase of that amplitude. Interestingly, subsequent measurements of these parameters has shown no energy dependence, but at the time of our story it was a "tantalizing" effect.

[38] Although a constant that varies with distance might seem to be an oxymoron, it is a useful shorthand.

[39] Unmodified gravity is such a long-range force, and that force is what Fischbach had used in his attempt to explain CP violation.

[40] Such a possibility had been mentioned by Fischbach *et al.* in their original paper (1986), and was discussed in a paper written at the time, but never published.

[41] The importance of the Moriond workshops cannot be overstated. It was at these meetings that many of those working on the fifth force gathered for both formal presentation of papers and informal discussion.

[42] Boynton would later refer to this as a marginal observation.

[43] The list of negative Fifth Force results includes; Fitch *et al.* (1988), Moore *et al.* (1988), Kuroda and Mio (1989a, 1989b, 1990), Speake and Quinn (1988), Cowsik *et al.* (1988, 1990), Bennett (1989), Newman, Graham, and Nelson (1989), Nelson *et al.* (1990), Stubbs (1989), Stubbs *et al.* (1989), Adelberger (1989), Bizzeti *et al.*(1989a, 1989b), Kasameyer *et al.* (1989), Thomas *et al.* (1989), Heckel *et al.* (1989), Muller *et al.* (1989), Boynton and Peters (1989), Speake *et al.*(1990), Kammeraad *et al.* (1990), Jekeli *et al.* (1990), and Eckhardt *et al.* (1990). The only positive experimental result reported in this period, from a measurement of gravity in a borehole in the Greenland icecap (Ander 1989), was later shown to be consistent with Newtonian gravity by members of the group (Zumberge 1988, Parker and Zumberge 1989).

[44] I was present at the discussion.

[45] This decison was based on an overwhelming preponderance of evidence. This was not just a group of eminent scientists declaring that the search was over, but rather the considered judgment of those working in the field. In particular, it included Paul Boynton who had reported one of the positive fifth-force results. By this time, Peter Thieberger, who had reported the other positive result, believed that he must have made an error (see discussion in the text). It also included Fischbach, Aronson, and Talmadge, who had originally proposed the fifth-force hypothesis. Although the opinion of the originators that a hypothesis has been refuted is not conclusive, it is certainly a significant indication of the evidential situation. The decision of this group was effectively final. When they stopped working on the fifth force, there was no one left working on it. Although some experimental work continues (See Franklin 1993a for details) it is concerned with looking for much smaller deviations from the law of gravity than the original fifth force, which had a strength of about one percent that of gravity.

[46] It is, however, a good reason not to regard them as good evidence.

[47] I am not, of course, denying that the factors discussed earlier enter into the decision to do an experiment (pursuit), or to publish its results. I deny, however, that in the long run they have any impact on the results presented. The theoretical presuppositions of scientists may, in the short run, influence the experimental results presented. For a discussion of the influence of theoretical presuppositions on experimental results see Galison (1987, Chapters. 2 and 5) and Franklin (1986, Chapter 5) and (1990, p. 196). I believe that these studies show that the presuppositions do not have any long-lasting effect. For a more extensive discussion of factors involved in experiment and its results see Franklin (1993a, Chapter 3).

CHAPTER 7

THE RESOLUTION OF DISCORDANT RESULTS

Experiments often disagree. How then can scientific knowledge be based on experimental evidence? In this paper I will examine four episodes from the history of recent physics: (1) the suggestion of a Fifth Force, a modification of Newton's Law of Gravitation; (2) the early attempts to detect gravitational radiation (gravity waves); (3) the claim that a 17-keV neutrino exists; and (4) experiments on atomic parity violation and on the scattering of polarized electrons and their relation to the Weinberg-Salam unified theory of electroweak interactions. In each of these episodes discordant results were reported, and a consensus was later reached that one result--or set of results-- was incorrect. I will examine the process of reaching that consensus. I will show that the decision was reached by reasoned discussion based on epistemological and methodological criteria. It then follows that we may use experimental evidence as the basis of scientific knowledge.

The late Richard Feynman, one of the leading theoretical physicists of the twentieth century remarked, "The principle of science, the definition, almost, is the following: *The test of all knowledge is experiment.* Experiment is the *sole judge* of scientific 'truth'" (Feynman, Leighton and Sands 1963, p. 1-1). Yet experiments often disagree. How then can scientific knowledge be based on experiment? Although in practice, the discord between experimental results is usually resolved within a reasonable time,[1] questions remain as to whether or not the method by which the resolution is achieved provides grounds for knowledge. Social constructivists imply, however much they may disclaim it,[2] that it does not. In their view the resolution of such disputes, as well as the acceptance of experimental results in general, is based on "negotiation" within the scientific community, which does not include epistemological or methodological criteria. Such negotiations do include considerations such as career interests, professional commitments, prestige of the scientists' institutions, and the perceived utility for future research. For example, Pickering states, "Quite simply, particle physicists accepted the existence of the neutral current because they could see how to ply their trade more profitably in a world in which the neutral current was real" (Pickering 1984b), p. 87). The emphasis on career interests and future utility is clear.

Part of the problem is that the constructivists conflate pursuit, the further investigation of a theory or of an experimental result, with justification, the process by which that theory or result becomes accepted as scientific knowledge. No one would deny that the considerations suggested by the constructivists enter into pursuit, along with other reasons such as the recycling of expertise, instrumental loyalty, and scientific interest (Franklin 1993b). I suggest that these considerations do not enter into justification. Scientific knowledge is too important for that. This is the reason why scientific fraud is so severely punished. For details see Franklin (1986, chap. 8).

Collins has summed up the argument against both experimental results and against reasoned resolution of discordant results in what he calls the "experimenters' regress." In his discussion of the early attempts to detect gravity waves he asks *"But what is the correct outcome?*

What the correct outcome is depends on whether or not there are gravity waves hitting the Earth in detectable fluxes. To find this out we must build a good gravity wave detector and have a look. But we won't know if we have built a good detector until we have tried it and obtained the correct outcome! But we don't know what the correct outcome is until... and so on ad infinitum.The existence of this circle, which I call the 'experimenters' regress' comprises the central argument of this book" (Collins 1985, p. 84). More succinctly, "Proper operation of the apparatus, parts of the apparatus and the experimenter are defined by the ability to take part in producing the proper experimental outcome. Other indicators cannot be found" (Collins 1985, p. 74).

I disagree. I believe that the discord between experimental results is resolved by reasoned argument, based on epistemological and methodological considerations. These are the other indicators. This does not preclude a joint decision concerning whether or not a detector works properly and whether or not the phenomenon in question exists. The disagreement between my view and that of the constructivists concerns the reasons for that decision.

What is at stake here is the status of science as knowledge. If we don't have good reasons for belief in experimental results or for our choice of one of a set of discordant results rather than another, then experimental evidence cannot provide grounds for scientific knowledge. Pickering would then be correct when he says, "There is no obligation upon anyone framing a view of the world to take into account what twentieth century science has to say" (Pickering 1984a, p. 413). I believe he is wrong.

Some commentators, as well as social constructivists themselves, have argued that constructivists do not claim that scientists don't provide reasons for their decisions--but rather that the reasons are insufficient. "Social constructivists do *not* say that experimental evidence is irrelevant to theory choice, confirmation, or refutation. Nor do they argue that there are no good reasons for belief in the validity of evidence" (Lynch 1991, pp. 476-77). Nevertheless in studies presented by constructivists evidence does not enter into such decisions, nor are good reasons for belief in evidence ever discussed. The contructivist claim is twofold. First, such reasons do not provide justification either for experimental evidence or for hypothesis testing on the basis of that evidence. The second point is that even if such reasons were sufficient within science, they do not have any standing beyond the scientific community.

I shall begin with a discussion of the second point. As discussed below, I believe that there is an epistemology of experiment, a set of strategies that provides grounds for reasonable belief in experimental results. I have further argued that these strategies have independent philosophical justification. One such strategy is based on the idea that observation of evidence entailed by a hypothesis should strengthen belief in that hypothesis. This would seem to be a foundation for reasonable thought. If such evidence shouldn't strengthen one's belief then what should? Similarly, two independent observations supporting the same hypothesis or experimental result, provide more support than do two repetitions of the same experiment.

Decisions between discordant results are made by the community of scientists, and are thus inherently social and dependent on historical context, particularly on what is accepted as scientific knowledge at a given time. I do not deny that scientists have the usual human motivations such as career advancement, or desire for credit, prestige, and economic gain. (Scientists also have an interest in producing scientific knowledge, as well as a career interest in producing correct results). I do claim, however, that such decisions are based on epistemological and methodological criteria, and that these criteria are not justified merely by their acceptance by the scientific community.

I believe that we have independent grounds for believing that science and its methodology provide us with reliable knowledge about the world. It is not just the successful practice of science, which is, after all, decided by scientists themselves, but, rather evidence from the "real" world that underlies this judgment. It is not mystical incantations by Faraday, Maxwell, or other scientists that causes a light to come on when a switch is thrown. Objects would not suddenly fall up rather than down if the American Physical Society voted to repeal Newton's Law of Universal Gravitation.[3] These, and other examples too numerous to mention, provide grounds for believing that science is actually telling us something reliable about the world. As Ian Hacking said, in the more limited context of discussing the reality of scientific entities such as electrons, "We are completely convinced of the reality of electrons when we regularly set out to build -- and often enough succeed in building -- new kinds of device that use various well-understood causal properties of electrons to interfere in other more hypothetical parts of nature" (Hacking 1983, p. 265). It is this practical intervention in the world that persuades us that we should take account of what twentieth century physics has to say when we formulate a worldview. It is possible that negotiations based on the considerations suggested by the constructivists might give us reliable knowledge about the world, but that seems rather unlikely. Why should the world be such that it benefits the career interests of scientists?

The first point, concerning whether or not reasons are sufficient to provide justification for evidence or theories, relies on two philosophical points (Nelson 1994). The first is the underdetermination of theory by evidence, the fact that one can always find an alternative explanation for a given experimental result. The second is the Duhem-Quine thesis: if an experiment seems to refute a theory, it in fact refutes the conjunction of both the theory and background knowledge; and one does not know where to place the blame for the failure. One may save a hypothesis from refutation by suitable changes in one's background knowledge. I believe that adequate answers have already been provided for these points (e.g., see Franklin (1990, pp. 144-161; 1993b, pp. 260-267)). An adequate discussion of these issues would take us too far from the central issue of this paper--namely, how is the discord between experimental results resolved.

In previous work I have argued for the existence of an epistemology of experiment, a set of strategies that can be used to argue for the correctness of an experimental result. These strategies include: 1) Experimental checks and calibration, in which the experimental apparatus reproduces known phenomena; 2) Reproducing artifacts that are known in advance to be present;[4] 3) Intervention, in which the experimenter manipulates the object under observation; 4) Independent confirmation using independent experiments; 5) Elimination of plausible sources of error and alternative explanations of the result (the Sherlock Holmes strategy); 6) Using the results themselves to argue for their validity; 7) Using an independently well-corroborated theory of the phenomena to explain the results; 8) Using an apparatus based on a well-corroborated theory; and 9) Using statistical arguments. The difficulty is that, in cases of discordant results, each of the discordant experiments has used such strategies. The resolution must proceed by demonstrating that, in at least some of the experiments, the strategies have been incorrectly applied.

Perhaps the most important method of arguing for this is to show that the Sherlock Holmes strategy has been incorrectly applied. One can argue that the experimental result can be explained by an alternative hypothesis or that a plausible source of error, a background that might either mask or mimic the correct result, has been overlooked. Another argument is to demonstrate that the use of a particular strategy generates a contradiction with accepted results. Similarly, one might examine the assumptions concerning the operation of the apparatus and show empirically that they are incorrect. Plausible interpretations of the results may also be shown to be incorrect.

Other criteria can also be used. In a particular experiment some epistemological strategies may have been applied successfully, whereas others had failed, casting doubt on the result. Sometimes the failure to reproduce an observation, despite numerous attempts to do so, might be legitimately regarded as casting doubt on the original observation, even when no error has been found in the original experiment. This would be a case of preponderance of evidence.

There are several different types of discordant experimental results. One may have experiments that measure the same quantity with the same, or similar, types of apparatus. Discordant results may also involve the measurement of the same quantity, but with different types of experimental apparatus. In this case one might worry that the difference in the results is due to some crucial difference in the apparatus. A third type of discord occurs when different experiments, measuring different quantities, both of which are predicted by the same theory, give results one of which confirms the theory whereas the other disagrees with the theoretical prediction. Each of these types of discord will be illustrated in the episodes discussed.

In this paper I will examine four episodes from the history of recent physics: the suggestion of a Fifth Force, a modification of Newton's Law of Gravitation; the early attempts to detect gravitational radiation (gravity waves); the claim that a 17-keV neutrino exists; and experiments on atomic parity violation and on the scattering of polarized electrons and their relation to the Weinberg-Salam unified theory of electroweak interactions. In each of these episodes discordant results were reported, and a consensus was later reached that one result or set of results was incorrect. I will examine the process of reaching that consensus. I will show that this process was based on the epistemological and methodological criteria I have suggested.

One might ask whether case studies can be used to demonstrate that scientists resolve the discord between experimental results by the application of epistemological and methodological criteria. The case studies show only that in the four episodes presented this was true. Nevertheless case studies do support the generalization. Although it is dangerous to generalize from only four instances, I believe that these four episodes provide a reasonable picture of the practice of modern physics. I note, in addition, that constructivists provide case studies to support their view of science. Two of the episodes considered here--namely those the early attempts to detect gravity waves and of atomic-parity violation experiments--have been used by constructivists to support their view that the resolution of such discordant results does cast doubt on the status of science as knowledge (Collins 1985; Pickering 1984a, 1991). It seems fair then to present an alternative account of those episodes and also to present other case studies. I have argued in detail elsewhere that their accounts are incorrect (Franklin 1990, 1993c; 1994). Constructivists such as Pickering and Collins seem to imply that epistemological criteria are never decisive in resolving the dispute between discordant results. In that case, the presentation of even one case study in which the criteria were decisive will cast doubt on their view. In none of the case studies presented by constructivists do methodological or epistemological criteria play an important, much less decisive, role.

1. THE FIFTH FORCE[5]

In January 1986, Sam Aronson, Ephraim Fischbach, and Carrick Talmadge proposed a modification of Newton's Law of Universal Gravitation (Fischbach *et al.* 1986). The Newtonian gravitational potential is $V = -Gm_1m_2/r$.[6] Their modification took the mathematical form $V = -Gm_1m_2/r [1 + \alpha e^{-r/\lambda}]$, where α was the strength of the new interaction, and λ was its range. This new interaction became known as the "Fifth Force." Their initial suggestion was that α was approximately 1 percent, and λ approximately 100 m. Unlike the gravitational force itself, the new force was composition dependent. The Fifth Force between a copper mass and a platinum mass would be different from that between a copper mass and an iron mass. By early 1990 the consensus was that such a force did not exist. The decision process was not simple. There were two different sets of discordant results; (1) from measurements of gravity using towers and mineshafts, which examined the distance dependence of the force, and (2) from experiments on the composition dependence of the force. For a reasoned decision to be reached concerning the existence of the Fifth Force, the discords had to be resolved.

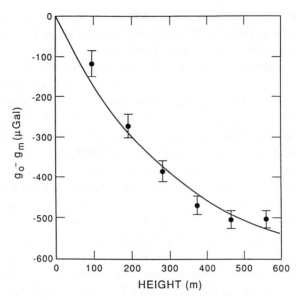

Figure 1. Eckhardt's experimental results fitted to a scalar Yukawa model. The difference between the predictions of Newtonian gravity and the measured values are plotted as a function of the height in the tower. From Fairbank (1988).

a) Tower Gravity Experiments

One way in which the presence of the Fifth Force could be tested was by investigating the distance dependence of the gravitational force, to see if there was a deviation from Newton's inverse-square law. This type of experiment measured the variation of gravity with position, usually in a tower, or in a mineshaft or borehole. All of the experiments used a standard device, a LaCoste-Romberg gravimeter, to measure gravity. The measurements were then compared with the values calculated using a model of the earth, surface gravity measurements, and Newton's law of gravitation.[7] In this case the experiments used the same type of apparatus to measure the same quantity.

Evidence from such measurements had provided some of the initial support for the existence of the Fifth Force. Geophysical measurements during the 1970s and 1980s had given values of G, the universal gravitational constant, that were consistently higher, by about 1 percent, than that obtained in the laboratory.[8] Because of possible local mass anomalies they were also "tantalizingly uncertain" (Stacey 1981).

After the proposal of the Fifth Force, further experimental work was done. At the Moriond workshop in January, 1988,[9] Eckhardt presented results from the first of the new tower gravity experiments (Eckhardt *et al.* 1988a).[10] The results differed from the predictions of the inverse square law, by -500 ± 35 μGal (1 μgal = 10^{-8} ms^{-2}) at the top of the tower (Figure 1). A second result was also presented at the workshop by the Livermore group (Thomas, Vogel, and Kasameyer 1988). They used gravity measurements from five boreholes and found a 2.5-percent discrepancy between their observed gravity gradient and that predicted by their Newtonian model. This result also differed in magnitude from the 0.52-percent discrepancy reported by Stacey and in both sign and magnitude from the 0.29-percent discrepancy reported by Eckhardt. They noted, however, that their measured free-air gradients disagreed with those calculated from their model and concluded, "that the model does not reflect the total mass distribution of the earth with sufficient accuracy to make a statement about Newtonian gravity" [or about the Fifth Force] (p. 591).

Further evidence for the existence of a Fifth Force was provided by a group that measured the variations in gravity in a borehole in the Greenland icecap (Ander *et al.* 1989). They found an unexplained difference of 3.87 mGal between the measurements taken ar a depth of 213 m and a depth of 1673 m. The experimental advantage of the Greenland experiment was the uniform density of the icecap. The disadvantages were the paucity of surface gravity measurements and the presence of underground geological features that could produce gravitational anomalies.

All of the evidence from tower and mineshaft experiments prior to 1988 supported the Fifth Force. There was, however considerable--although not unambiguous--negative evidence from other types of experiment. Negative evidence from tower experiments would, however, be forthcoming, and it is the discrepancy between the tower results that I will address here. (The discord between the other experimental results on the composition dependence of the Fifth Force will be addressed in the next section).

Even before those negative results appeared questions and doubts were raised concerning the positive results. It was not, in fact, the gravity measurements themselves that were questioned. These were all obtained with a standard and reliable apparatus. It was, rather, the theoretical calculations that were used for the theory-experiment comparison that were criticized. One of the important features needed in these calculations was an adequate model of the earth. Recall that the Livermore group had doubted their own comparison because their model had not given an adequate account of the measured free-air gradients.

The Greenland group's calculation was the first to be criticized. It was subjected to severe criticism, particularly for the paucity of surface gravity measurements near the location of their experiment, (their survey included only 16 such points), and for the inadequacy of their model of the earth. It was pointed out that in Greenland there were underground features of the type that could produce such gravitational anomalies. The Greenland group was criticized both for having overlooked plausible sources of error in their experiment-theory comparison and for overlooking plausible alternative explanations of their result.

When this result was later presented the group stated that their result could be interpreted either as evidence for non-Newtonian gravity (a Fifth Force), or explained by local density variations. "We cannot unambiguously attribute it to a breakdown of Newtonian gravity because we have shown that it might be due to unexpected geological features below the ice" (Ander *et al.* 1989, p. 985). Parker, a member of the Greenland group, as well as Bartlett and Tew, suggested that both the positive evidence for the Fifth Force of Eckhardt and collaborators and that of Stacey and collaborators could be explained by either local density variations or by inadequate modeling of the local terrain. Bartlett and Tew gave more details of their criticism at the 1989 Moriond workshop (Bartlett and Tew 1989a). Bartlett and Tew admitted that it was still an open question as to whether or not the models of Stacey and Eckhardt properly accounted for local terrain, and presented a calculation arguing that 60 - 65 of Eckhardt's tower residuals could be explained by local terrain.

Eckhardt disagreed. His group presented a revised value for the deviation from Newtonian gravity at the top of their tower of 350 ± 110 µGal. They attributed this change, a reduction of about one-third, to better surface gravity data, and to finding an elevation bias in their previous survey.[11] "We also had the help of critics who found our claims outrageous"(Eckhardt 1989), p. 526). They concluded, "Nevertheless the experiment and its reanalysis are incomplete and we are not prepared to offer a final result" (p. 526).

The Livermore group presented a definite result from their gravity measurements at the BREN tower at the Nevada test site (Kasameyer *et al.* 1989). To overcome the difficulties with their previous calculations, they had extended their gravity survey to include 91 of their own gravity measurements, within 2.5 km of the tower, supplemented with 60,000 surface gravity measurements within 300 km, that were done by others.[12] They presented preliminary results in agreement with Newtonian gravity with a difference between the measured and predicted values of 93 ± 95 µGal at the top of the tower.[13]

Bartlett and Tew continued their work on the effects of local terrain. They argued that the Hilton mine results of Stacey and his collaborators could also be due to a failure to include local terrain in their theoretical model (Bartlett and Tew 1989b). They communicated their concerns to Stacey privately. Their view was confirmed when, at the General Relativity and Gravitation Conference in July 1989, Tuck reported that their group had incorporated a new and more extensive surface gravity survey into their calculation. "Preliminary analysis of these data indicates a regional bias that reduces the anomalous gravity gradient to two thirds of the value that we had previously reported (with a 50% uncertainty)..." (Tuck 1989). With such a large uncertainty, the results of Stacey and his collaborators could no longer be considered as support for the concept of a Fifth Force.

Parker and Zumberge, two members of the Greenland group, offered a general criticism of tower experiments (Parker and Zumberge 1989). They argued, in some detail, that they could explain the anomalies reported in both Eckhardt's tower experiment and in their own ice cap experiment, using conventional physics and plausible local density variations.[14] They concluded that there was "no compelling evidence for non-Newtonian long-range forces in the three most widely cited geophysical experiments [those of Eckhardt, of Stacey, and their own]... and that the case for the failure of Newton's Law could not be established "(Parker and Zumberge 1989, p. 31).

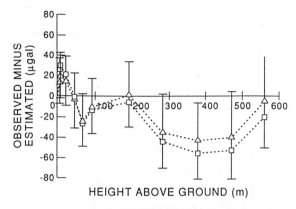

Figure 2. Difference between measured and calculated values of *g* as a function of height. No substantial difference is seen. From Jekeli *et al.* (1990).

The last hurrah for tower gravity experiments that supported the concept of a Fifth Force was signalled in the paper, "Tower Gravity Experiment: No Evidence for Non-Newtonian Gravity" (Jekeli, Eckhardt and Romaides 1990). In this paper Eckhardt's group presented their final analysis of their data, which included a revised theoretical model, and concluded that there was, in fact, no deviation from Newtonian gravity. (See Figure 2, and contrast this with their initial positive result shown in Figure 1). Two subsequent tower results also supported Newton's Law (Kammeraad *et al.* 1990), Speake *et al.* 1990).

The discord had been resolved. The measurements were correct. It was the comparison between theory and experiment that had led to the discord. It had been shown that the results supporting the existence of a Fifth Force could be explained by inadequate theoretical models: either failure to account adequately for local terrain or the failure to include plausible local density variations. In other words, the Sherlock Holmes strategy had been incorrectly applied. The experimenters had overlooked plausible alternative explanations of the results or possible sources of error.

The careful reader will have noted that it had not been demonstrated that the original theoretical models were incorrect. It had only been shown that the measurements agreed with the calculations when plausible sources of error were eliminated. Although this made the positive Fifth Force results very questionable, it was not an airtight argument. The new calculations could have been wrong. I note, however, that the experimenters themselves agreed that the newer models were better.

Scientists make decisions in an evidential context. The Fifth Force was a modification of Newtonian gravity. Newtonian gravity, and its successor, general relativity, were strongly supported by existing evidence. In addition, there were other credible negative tower gravity results that did not suffer from the same difficulties as did the positive results. There was also, as discussed in the next section, an overwhelming preponderance of evidence against the existence of a Fifth Force, from

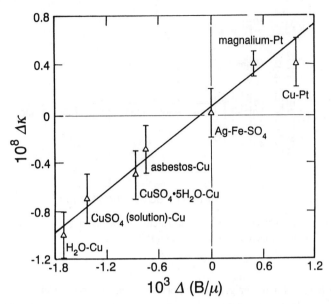

Figure 3. Δk, the fractional change in g, as a function of $\Delta(B/\mu)$. A substantial composition dependence is clearly seen. From Fischbach et al. (1986).

other types of experiment. The decision as to which theory-experiment comparison was correct was not made solely on the basis of the experiments and calculations themselves, although one could have justified this. Scientists examined all of the available evidence, and came to a reasoned decision about which were the correct results, and that a Fifth Force did not exist.

b) The Search for a Composition Dependent Force[15]

The other strand of experimental investigation of a Fifth Force was the search for composition dependence of the gravitational force. The strongest piece of evidence cited when the Fifth Force was originally proposed came from a reanalysis of the Eötvös experiment. The original Eötvös experiment was designed to demonstrate the equality of gravitational and inertial mass for all substances. Eötvös reported equality to about one part in a million. Fischbach and collaborators had reanalyzed the Eötvös data and reported a large and surprising composition-dependent effect (Figure 3).

This was the effect that was subsequently investigated. Two types of composition-dependence experiments are shown in Figure 4. In order to observe the effect of a short-range force such as the Fifth Force, one needs a local mass asymmetry. This asymmetry was provided by either a terrestrial source--a hillside or a cliff--or by a large, local, laboratory mass. If there were a composition-dependent, short-range force the torsion pendulum would twist. A variant of this experiment was the float experiment, in which an object floated in a fluid and in which the difference in gravitational force on the float

THE RESOLUTION OF DISCORDANT RESULTS

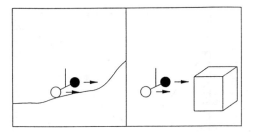

Figure 4. Types of composition-dependence experiments used to search for the Fifth Force. From Stubbs (1990).

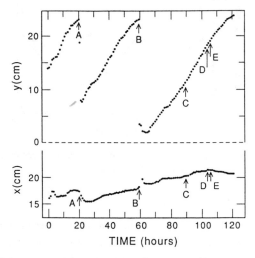

Figure 5. Position of the center of the sphere as a function of time. The *y* axis points away from the cliff. From Thieberger (1987a).

and on the fluid would be detected by the motion of the float. These were done with terrestrial sources.

The first results of tests for a composition-dependent force appeared in January, 1987, a year after the concept of a Fifth Force was first proposed. They disagreed. Thieberger, using a float experiment, found results consistent with the presence of such a force (Thieberger 1987a). A group at the University of Washington, headed by Eric Adelberger and whimsically named the "Eöt-Wash group," found no evidence for such a force and set rather stringent limits on its presence (Adelberger et al. 1987).

The results of Thieberger's experiment, done on the Palisades cliff in New Jersey, are shown in Figure 5. One can see that the float moves quite consistently and steadily away from the cliff (the y-direction) as one would expect if there were a Fifth Force. Thieberger eliminated other possible causes for the observed motions; these causes included magnetic effects, thermal gradients, and leveling errors. He also rotated his apparatus by 90°, to check for possible instrumental asymmetries, and obtained the

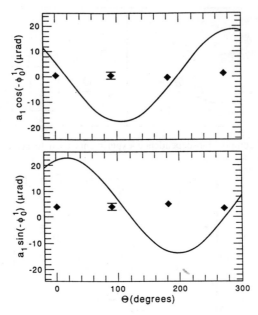

Figure 6. Deflection signal as a function of θ. The theoretical curves correspond to the signal expected for $\alpha = 0.01$ and $\lambda = 100$ m. From Raab (1987).

same positive result. In addition, he performed the same experiment at another location, one without a local mass asymmetry or cliff, and found no effect, as expected. He concluded, "The present results are compatible with the existence of a medium-range, substance-dependent force which is more repulsive (or less attractive) for Cu than for H_2O.... Much work remains before the existence of a new substance-dependent force is conclusively demonstrated and its properties fully characterized" (Thiberger 1987a, p. 1068).

The Eöt-Wash experiment used a torsion pendulum, shown schematically in Figure 4. It was located on the side of a hill on the University of Washington campus. If the hill attracted the copper and beryllium test bodies differently, then the torsion balance would experience a net torque. None was observed (Figure 6). The group minimized asymmetries that might produce a spurious effect, by machining the test bodies to be identical to within very small tolerances. The bodies were coated with gold to minimize electrostatic forces. Magnetic, thermal, leveling, and gravity gradient effects were shown to be negligible.

The discordant results were an obvious problem for the physics community. Both experiments appeared to be carefully done, with all plausible and significant sources of possible error and background adequately accounted for. Yet the two experiments disagreed.[16]

In this case we are dealing with attempts to observe and measure the same quantity--a composition-dependent force--with very different apparatuses, a float experiment and a

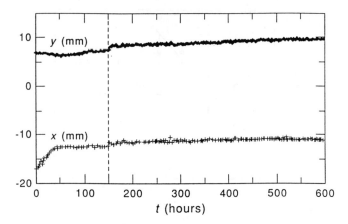

Figure 7. Position of the sphere completely immersed in liquid as a function of time. The vertical line marks the time at which restraining wires were removed. From Bizzeti et al.(1988).

torsion pendulum. Was there some unknown but crucial background in one of the experiments that produced the wrong result? To this day, no one has found an error in Thieberger's experiment, but the consensus is that the Eöt-Wash group is correct and that Thieberger is wrong--that there is no Fifth Force. How was the discord resolved?

In this episode it was resolved by an overwhelming preponderance of evidence. The torsion pendulum experiments were repeated by others including Fitch, Cowsik, Bennett, and Newman, and by the Eöt-Wash group itelf (for details and references see Franklin, 1993a). These repetitions, in different locations and with different substances, gave consistently negative results. Evidence against the concept of a Fifth Force was also provided by modern versions of Galileo's Leaning Tower of Pisa experiment by Kuroda and by Faller. For a graphical illustration of how the evidence concerning a composition-dependent force changed with time see Figures 3, 8, and 9. As more evidence was provided the initial--and startling--effect claimed by Fischbach and collaborators became far less noticeable. In addition, Bizzeti, using a float apparatus similar to that used by Thieberger, also obtained results showing no evidence of a Fifth Force (Bizzeti et al. 1989b). (Compare Bizzeti's results [Figure 7] with those of Thieberger (1987a) [Figure 5]). Bizzeti's result was quite important. Had he agreed with Thieberger, then one might well have wondered whether between torsion balance experiments and float experiments, there was some systematic difference that gave rise to the conflicting results. This did not happen. There was an overwhelming preponderance of evidence against composition dependence of a Fifth Force. Even Thieberger, although he had not found any error in his own experiment remarked, "Unanticipated spurious effects can easily appear when a new method is used for the first time to detect a weak signal... Even though the sites and the substances vary, effects of the magnitude expected have not been observed.... It now seems likely that

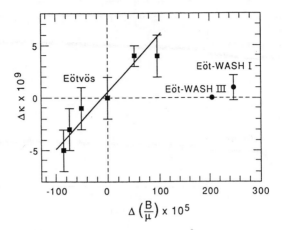

Figure 8. Comparison of the Eötvös reanalysis of Fischbach *et al.* (1986) with the results of the Eöt-Wash I and III experiments. The error bar on the Eöt-Wash III datum is smaller than the dot. From Adelberger (1989).

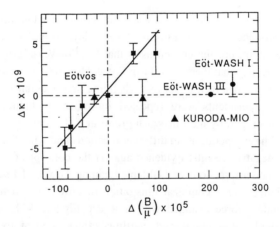

Figure 9. The data of Kuroda and Mio (1989b) added to Figure 8.

some other spurious effect may have caused the motion observed at the Palisades cliff" (Thieberger 1989, p. 810).[17]

2. GRAVITY WAVES[18]

Beginning in the late 1960s and continuing through the early 1970s, there were numerous attempts to detect gravitational radiation (gravity waves). Gravity waves were predicted by the general theory of relativity but had not been detected. Each of these early attempts used variants of a standard detector developed by Joseph Weber. Weber had used a massive aluminum alloy bar, or antenna, (known as a Weber bar), which was

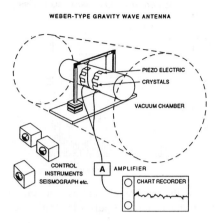

Figure 10. A Weber-type gravity wave detector. From Collins (1985).

supposed to oscillate when struck by gravitational radiation (Figure 10). The oscillation was to be detected by observing the amplified signal from piezo-electric crystals attached to the antenna. The expected signals were quite small (the gravitational force is quite weak in comparison to the electromagnetic force) and the bar had to be insulated from sources of noise such as electrical, magnetic, thermal, acoustic, and seismic forces. Because the bar was at a temperature different from absolute zero, thermal noise could not be avoided, and to minimize its effect Weber set a threshold for pulse acceptance. Weber claimed to have observed above-threshold pulses, in excess of those expected from thermal noise.[19] In 1969 Weber claimed to have detected approximately seven pulses/day that were due to gravitational radiation.

The problem was that Weber's reported rate was far greater (by a factor > 1000) than that expected from calculations of cosmic events and his early claims were met with skepticism. During the late 1960s and early 1970s, however, Weber introduced several modifications and improvements that increased the credibility of his results. He claimed that above-threshold peaks had been observed simultaneously in two detectors separated by 1,000 miles. Such coincidences were extremely unlikely if they were due to random thermal fluctuations. In addition, he reported a 24 hour periodicity in his peaks, the sidereal correlation, that indicated a single source for the radiation, perhaps near the center of our galaxy. These results increased the plausibility of his claims sufficiently so that others attempted to replicate his findings. By 1975, six other experimental results had been reported. None were in agreement with Weber. The consensus was that Weber was wrong and that gravity waves had not been detected.

How was this agreement reached? This is a case in which the same apparatus or a slight variant of that apparatus was used to search for the same quantity. What complicated the search was that the phenomenon had not been previously observed. Although questions might have been raised concerning what constituted a good gravity

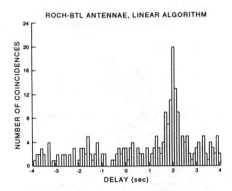

Figure 11. A plot showing the calibration pulses for the Rochester-Bell Laboratory collaboration. The peak due to the calibration pulses is clearly seen. From Shaviv and Rosen (1975).

wave detector, there was no such dispute. Everyone agreed that the Weber bar was a suitable detector. The questions were, rather, whether or not the detector was operating properly and whether or not the data were being analyzed correctly.

The first issue raised was the question of the calibration of the apparatus, the use of a surrogate signal to standardize an instrument. In this case, scientists injected pulses of electromagnetic energy into the antenna and determined whether their apparatus could detect such pulses. Weber's apparatus failed to detect the pulses, whereas each of the six discordant experiments performed by his critics detected them with high efficiency. This difference was due to a difference in the analysis procedures used by Weber and that used by his critics.

The question of determining whether there is a signal in a gravitational wave detector or whether two such detectors have fired simultaneously is not easy to answer. There are several problems. One is that there are energy fluctuations in the bar due to thermal, acoustic, electrical, magnetic, and seismic noise, etc. When a gravity wave strikes the antenna, its energy is added to the existing energy. This may change either the amplitude or the phase, or both, of the signal emerging from the bar. It is not just a simple case of observing a larger signal from the antenna after a gravitational wave strikes it. This difficulty informs the discussion of which is the best analysis procedure to use.

The nonlinear, or energy, algorithm preferred by Weber was sensitive only to changes in the amplitude of the signal. The linear algorithm, preferred by everyone else, was sensitive to changes in both the amplitude and the phase of the signal. Weber preferred the nonlinear procedure because it resulted in proliferation, several pulses exceeding threshold for each input pulse to his detector.[20] Weber admitted that the linear algorithm, preferred by his critics, was more efficient at detecting calibration pulses. Similar results on the superiority of the linear algorithm for detecting calibration pulses were reported by both Kafka (pp. 258-9) and Tyson (pp. 281-2). Tyson's results

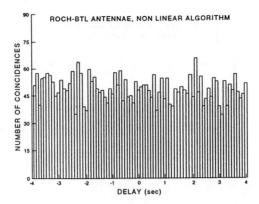

Figure 12. A time-delay plot for the Rochester-Bell Laboratory collaboration, using the non-linear algorithm. No sign of a zero-delay peak is seen. From Shaviv and Rosen (1975).

for calibration pulse detection are shown here for the linear algorithm (Figure 11), and for the nonlinear algorithm (Figure 12). There is a clear peak for the linear algorithm, whereas no such peak is apparent for the nonlinear procedure. (The calibration pulses were inserted periodically during data-taking runs. The peak was displaced by two seconds by the insertion of a time delay, so that the calibration pulses would not mask any possible real signal, which was expected at zero time delay).

Nevertheless, Weber preferred the nonlinear algorithm. His reason for this was that this procedure gave a more significant signal than did the linear one. This is illustrated in Figure 13, in which the upper panel shows the data analyzed with the non-linear algorithm and the lower panel shows the data analyzed with the linear procedure. Weber was, in fact, using the positive result to decide which was the better analysis procedure. If anyone was "regressing," it was Weber.

Weber's failure to calibrate his apparatus was criticized by others. "Finally, Weber has not published any results in calibrating his system by the impulsive introduction of known amounts of mechanical energy into the bar, followed by the observation of the results either on the single detectors or in coincidence"(Levine and Garwin 1973, p. 177).

His critics, however, analyzed their own data by using both algorithms. If it was the case that, unlike the calibration pulses where the linear algorithm was superior, using the linear algorithm either masked or failed to detect a real signal, then using the nonlinear algorithm on their own data should produce a clear signal. None appeared. Typical results are shown in Figures 12 and 14. Figure 12, which shows Tyson's data analyzed with the nonlinear algorithm, not only shows no calibration peak, but it does not show a signal peak at zero time delay. It is quite similar to the data analyzed with the linear algorithm (Figure 14).

Weber had an answer. He admitted that the linear algorithm was better for detecting calibration pulses, which were short. He claimed, however, that the real signal for

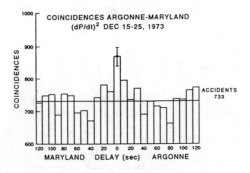
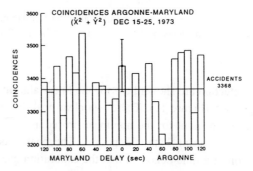

Figure 13. Weber's time-delay data for the Maryland-Argonne collaboration for the period 15-25 December, 1973. The top graph uses the nonlinear algorithm, whereas the bottom uses the linear algorithm. The zero-delay peak is seen only with the nonlinear algorithm. From Shaviv and Rosen (1975).

gravitational waves was a longer pulse than most investigators believed. He argued that the nonlinear algorithm was better for detecting these long pulses. If the gravity wave signal was longer than expected, then one would have expected it to show up when the critics' data was processed with the nonlinear algorithm. It did not. (See Figure 12).[21] Weber's experiment had failed the calibration test.

The various experimental groups cooperated, exchanging both data and analysis programs. This led to the first of several questions concerning possible serious errors in Weber's analysis of his data. Douglass pointed out that Weber's analysis program contained an error. It generated coincidences between detectors even when none were present. Douglass also pointed out that this error accounted for all of the coincidences observed in the tape of Weber's data that he had examined. Weber admitted the error, but did not agree with the claim that it accounted for all of his observed coincidences. At the very least, this error raised legitimate doubts about Weber's results.

Another question was raised concerning Weber's analysis of his data. This was the question of selectivity and bias. The problem was with setting the threshold for a gravity-wave pulse. Weber's critics adopted a single threshold and used it consistently in their analysis procedure. The critics claimed that Weber varied his threshold for

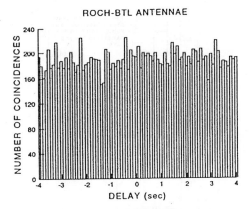

Figure 14. A time-delay plot for the Rochester-Bell Laboratory collaboration, using the linear algorithm. No sign of a zero-delay peak is seen. From Shaviv and Rosen (1975).

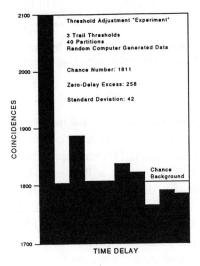

Figure 15. The result of selecting thresholds that maximized the zero-delay signal, for Levine's computer simulation. Such selectivity can produce a spurious signal at zero time delay. From Garwin (1974).

different segments of his data so as to produce a maximum signal. Garwin also presented a computer simulation that showed that such selectivity could produce a positive result (Figure 15).[22] Weber denied the charges. He admitted that his data were analyzed with 31 different thresholds, but claimed that he exercised no selectivity and that his results were robust against changes in the threshold. He did not, however, specify how his data were selected. In particular he did not state that all of the results presented in his histograms were obtained with the same threshold.

There was also a rather odd result reported by Weber.

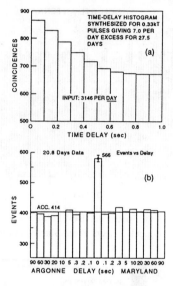

Figure 16. (a) Computer simulation result obtained by Levine for signals passing through Weber's electronics. (b) Weber's reported result. The difference is clear. From Levine and Garwin (1974).

First, Weber has revealed at international meetings (Warsaw, 1973. etc.) that he had detected a 2.6-standard deviation excess in coincidence rate between a Maryland antenna [Weber's apparatus] and the antenna of David Douglass at the University of Rochester. Coincidence excess was located not at zero time delay but at "1.2 seconds," corresponding to a 1-sec intentional offset in the Rochester clock and a 150-millisecond clock error. At CCR-5, Douglass revealed, and Weber agreed, that the Maryland Group had mistakenly assumed that the two antennas used the same time reference, whereas one was on Eastern Daylight Time and the other on Greenwich Mean Time. Therefore, the "significant" 2.6 standard deviation excess referred to gravity waves that took four hours, zero minutes and 1.2 seconds to travel between Maryland and Rochester (Garwin 1974, p. 9).

Weber answered that he had never claimed that the 2.6 standard-deviation effect he had reported was a positive result. Nevertheless, by producing a positive result where none was expected--or even possible--Weber had cast doubt on his own analysis procedures.

Garwin (1974) and Levine and Garwin (1974) raised yet another question about Weber's results. They used a computer simulation to show that, if Weber's apparatus was as he described it, then it could not have produced the result he claimed. In particular, they argued that the narrow signal seen by Weber should have been broader (Figure 16).

Let us summarize the evidential situation, concerning gravity waves, that obtained at the beginning of 1975. There were discordant results. Weber had reported positive results on gravitational radiation, whereas six other groups had reported no evidence for such radiation. The critics' results were not only more numerous, but had also been carefully cross-checked. The groups had exchanged both data and analysis programs and had confirmed the results. The critics had also investigated whether their analysis procedure, the use of a linear algorithm, could account for their failure to observe Weber's reported results. They had used Weber's preferred procedure, a nonlinear algorithm, to analyze their data, and still found no sign of an effect. They had also calibrated their experimental apparatuses by inserting electrostatic pulses of known energy and finding that they could detect a signal. Weber, on the other hand, as well as his critics using his analysis procedure, could not detect such calibration pulses. Under ordinary circumstances Weber's calibration failure would have been decisive; it was because this episode is atypical, one in which a new type of apparatus was used to search for a hitherto unobserved phenomenon, that it was not. Other arguments were both needed and provided.

There were, in addition, several other serious questions raised about Weber's analysis procedures. These included an admitted programming error that generated spurious coincidences between Weber's two detectors; Weber's report of coincidences between two detectors, when the data had been taken four hours apart, and thus could not have produced real coincidences; the question of selectivity in setting signal thresholds; and whether or not Weber's experimental apparatus could produce the narrow coincidences claimed.

It seems clear that, according to the epistemological criteria discussed above, the critics' results were far more credible than Weber's. They had checked their results by independent confirmation, which included the sharing of data and analysis programs. They had also eliminated a plausible source of error, that of the pulses being longer than expected, by analyzing their results using the nonlinear algorithm and by looking for such long pulses. They had also calibrated their apparatuses by injecting known pulses of energy and observing the output.

In addition, Weber's reported result failed several tests suggested by these criteria. Weber had not eliminated the plausible error of a mistake in his computer program. It was, in fact, shown that this error could account for his result. It was also argued that Weber's analysis procedure, which varied the threshold accepted, could also have produced his result. Having increased the credibility of his result when he showed that it disappeared when the signal from one of the two detectors was delayed, he then undermined his result by obtaining a positive result when he thought two detectors were simultaneous, when, in fact, one of them had been delayed by 4 hours. As Garwin also argued, Weber's result itself also argued against its credibility. The coincidence in the time delay graph was too narrow to have been produced by Weber's apparatus. Weber's analysis procedure also failed to detect calibration pulses.

I believe that the evidence against Weber's result was overwhelming, and did use epistemological criteria. Although no formal rules were applied, the procedure was reasonable.

3. THE 17-KEV NEUTRINO[23]

Another interesting case of discordant results is the recent history of experiments concerning the existence of a heavy, 17-keV neutrino.[24] What makes this episode so intriguing is that both the original positive claim, and all subsequent positive claims were obtained in experiments using one type of apparatus, namely those incorporating a solid-state detector, whereas the initial negative evidence resulted from experiments using another type of detector, a magnetic spectrometer.[25] These were both seemingly reliable types of experimental apparatus. Solid state detectors had been in wide use since the early 1960s, and their use was well understood. Magnetic spectrometers had been used in nuclear β-decay experiments since the 1930s, and both the problems and advantages of using this technique had been well studied.[26] This is an illustration of discordant results obtained using different types of apparatus. One might worry that the discordant results were due to some crucial difference between the types of apparatus or to different sources of background that might mimic or mask the signal.

Simpson first reported evidence for the 17-keV neutrino April, 1985 (Simpson 1985a).[27] He had searched for a heavy neutrino by looking for a kink in the energy spectrum of tritium, or in the Kurie plot,[28] at an energy equal to the maximum allowed decay energy minus the mass of the heavy neutrino in energy units. Simpson's result is shown in Figure 17. A kink is clearly visible at an energy of 1.5 keV, corresponding to a 17 keV neutrino.[29] "In summary, the β spectrum of tritium recorded in the present experiment is consistent with the emission of a heavy neutrino of mass about 17.1 keV and a mixing probability of about 3%" (Simpson 1985a, p. 1893).

Simpson had been using the apparatus for some time.[30] In 1981 he had attempted to measure, or to set an upper limit on, the mass of the neutrino by a precise measurement of the endpoint energy of the β-decay spectrum of tritium (Simpson 1981a).[31] If the neutrino had mass then the measured endpoint energy would be lower than that predicted, by an amount equal to the mass of the neutrino. In addition, the shape of the energy spectrum near the endpoint was sensitive to the mass of the neutrino. Earlier measurements on tritium had been made with magnetic spectrometers, whereas Simpson used a different type of experimental apparatus, in which the tritium was implanted in a Si(Li) x-ray detector, a solid-state device. Simpson provided additional confirmation for those negative searches by using a different type of experimental apparatus.[32] He continued his work with an unsuccessful search for a heavy neutrino with mass less than 10 keV (Simpson 1981b).[33] These previous studies had provided Simpson with expertise in the use of the apparatus, particularly in the energy calibration of the detector, an important factor in performing such experiments.

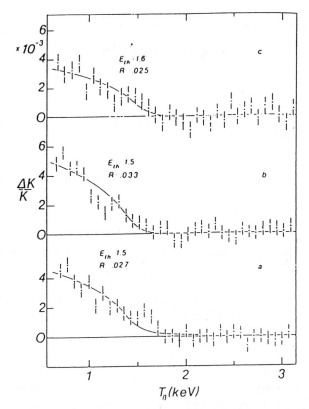

Figure 17. The data of three runs presented as ΔK/K (the fractional change in the Kurie plot) as a function of the kinetic energy of the β particles. E_{th} is the threshold energy, the difference between the endpoint energy and the mass of the heavy neutrino. A kink is clearly seen at E_{th} = 1.5 keV, or at a mass of 17.1 keV. Run *a* included active pileup rejection, whereas runs *b* and *c* did not. No difference is apparent. From Simpson (1985).

By the end of 1985 the results of five other experimental searches for the particle had appeared in the published literature (Altzitzoglou *et al.* 1985; Apalikov *et al.* 1985; Datar *et al.* 1985; Markey and Boehm 1985; Ohi *et al.* 1985). All of these results were negative. The experiments set limits of less than one percent for a 17-keV branch of the decay, in contrast to Simpson's value of three percent. A typical result is shown in Figure 18 and should be compared to Simpson's result shown in Figure 17. No kink of any kind is apparent.

Each of the experiments had examined the β-decay spectrum of ^{35}S, and searched for a kink at an energy of 150 keV, 17 keV below the endpoint energy of 167 keV. Three of the experiments, those of Altzitzoglou *et al.*, of Apalikov *et al.*, and of Markey and Boehm, used magnetic spectrometers. Those of Datar *et al.* and of Ohi *et al.* used Si(Li) detectors, the same type used by Simpson. In the latter two cases, however, the source was not implanted in the detector, as Simpson had done, but was separated from it.

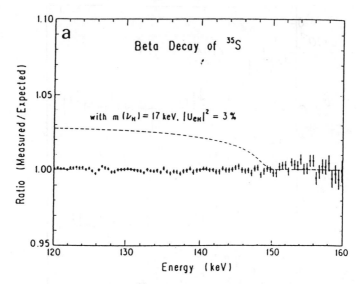

Figure 18. The ratio of the measured ^{35}S beta-ray spectrum to the theoretical spectrum. A three percent mixing of a 17-keV neutrino should distort the spectrum as indicated by the dashed curve. From Ohi et al. (1985).

Such an arrangement would change the atomic physics corrections to the spectrum.[34] In addition, the ^{35}S β-decay sources used in the experiments had a higher endpoint energy than did the tritium used by Simpson (167 keV, in contrast to 18.6 keV). This higher endpoint energy made atomic physics corrections to the beta-decay spectrum less important.

These experiments were an attempt to provide independent confirmation of Simpson's result using different experiments. By using a ^{35}S source, as opposed to tritium, one could check on whether Simpson's observed effect might be due to some atomic physics phenomena peculiar to his choice of decay source and detector. Had positive results been found, then one would have concluded that such effects were negligible and the experiments would have provided more support for Simpson's original result than would have been the case if the new experiments had used the same source and detector. The difficulty is that, although different experiments which agree do provide greater support for a result,[35] when different experiments disagree we don't know which result is correct and will suspect that the different results are caused by some difference either in the experimental apparatus or in the data analysis.

Simpson's first report of the 17-keV neutrino was unexpected. It was not predicted-- or even suggested--by any existing theory. Faced with such an unexpected result, the physics community took a reasonable approach. Some scientists tried to explain the result within the context of accepted theory. They argued that a plausible alternative explanation of the result had not been considered. This involved the question of

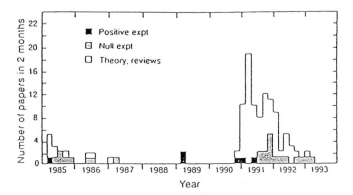

Figure 19. Preprints on the 17-keV neutrino received at CERN as a function of time. From Morrison (1993).

whether the theory used in the analysis of the data--and to compare the experimental result with the theory of the phenomenon--was correct. This is an important point. An experimental result is not immediately given by an examination of the raw data, but requires considerable analysis. In this case the analysis included atomic-physics corrections, needed for the comparison of the theoretical spectrum and the experimental data. Everyone involved agreed that such corrections had to be made. The question was what were the proper corrections. The atomic physics corrections used by Simpson in his analysis, particularly the screening potential, were questioned by other scientists (Haxton 1985; Kalbfleisch and Milton 1985; Drukarev and Strikman 1986; Eman and Tadic 1986; Lindhard and Hansen 1986).[36] These suggestions were aimed at accommodating the unexpected result. Several calculations indicated, at least qualitatively, that Simpson's result could be accommodated within accepted theory, and that there was no need for the suggestion of a new particle. "A detailed account of the decay energy and Coulomb-screening effects raises the theoretical curve in precisely this energy range so that little, if any, of the excess remains" (Lindhard and Hansen 1986, p. 965).

The combination of negative experimental searches combined with plausible theoretical explanations of Simpson's result had a chilling effect on the field (see Figure 19).[37] Simpson, however, continued his work. He presented further evidence in support of the 17-keV neutrino using a somewhat modified experimental apparatus (Simpson 1986b). He also took the criticism of his work seriously and for these new data presented an analysis that incorporated the screening potential suggested by his critics. Although this reduced the size of his effect by approximately 20 percent, the effect was still clearly present. He also questioned whether the analysis procedures used in the five negative searches were adequate to set the upper limits they had reported. He argued that the wide energy range used to fit the β-decay spectrum tended to minimize any

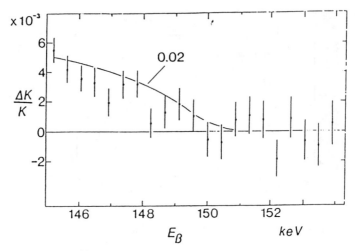

Figure 20. ΔK/K for the ^{35}S spectra of (Ohi *et al.* 1985) as recalculated by Simpson. From Simpson (1986a).

possible effect of a heavy neutrino, which would appear primarily in a narrow energy band near threshold. He also questioned the procedure of merely adding the contribution of a 17-keV neutrino to the already fitted spectrum, a point with which others agreed (Borge *et al.* 1986). Simpson further questioned whether or not the "shape-correction" factor needed to fit the spectra in magnetic spectrometer experiments could mask a kink due the presence of a heavy neutrino. He also presented a reanalysis of Ohi's data using his own preferred analysis procedure. He found a positive effect (Figure 20; compare this with Ohi's own reported result shown in Figure 18).

Further negative evidence was provided by (Borge *et al.* 1986; Hetherington *et al.* 1986; Hetherington *et al.* 1987; Zlimen, Kaucic and Ljubicic 1988). Hetherington *et al.* urged caution concerning Simpson's method of data analysis. They pointed out that "concentrating on too narrow a region can lead to misinterpretation of a local statistical anomaly as a more general trend..." (1987, p. 1512). At the end of 1988 the situation was much as it was at the end of 1985. Simpson had presented positive results on a 17-keV neutrino. There were nine negative experimental reports, as well as plausible theoretical explanations of his result.

In 1989 Simpson and Hime presented two additional positive results, using both tritium and ^{35}S, the spectrum used in the original negative searches. In these reports the value of the mixing fraction of 17-keV neutrinos had been reduced to approximately 1percent. The new atomic physics corrections had reduced the originally reported effect from 3 percent to 1.6 percent (Hime and Simpson 1989; Simpson and Hime 1989).

The situation changed dramatically in 1991. New positive results were reported at both conferences and in the published literature by groups at Oxford and at (Hime and

Jelley 1991; Sur et al. 1991).[38] These results were quite persuasive. Hime and Jelley had incorporated anti-scatter baffles into their apparatus, to guard against a distortion of the spectrum caused by scattering of the decay electrons, a possible problem in the earlier experiments. The Berkeley group had embedded their ^{14}C source in a solid-state detector and included a guard ring veto to reject decays occurring near the boundary, which might not deposit their full energy and thus distort the spectrum. They claimed that their result "supports the claim by Simpson that there is a 17-keV neutrino emitted with ~1% probability in β decay" (Sur et al., p. 2447). They also claimed to rule out the null hypothesis (no heavy neutrino) at the 99-percent confidence level. A further positive result was reported by Zlimen et al. (1991). The revitalizing effect of these new results on work done on the 17-keV neutrino is clearly seen in Figure 19. It generated considerable new experimental and theoretical work. Sheldon Glashow, a Nobel-Prize winning theorist, remarked, "Simpson's extraordinary finding proves that Nature's bag of tricks is not empty, and demonstrates the virtue of consulting her, not her prophets" (Glashow 1991, p. 257).

This marked the high point in the life of the 17-keV neutrino. From this time forward only negative results would be reported, and errors would be found in the most persuasive positive results. In 1992 Piilonen and Abashian suggested that Hime and Jelley had overlooked a background effect that might have simulated the effect of a 17-keV neutrino in their experiment (Piilonen and Abashian 1992). The appearance of several negative results (discussed below) encouraged Hime to consider the Piilonen-Abashian suggestion seriously and to reanalyze his own result (Hime 1993). He found, using an experimentally checked Monte Carlo calculation, that the scattering of the decay electrons in the experimental apparatus could explain the result without the need for a 17-keV neutrino. "It will be shown that scattering effects are sufficient to describe the Oxford β-decay measurements and that the model can be verified using existing calibration data. Surprisingly, the β spectra are very sensitive to the small corrections considered" (p. 166). He also suggested that similar effects might explain his earlier positive results obtained in collaboration with Simpson.

Hime briefly reviewed the evidential situation, noting that the major evidence against the existence of the 17-keV neutrino came from magnetic-spectrometer experiments in which questions had been raised concerning the shape corrections. He commented that Bonvicini (in a CERN report, CERN-PPE/92-54) had shown that nonlinear distortions could mask the presence of a heavy neutrino signature and still be described by a smooth shape correction.[39] He remarked, however, that "A measurement of the ^{63}Ni spectrum (Kawakami et al. 1992) has circumvented this difficulty. The sufficiently narrow energy interval studied, and the very high statistics accumulated in the region of interest, makes it very unlikely that a 17-keV threshold has been missed in this experiment" (Hime 1993, p. 165). He also cited a new result from a group at Argonne National Laboratory (Mortara et al. 1993, discussed in detail below), that provided "convincing evidence against a 17-keV neutrino" (Hime 1993, p. 165). In particular, the Argonne group had demonstrated the sensitivity of their magnetic spectrometer experiment to a possible 17-keV neutrino by admixing a small component of ^{14}C in

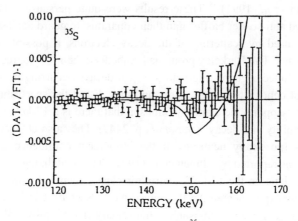

Figure 21. Residuals from a fit to the pile-up corrected ^{35}S data assuming no massive neutrino; the reduced χ^2 for the fit is 0.88. The solid curve represents the residuals expected for decay with a 17-keV neutrino and $\sin^2\theta = 0.85$ percent; the reduced χ^2 of the data is 2.82. From Mortara et al. (1993).

their ^{35}S source and detecting the resulting kink in their composite spectrum (see Figure 22 below). These negative results provided the impetus for Hime's reexamination of his result.

Further evidence against the 17-keV neutrino was provided by the Argonne group (Mortara et al. 1993). This experiment used a solid-state, Si(Li) detector (the same type used originally by Simpson), an external ^{35}S source, and a solenoidal magnetic field to focus the decay electrons. The field also had the effect of reducing the backscattering of the decay electrons, a possible problem. Their final result, shown in Figure 21, was $\sin^2\theta = -0.0004 \pm 0.0008$ (statistical) ± 0.0008 (systematic), for the mixing probability of the 17-keV neutrino. They had found no evidence for a 17-keV neutrino.

The experimenters demonstrated the sensitivity of their apparatus to a possible 17-keV neutrino.

> To assess the reliability of our procedure, we introduced a known distortion into the ^{35}S beta spectrum and attempted to detect it. A drop of ^{14}C-doped valine ($E_o - m_e \sim 156$ keV) was deposited on a carbon foil and a much stronger ^{35}S source was deposited over it. The data from the composite source were fitted using the ^{35}S theory, ignoring the ^{14}C contaminant. The residuals are shown in Figure [22]. The distribution is not flat; the solid curve shows the expected deviations from the single component spectrum with the measured amount of ^{14}C. The fraction of decays from ^{14}C determined from the fit to the beta spectrum is $(1.4 \pm 0.1)\%$. This agrees with the value of 1.34% inferred from measuring the total decay rate of the ^{14}C alone while the source was being prepared. This exercise demonstrates that our method is sensitive to a distortion at

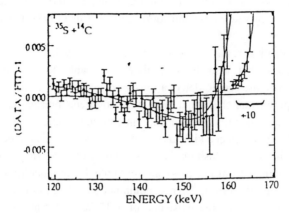

Figure 22. Residuals from fitting the beta spectrum of a mixed source of ^{14}C and ^{35}S with a pure ^{35}S shape; the reduced χ^2 of the data is 3.59. The solid curve indicates residuals expected from the known ^{14}C contamination. The best fit yields a mixing of (1.4 ± 0.1) percent and reduced χ^2 of 1.06. From Mortara et al. (1993).

the level of the positive experiments. Indeed, the smoother distortion with the composite source is more difficult to detect than the discontinuity expected from the massive neutrino.

In conclusion, we have performed a solid-state counter search for a 17 keV neutrino with an apparatus with demonstrated sensitivity. We find no evidence for a heavy neutrino, in serious conflict with some previous experiments (Mortara et al. 1993, p. 396).

The Berkeley group continued to work on trying to find the reason for the artifact in their ^{14}C data. The cause, found in 1993, was quite subtle. The way in which the center detector was separated from the guard ring was by cutting a groove in the detector. "The n$^+$ is divided by a 1-mm-wide circular groove into a 'center region' 3.2 cm in diameter, and an outer 'guard ring.' By operating the guard ring in anticoincidence mode, one can reject events occurring near the boundary which are not fully contained within the center region" (Sur et al. 1991, p. 2444). Such events would not give a full energy signal and would thus distort the observed spectrum.

What the Berkeley group found was that ^{14}C decays occurring under the groove shared the energy between both regions without necessarily giving a veto signal and thus gave an incorrect event energy, distorting the spectrum. They also found that, although their earlier tests had indicated that the ^{14}C was uniformly distributed in the detector, their new tests showed that between one third and one half of the ^{14}C was localized in grains. They also found that approximately 1 percent of the grains were located under the groove. Thus, the localization of the ^{14}C, combined with the energy sharing, gave rise to a spectrum distortion that simulated that expected from a 17-keV neutrino (Norman, private communication, LBL-36136 (1994) and (Wietfeldt *et al.* 1993b)).

The newer negative results were persuasive not only because of their improved statistical accuracy, but also because they were able to demonstrate that their experimental apparatuses could detect a kink in the spectrum if one were present. This was a direct experimental check that there were no effects present that would mask the presence of a heavy neutrino. These experiments met Hime's suggested criteria--(a) demonstrated ability to detect a kink and (b) high statistics so that a local analysis of the spectrum could be done.[40] Ohshima had also shown that the shape correction factors used in their experiment did not mask any possible 17-keV neutrino effect (Ohshima 1993; Ohshima *et al.* 1993).[41] This combination of almost overwhelming and persuasive evidence against the existence of a 17-keV neutrino, combined with the demonstrated and admitted problems with the positive results, decided the issue. There was no 17-keV neutrino.

It seems clear that this decision was based on experimental evidence, discussion, and criticism or, in other words, epistemological criteria. It had been shown that the two most persuasive positive results had overlooked effects that mimicked the presence of a 17-keV neutrino. The Sherlock Holmes strategy had been incorrectly applied. In addition, the new negative results had answered the criticisms made previously concerning the "shape-correction" factor and had demonstrated that they could detect a kink in the spectrum if one were present.

The process of designing a good "17-keV neutrino" detector was not simply a matter of deciding whether the particle existed, and then asserting that a good detector was one that gave the correct answer. The community decided which were the good detectors, based on epistemological criteria, and then decided that the particle did not exist.[42]

4. ATOMIC PARITY VIOLATION, SLAC E122, AND THE WEINBERG-SALAM THEORY

The final episode I will consider involves discordant results from experiments that tested the Weinberg-Salam (W-S) unified theory of electroweak interactions. This is the most complex of the episodes considered because it involved two different, and intertwined, instances of discordant experimental results. First, there was a disagreement between different experiments that measured atomic parity violation. Some of these experiments seemed to refute the W-S theory, whereas others supported it. In addition, the question of whether or not these experimental results confirmed the W-S theory was dependent on atomic physics calculations, which were themselves uncertain. The second discordance was between the early negative results from atomic-parity violation experiments and the supportive result produced by the SLAC (Stanford Linear Accelerator Center) E122 experiment, for the W-S theory, which measured an asymmetry in the scattering of polarized electrons from deuterium. Thus we have discordant results produced in experiments which measured the same. or similar, quantities, by similar techniques. We also have discordant results between experiments that measured very different quantities by very different experimental methods, in which the discord was whether or not the experiments supported or refuted the same theory.

The early history of the episode may be summarized as follows. In 1957, it had been experimentally demonstrated that parity, or left-right symmetry, was violated in the weak interactions. This feature of the weak interactions had been incorporated into the W-S unified theory of electroweak interactions. The theory predicted that one would see weak neutral-current effects in the interactions of electrons with hadrons, the strongly interacting particles. The effect would be quite small when compared to the dominant electromagnetic interaction, but could be distinguished from it by the fact that it violated parity conservation. A demonstration of such a parity-violating effect and a measurement of its magnitude would test the W-S theory. One such predicted effect was the rotation of the plane of polarization of polarized light when it passed through bismuth vapor. Such a rotation is possible only if parity is violated.

In 1976 and 1977, experimental groups at Oxford University and at the University of Washington reported results from atomic parity-violation experiments that disagreed with the predictions of the Weinberg-Salam theory. At the time, the theory had other experimental support, but was not universally accepted. In 1978 and 1979 a group at the Stanford Linear Accelerator (the SLAC E122 experiment) reported results on the scattering of polarized electrons from deuterium, which confirmed the W-S theory. By 1979 the Weinberg-Salam theory was regarded by the high-energy physics community as established, despite the fact that as Pickering states, "there had been no *intrinsic* change in the status of the Washington-Oxford experiments" (Pickering 1984a, p. 301).[43]

In Pickering's view "particle physicists *chose* to accept the results of the SLAC experiment, *chose* to interpret them in terms of the standard model (rather than some alternative which might reconcile them with the atomic physics results) and therefore *chose* to regard the Washington-Oxford experiments as somehow defective in performance or interpretation" (Pickering, 1984a, p. 301). The implication seems to be that these choices were made so that the experimental evidence would be consistent with the standard model--and that there were not good, independent reasons for the decision. In other words, the disagreement was not resolved on epistemological or methodological grounds, but rather by loyalty to existing community commitments.

My view is quite different. I regard the two experimental results as having quite different evidential weights. The initial Washington-Oxford results (there were later ones) used new and untested experimental apparatus and had large systematic uncertainties (as large as the predicted effects). In addition, their initial results, reported in 1976 and 1977, were internally inconsistent, and by 1979 there were other atomic parity-violation results, discussed below, that confirmed the W-S theory. Thus, by the end of 1979, the overall situation with respect to the atomic parity results was quite uncertain. The SLAC experiment, on the other hand, although also using new techniques, had been very carefully checked, and had far more evidential weight.

Faced with this situation the physics community chose to accept the SLAC results, which supported the Weinberg-Salam theory, and to await further developments on the uncertain atomic-parity violation results. The experiment-theory disagreement and the discord between the two sets of experimental results on atomic parity violation was later resolved, as was the disagreement between the early atomic parity violation results and that of SLAC E122. Both disagreements were resolved by reasoned argument, based on experimental evidence and on epistemological and methodological criteria. I will argue for this by a detailed examination of the history of this episode.

a) Early Atomic Parity Violation Experiments

The first experimental tests of the W-S theory were performed by groups at Oxford and Washington. They looked for a parity-violating rotation of the plane of polarization of light when it passed through bismuth vapor. They both used bismuth vapor, but they used light corresponding to different transitions in bismuth: λ = 648 nm (Oxford) and λ = 876 nm (Washington). They published a joint preliminary report noting that, "we feel that there is sufficient interest to justify an interim report." (Baird *et al.* 1976), p. 528). They reported values for R, the parity violating parameter, of R = $(-8 \pm 3) \times 10^{-8}$ (Washington) and R = $(+10 \pm 8) \times 10^{-8}$ (Oxford). "We conclude from the two experiments that the optical rotation, if it exists, is smaller than the values -3×10^{-7} and -4×10^{-7} predicted by the Weinberg-Salam model plus the atomic central field approximation" (Baird *et al.,* 1976, p. 529).[44]

The experimental results were quite uncertain, and included systematic uncertainties, which were not fully understood, of the order of $\pm 10 \times 10^{-8}$. These systematic experimental uncertainties were of the same order of magnitude as the expected effect. These were also novel experiments, using new and previously untried techniques, which also tended to make the experimental results uncertain.

In September, 1977, both the Washington and Oxford groups published more detailed accounts of their experiments with somewhat revised results (Baird *et al.* 1977; Lewis *et al.* 1977). Both groups again reported results in substantial disagreement with the predictions of the Weinberg-Salam theory, although the Washington group stated that, "more complete calculations that include many-particle effects are clearly desirable" (Lewis *et al.*, 1977, p. 795). The Washington group reported a value of R = $(-0.7 \pm 3.2) \times 10^{-8}$, which was in disagreement with the prediction of approximately -2.5×10^{-7} (see Table 1). This value was also inconsistent with their earlier result of $(-8 \pm 3) \times 10^{-8}$.[45] The difference between the two values is $(7.3 \pm 2.5) \times 10^{-8}$, a 2.9 standard-deviation effect, which has a 0.37-percent probability of being equal to 0, an unlikely occurrence. This inconsistency was not discussed by the experimenters in the published paper, but it was discussed by others within the atomic physics community and lessened the credibility of the result.[46] The Oxford result was R = $(+2.7 \pm 4.7) \times 10^{-8}$, again in disagreement with the Weinberg-Salam prediction of approximately -2.5×10^{-7}. They noted, however, that there was a systematic effect in their apparatus. They found a

Table 1. Calculated Parity-Violation Effect in Bismuth

Method	R(10^{-7})	Reference
Hartree-Fock	-2.3	Brimicombe and others (1976)
Hartree-Fock	-3.5	Henley and Wilets (1976)
Semiempirical	-1.7	Novikov and others (1976)
Multiconfiguration	-2.4	Grant (private communication)

change in ϕ, the rotation angle, due to slight misalignment of the polarizers, optical rotation in the windows, etc., of order 2×10^{-7} radians. "Unfortunately, it varies with time over a period of minutes, and depends sensitively on the setting of the laser and the optical path through the polarizer. While we believe we understand this effect in terms of imperfections in the polarizers combined with changes in laser beam intensity distribution, we have been unable to reduce it significantly"(Baird et al. 1977, p. 800). A systematic effect of this size, the same as that of the theoretically predicted effect, cast doubt on the result, and on the comparison between experiment and theory.

The theoretical calculations of the expected effect were also uncertain. The problem was that, for an atom with few electrons, where the electron wavefunctions could be calculated quite reliably, the predicted effect was small. For a multi-electron atom such as bismuth, in which the predicted effect was much larger, the wavefunctions could be calculated only approximately and with a fair amount of uncertainty. There were, at the time, four different calculations of the expected effect, which agreed with one another to within ± 25 percent. This made the largest and smallest calculated values of R differ by almost a factor of 2. The experimenters thought that this rough agreement was encouraging, although they could not say that the many-body effects which had been neglected in the calculation could resolve the discrepancy between theory and experiment.

How were these results viewed at the time by the physics community? In the same issue of *Nature* in which the original joint Oxford-Washington paper was published, Frank Close, a particle theorist, summarized the situation. "Is parity violated in atomic physics? According to experiments being performed independently at Oxford and the University of Washington the answer may well be no.... This is a very interesting result in light of last month's report... claiming that parity is violated in high energy 'neutral current' interactions between neutrinos and matter" (Close 1976, p. 505). The experiment that Close referred to had concluded, "Measurements of R^υ and $R^{\bar{\upsilon}}$, the ratios of neutral current to charged current rates for υ and $\bar{\upsilon}$ [neutrino and antineutrino] cross sections, yield neutral current rates for υ and $\bar{\upsilon}$ that are consistent with a pure V-A interaction but 3 standard deviations from pure V or pure A, indicating the presence of parity nonconservation in the weak neutral current"(Benvenuti et al. 1976, p. 1039).

Close noted that, as the atomic physics results stood, they appeared to be inconsistent with the predictions of the Weinberg-Salam model supplemented by atomic-physics calculations. He also remarked that, "At present the discrepancy can conceivably be the combined effect of systematic effects in atomic physics calculations and systematic uncertainties in the experiments" (Close 1976, pp. 505-6). Close discussed the possibility that neutral current effects might violate parity in neutrino interactions and conserve parity in electron interactions. He also discussed an alternative that had an unexpected (on the basis of accepted theory) energy dependence, so that the high energy experiments (the neutrino interactions) showed parity nonconservation whereas the low energy atomic physics experiments would not. "Whether such a possibility could be incorporated into the unification ideas is not clear. It also isn't clear, yet, if we have to worry. However, the clear blue sky of summer now has a cloud in it. We wait to see if it heralds a storm" (Close 1976, p. 506).

The uncertainty caused by these atomic parity violation results is shown in a summary of the Symposium on Lepton and Photon Interactions at High Energies, held in Hamburg August 25-31, 1977, given by David Miller. Miller noted that Sandars had reported that neither his group at Oxford nor the Washington group had seen any parity-violating effects and that "they have spent a great deal of time checking both their experimental sensitivity and the theory in order to be sure" (Miller 1977, p. 288). Miller went on to state that "S. Weinberg and others discussed the meaning of these results. It seems that the $SU(2) \times U(1)$ is to the weak interaction what the naive quark-parton model has been to QCD, a first approximation which has fitted a surprisingly large amount of data. Now it will be necessary to enlarge the model to accommodate the new quarks and leptons, the absence of atomic neutral currents, and perhaps also whatever it is that is causing trimuon events" (Miller 1977, p. 288). I believe, however, that the uncertainty in these experimental results made the disagreement with the W-S theory only a worrisome situation and not a crisis. In any event, the monopoly of Washington and Oxford was soon broken.

The evidential situation changed in 1978 when Barkov and Zolotorev (1978a, 1978b, 1979a), two Soviet scientists from Novosibirsk, reported measurements on the same transition in bismuth as did the Oxford group. Their results agreed with the predictions of the W-S model. They gave a value for $\psi_{exp}/\psi_{W-S} = (+1.4 \pm 0.3)$ k, where ψ was the angle of rotation of the plane of polarization by the bismuth vapor. "The factor k was introduced because of inexact knowledge of the bismuth vapor, and also because of some uncertainty in the estimate, the factor lies in the interval from 0.5 to 1.5" (Barkov and Zolotorev, 1978a, p.360). They concluded that their result "does not contradict the predictions of the Weinberg-Salam model." A point to be emphasized here is that agreement with theoretical prediction depended (and still does depend) on which method of calculation one chose, as discussed earlier. A somewhat later paper changed the result to $\psi_{exp}/\psi_{W-S} = 1.1 \pm 0.3$ (Barkov and Zolotorev 1978b).

Subsequent papers, in 1979 and 1980 (Barkov and Zolotorev 1979a, 1980a, 1980b) reported more extensive data and found a value for $R_{exp}/R_{theor} = 1.07 \pm 0.14$. They also reported that the latest unpublished results from the Washington and Oxford groups, which had been communicated to them privately, showed parity violation, although "the results of their new experiments have not reached good reproducibility" (Barkov and Zolotorev 1979a, p. 312). These later results were also presented at the 1979 conference, discussed below, at which Dydak reviewed the situation.

During September, 1979 an international workshop devoted to neutral-current interactions in atoms was held in Cargese. This workshop was attended by representatives of virtually all of the groups actively working in the field, including Oxford, Washington, and Novosibirsk. At that workshop the Novosibirsk group presented a very detailed account of their experiment (Barkov and Zolotorev 1979b). C. Bouchiat remarked in his workshop summary paper, "Professor Barkov, in his talk, gave a very detailed account of the Novosibirsk experiment and answered many questions concerning possible systematic errors" (Bouchiat 1980, p.364). There was also communication between the Soviet and Oxford groups. The Soviets reported that they had been able to uniquely identify the hyperfine structure of the 6477 A° [648 nm] line of atomic bismuth and that "the results of these measurements agree also with the results in Oxford (P. Sandars, private communication)" (Barkov and Zolotorev 1978a, p. 359).

In early 1979, a Berkeley group reported an atomic-physics result for thallium that agreed with the predictions of the W-S model (Conti *et al.* 1979). They investigated the polarization of light passing through thallium vapor and found a circular dichroism $\delta = (+5.2 \pm 2.4) \times 10^{-3}$ in comparison with the theoretical prediction of $(+2.3 \pm 0.9) \times 10^{-3}$. Although these were not definitive results--they were only two SD from zero--they did agree with the model, in both sign and magnitude.

It seems fair to say that in mid-1979 the atomic physics results concerning the W-S theory were inconclusive. The Oxford group and the Washington group had originally reported a discrepancy between their experimental results and the theory, but their more recent results, although preliminary, showed the presence of the predicted parity-nonconserving effects. In addition, the Soviet and Berkeley results agreed with the model. Dydak summarized the situation in a talk at a 1979 conference. "It is difficult to choose between the conflicting results in order to determine the *eq* [electron-quark] coupling constants. Tentatively, we go along with the positive results from Novosibirsk and Berkeley groups and hope that future development will justify this step (it cannot be justified at present, on clear-cut experimental grounds)" (Dydak 1979), p. 35)."[47]

Bouchiat's summary paper at the Cargese Workshop was more positive. After reviewing the Novosibirsk experiment as well as the conflict between the earlier and later Washington and Oxford results he remarked that, "*As a conclusion on this Bismuth session, one can say that parity violation has been observed roughly with the magnitude predicted by the Weinberg-Salam theory* (emphasis in original)" (Bouchiat 1980, p. 365). Even this more positive statement does not conclude that the results agree with the predictions of the theory; it states only that the experimental results were of the correct order of magnitude.[48]

b) The SLAC E122 Experiment

The evidential situation was made even more complex when a SLAC group reported a result on the scattering of polarized electrons from deuterium, which agreed with the W-S model (Prescott et al. 1978, 1979). This was the SLAC E122 experiment. They found not only the predicted scattering asymmetry but also obtained a value for $\sin^2\theta_W$ = 0.20 ± 0.03 (1978) and 0.224 ± 0.020 (1979), in agreement with other measurements at the time. ($\sin^2\theta_W$ is an important parameter in the W-S theory.) "We conclude that within experimental error our results are consistent with the W-S model, and furthermore our best value of $\sin^2\theta_W$ is in good agreement with the weighted average for the parameter obtained from neutrino experiments" (Prescott et al. 1979, p. 528).

Let us examine the arguments presented by the SLAC group in favor of the validity and reliability of their measurement. I agree with Pickering that, "In its own way E122 was just as innovatory as the Washington-Oxford experiments and its findings were, in principle, just as open to challenge" (Pickering 1984a, p. 301). For this reason, the SLAC group presented a very detailed analysis of their experimental apparatus and result and performed many checks on their experiment.

The experiment depended, in large part, on a new high-intensity source of longitudinally polarized electrons. The polarization of the electron beam could be varied by changing the voltage on a Pockels cell. "This reversal was done randomly on a pulse to pulse basis. The rapid reversals minimized the effects of drifts in the experiment, and the randomization avoided changing the helicity synchronously with periodic changes in experimental parameters" (Prescott et al. 1978, p. 348). It had been demonstrated, in an earlier experiment, that polarized electrons could be accelerated with negligible depolarization. In addition, both the sign and magnitude of the beam polarization were measured periodically by observing the known asymmetry in elastic electron-electron scattering from a magnetized iron foil.

The experimenters also checked whether or not the apparatus produced spurious asymmetries. They measured the scattering using the unpolarized beam from the regular SLAC electron gun, for which the asymmetry should be zero. They assigned polarizations to the beam using the same random number generator that determined the sign of the voltage on the Pockels cell. They obtained a value for $A_{exp}/P_e = (-2.5 \pm 2.2) \times 10^{-5}$, where A_{exp} was the experimental asymmetry and P_e was the beam polarization for the polarized source, $P_e = 0.37$. This was consistent with zero and demonstrated that the apparatus could measure asymmetries of the order of 10^{-5}.

They also varied the polarization of the beam by changing the angle of a calcite prism, thereby changing the polarization of the light striking the Pockels cell. They expected that $A_{exp} = |P_e| A \cos(2\phi_p)$, where ϕ_p was the prism angle. The results are shown in Figure 23. Not only do the data fit the expected curve, but the fact that the results at 45° are consistent with zero indicates that other sources of error in A_{exp} are small. The graph shows the results for two different detectors, a nitrogen-filled Cerenkov counter and a lead glass shower counter. The consistency of the results increases the belief in the validity of the measurements. "Although these two separate counters are not statistically independent, they were analyzed with independent

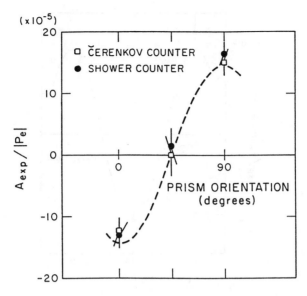

Figure 23. Experimental asymmetry as a function of prism angle for both the Cerenkov counter and the shower counter. The dashed line is the predicted behavior. From Prescott et al. (1978).

electronics and respond quite differently to potential backgrounds. The consistency between these counters serves as a check that such backgrounds are small" (Prescott et al. 1978, p. 350).

The electron beam helicity also depended on E_o, the beam energy, because of the g-2 precession of the spin as the electrons passed through the beam transport magnets. The expected distribution as well as the experimental data for $A_{exp}/|P_e|Q^2$ is shown in Figure 24. Q^2 is the square of the momentum transfer. "The data quite clearly follow the g-2 modulation of the helicity," and the fact that the value at 17.8 GeV is close to zero demonstrated that any transverse spin effects were small.

A serious source of potential error came from small systematic differences in the beam parameters for the two helicities. Small changes in beam position, angle, current, or energy could influence the measured yield and, if correlated with reversals of beam helicity, could cause apparent, but spurious, parity violating asymmetries. These quantities were carefully monitored and a feedback system used to stabilize them. "Using the measured pulse to pulse beam information together with the measured sensitivities of the yield to each of the beam parameters, we made corrections to the asymmetries for helicity dependent differences in beam parameters. For these corrections, we have assigned a systematic error equal to the correction itself. The most significant imbalance was less than one part per million in E_o [the beam energy] which contributed -0.26 × 10^{-5} to A/Q^2" (Prescott et al. 1978, p. 351). This is to be compared to their final result of $A/Q^2 = (-9.5 \pm 1.6) \times 10^{-5}$ GeV/c^2. This was regarded by the physics community as a reliable and convincing result.[49]

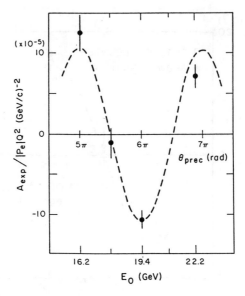

Figure 24. Experimental asymmetry as a function of beam energy. The expected behavior is the dashed line. From Prescott et al. (1978).

Hybrid models, which might have reconciled the discordant atomic parity results and the results of SLAC E122, were both considered and tested by the E122 group. In their first paper (Prescott et al. 1978) they pointed out that the hybrid model was consistent with their data only for values of $\sin^2\theta_W < 0.1$, which was inconsistent with the measured value of approximately 0.23. In their second paper (Prescott et al. 1979) they plotted their data as a function of $y = (E_o - E')/E_o$, where E' is the energy of the scattered electron. Both models, the W-S theory and the hybrid model, made definite predictions for this graph. The results are shown in Figure 25 and the superiority of the W-S model is obvious. For W-S they obtained a value of $\sin^2\theta_W = 0.224 \pm 0.020$ with a χ^2 probability of 40 percent. The hybrid model gave a value of 0.015 and a χ^2 probability of 6×10^{-4}, "which appears to rule out this model" (Prescott et al. 1979, p. 527).

The physics community chose to accept an extremely carefully done and carefully checked experimental result (SLAC E122) that confirmed the W-S theory and to await further developments in the atomic parity-violating experiments, which, as I have shown, were uncertain. This view is supported by Bouchiat's 1979 summary. After hearing a detailed account of the SLAC experiment by Prescott, he stated, "To our opinion, this experiment gave the first truly convincing evidence for parity violation in neutral current processes" (Bouchiat, 1980, p. 358). "I would like to say that I have been very much impressed by the care with which systematic errors have been treated in the experiment. It is certainly an example to be followed by all people working in this very difficult field" (pp. 359-60). This decision was based on evidential weight, determined by epistemological criteria.

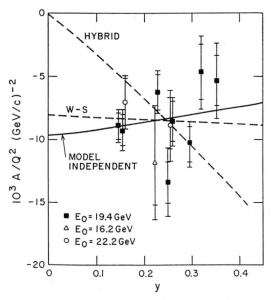

Figure 25. Asymmetries measured at three different energies are plotted as a function of y; (y = $(E_0 - E')/E_0$). The predictions of the hybrid model, the Weinberg-Salam theory, and a model independent calculation are shown. "The Weinberg-Salam model is an acceptable fit to the data; the hybrid model appears to be ruled out" (Prescott et al. 1979).

c) Later Atomic parity-Violation Experiments and the Resolution of the Discord

I discussed earlier the uncertainty in the 1977 Washington and Oxford results caused by systematic effects. This uncertainty is made even more evident when we examine the 1980 report by a group from Moscow. They reported measurements on the same transition in bismuth (λ = 648 nm) that the Oxford group had used (Bogdanov et al. 1980a, 1980b).[50] Their measurement was also in disagreement with the predictions of the W-S theory. They reported an optical rotation due to the parity nonconserving interaction of ϕ_{PNC} = (-0.22 ± 1.0) × 10^{-8} rad, in disagreement with the theoretical prediction of 10^{-7} rad. They discussed two sources of systematic errors that could give rise to effects of the same size as those expected from parity nonconservation: (1) variation in laser intensity due to scanning the laser frequency, and (2) interference between the main laser beam and scattered light. They also discussed the measures taken to reduce the errors due to these effects. Even so, they reported the following results for their six measurement series. (See Table 3 and discussion below).

They remarked that the spread in these individual series substantially exceeded the error in their quoted result and attributed that to time-dependent instrumental errors. Once again, there were systematic errors in this type of experiment that were approximately the same size as the effects predicted by the W-S theory.

A later paper by the Washington group, emphasized the discord between the various experimental results. "Our experiment and the bismuth optical-rotation experiments by

Table 2: Bismuth Optical Activity

λ Value and Group	$R \times 10^8$
(λ = 648 nm)	
Experiment:	
Novosibirsk	-20.6 ± 3.2
Oxford I	-10.3 ± 1.8
Oxford II	-11.2 ± 4.1
Old Oxford	2.7 ± 4.7
Theory:	
Novosibirsk	-19
Oxford	-14
(λ = 876 nm)	
Experiment:	
Washington	-10 ± 2
Old Washington	-0.7 ± 3.2
Theory:	
Novosibirsk	-14
Oxford	-12

Table 3. $\Delta\phi_{PNC}$ (10^{-8} rad

Experimental Series	$\Delta\phi_{PNC}$ in 10^{-8} rad
1	(-1.52 ± 2.1)
2	(+5.41 ± 3.5)
3	(-0.16 ± 2.4)
4	(-1.96 ± 2)
5	(-6.76 ± 2.6)
6	(+3.70 ± 2.4)

three other groups [Oxford, Moscow, and Novosibirsk] *have yielded results with significant mutual discrepancies far larger than the quoted errors*" (emphasis added) (Hollister et al. 1981, p. 643). They also pointed out that their earlier measurements "were not mutually consistent" (p. 643), empasizing the uncertainty in the results.

The moral of the story is clear. These were extremely difficult experiments, beset with systematic errors of approximately the same size as the predicted effects.

Let us briefly examine the subsequent history of the bismuth experiments, along with a brief treatment of the other atomic-physics parity-violation experiments that have relevance for the W-S theory. I begin with C. Bouchiat's 1980 summary of the situation. He presented the the summary shown in Table 2.

He remarked, that whereas the Novosibirsk result had been published, both the Washington and Oxford results were in the nature of progress reports on recent trends in their experiments, not definite results. He also noted that there was no explanation of the large difference between the old and new Washington and Oxford results, and that there was a factor-of-two discrepancy between the Novosibirsk and Oxford results, at λ = 648 nm. The difference in both theoretical approach and the numerical value of the calculation between the two groups was also mentioned.

In 1981, the Washington group published another measurement of the optical rotation in bismuth at λ = 876 nm (Hollister et al. 1981). The theoretical calculations they used to compare with their data ranged from $R = -8 \times 10^{-8}$ to -17×10^{-8}. Their value of $R = (-10.4 \pm 1.7) \times 10^{-8}$ agreed in "sign and approximate magnitude with recent calculations of the effect in bismuth based on the Weinberg-Salam theory" (Hollister et al. 1981, p. 643). They pointed out that since their earliest measurements they had "added a new laser, improved the optics, and included far more extensive experimental checks" (p. 643). They excluded these first three measurements from their average because they were made without the new systematic checks and controls, which were to be discussed in detail in a forthcoming paper.

This discussion appeared in an extensive review of atomic parity-violation experiments by Fortson and Lewis (1984), two members of the Washington group. They reported experimental controls on both the polarizer angle and the laser frequency. They also used alternate cycles, in which their bismuth oven was turned off, to avoid a spurious effect that could mimic the expected parity-nonconserving effect. They also examined their data for any correlations between the measured values of the parity-nonconserving parameter and any other experimental variables. This procedure set limits on known sources of systematic error and initially helped to uncover some errors and eliminate them. Included in these errors were those due to wavelength dependent effects and to beam movement. These were problems in all atomic-parity experiments.

The Novosibirsk group's results did not change very much from the value cited above by C. Bouchiat. Their last published measurement (Barkov and Zolotorev 1980b) gave $R = (-20.2 \pm 2.7) \times 10^{-8}$, which was approximately twice the value obtained by the Oxford group for the same, $\lambda = 648$ nm, transition. The Moscow group had originally reported (Bogdanov et al. 1980a) a value for R in disagreement with both the theoretical predictions and with the experimental results of both the Oxford group and the Novosibirsk group. Their value was $R_{exp}/R_{th} = -0.02 \pm 0.1$. A second publication (Bogdanov et al. 1980b) reported a value of $R = (-2.4 \pm 1.3) \times 10^{-8}$, still in disagreement with both theory and the other experimental measurements. They noted, however, that the errors within an individual series of measurements (see Table 3) exceeded the standard deviation in some cases, indicating that there were additional systematic errors present that varied comparatively slowly with time. The Moscow group continued their investigation of the sources and magnitudes of these systematic errors and found two principal sources of difficulty: (1) a change in the spatial distribution of light intensity as the frequency of the laser was changed and (2) interference between the signal beam and scattered light in the experimental apparatus (Birich et al. 1984). They took steps to minimize these effects and to measure any residual effects, and noted that their earlier results had not included all of these controls and corrections. Their final value was $R = (-7.8 \pm 1.8) \times 10^{-8}$ and they concluded that, "It is clear that our latest results and the results of the Oxford group [Oxford was reporting a value of approximately $(-9 \pm 2) \times 10^{-8}$ at this time] are in sufficient agreement with one another and with the results of the most detailed calculations" (p. 448).

The Oxford group continued their measurements on the $\lambda = 648$ nm line in bismuth through 1987. They had presented intermediate reports at conferences in 1982 and 1984, that were consistent with the W-S theory. Their 1987 result $R = (-9.3 \pm 1.4) \times 10^{-8}$ (Taylor et al. 1987) is consistent with the standard model (the uncertainty is now, primarily, in the theoretical calculations) and with the measurements by Birich et al. (1984) but inconsistent with those of Barkov and Zolotorev (1980a, b). During the 1980s the Oxford group devoted considerable effort to searching for systematic effects and trying to eliminate or correct for them. The difficulties of this type of experiment are severe. They noted that their method depends on changing the wavelength of the laser and that wavelength-dependent angles (WDA), comparable to the expected parity nonconserving optical rotation angle, are seen in their apparatus, even in the absence of bismuth.[51] This WDA varied with time in an apparently random way and affected the group's ability to make measurements and carry out diagnostic tests. They did not expect the WDA to give rise to systematic error in the bismuth measurements, because of its random nature; but its presence does indicate the possibility of angle effects of similar size to the expected effect and the need for precautions and checks. Their paper lists the following experimental checks: a) Angle sensitivity, b) Angle lock, c) Polarizer reversal, d) Faraday contamination, e) Pickup, f) Cross-modulation between laser and magnetic field, g) Transverse magnetic field effects, and h) Oven reversal. As one can see, making a valid measurement demands considerable care. The paper noted that their present result disagreed with their earlier published value of $R = (+2.75 \pm 4.7) \times 10^{-8}$. Because their new result involved an improved apparatus, considerably more data, and numerous checks against possible systematic error, they preferred their latest result. They concluded that their earlier result was in error. They admitted, however, that they did not have any explanation of what the error was or for the difference between the measurements. The rebuilding of the apparatus precluded testing many of the likely explanations.

The present situation is virtually the same as when M. Bouchiat and Pottier (Bouchiat and Pottier 1984) presented their summary (Table 4). The bismuth results are in approximate agreement with the Weinberg-Salam theory, although the discrepancy with the Novosibirsk measurement remains slightly worrisome. S. Blundell (Private communication) has told me that there are recent reports of a new Novosibirsk experiment whose results agree with those of the Oxford group.

Atomic parity violation has also been observed in elements other than bismuth. I mentioned earlier an experiment on thallium (Conti et al. 1979), which had given a result in approximate agreement with the W-S theory. This experiment had been part of the context in which the early bismuth experiments had been evaluated. The experiment had measured the circular dichroism, δ, and had found $\delta = (+5.2 \pm 2.4) \times 10^{-3}$ in agreement with the theoretical value $\delta_{th} = (+2.3 \pm 0.9) \times 10^{-3}$. Commins and his collaborators continued this series of experiments through the 1980s using the same basic method, although they made improvements in the experimental apparatus and carried out more thorough investigations of possible sources of systematic error. In 1981 they reported a value $\delta = (+2.8^{+1.0}_{-0.9}) \times 10^{-3}$ in comparison with the theoretical value $(+2.1 \pm 0.7) \times 10^{-3}$ (Bucksbaum, Commins and Hunter 1981a, b). The change in the theoretical value of δ was caused by a change in the experimentally measured value of $\sin^2\theta_W$ from 0.25 to 0.23.

Table IV. Bismuth Optical Activity

Bismuth Value and Group	Bismuth Optical Activity
($R \times 10^8$)	
648 nm:	
Experiment:	
Novosibirsk (1979)	-20.2 ± 2.7 -13
Oxford (1984)	- 9.3 ± 1.5
Moscow (1984)	- 7.8 ± 1.8
Theory:	
Sandars (1980)	-13
Novikov (1976)	-17
Barkov (1980)	-18.8
Martensson et al. (1981)	-10.5
876 nm	
Experiment:	
Washington (1981)	-10.4 ± 1.7
Theory:	
Martensson (1981)	-8
Sandars (1980)	-11
Novikov (1976)	-13

Parity-nonconserving optical rotation has also been observed in lead by the Washington group (Emmons, Reeve and Fortson 1983). Their experimental value of $R = (-9.9 \pm 2.5) \times 10^{-8}$ agrees, to within the uncertainties of both the measurement and the atomic theory calculation, with the theoretical prediction of $R = -13 \times 10^{-8}$. A series of measurements has also been done on cesium. As early as 1974, even before the existence of the weak neutral currents predicted by the Weinberg-Salam theory had been established, Bouchiat and Bouchiat (1974) had calculated the expected effect of such neutral currents in atomic parity-violation experiments. They had found that the effect would be enhanced in heavy atoms: going from hydrogen to cesium, one gets an enhancement of the order of 10^6.

The first experimental result on cesium was reported by M. Bouchiat and her collaborators (Bouchiat *et al.* 1982). They found that the parity nonconserving parameter $\text{Im}(E_1^{PNC}/\beta)\exp = (-1.34 \pm 0.22 \pm 0.11)$ mV/cm, where the theoretical value was (-1.73 ± 0.07) mV/cm. They concluded that, "in view of the experimental and theoretical uncertainties, this is quite consistent with the measured value" (Bouchiat *et al.* 1982, p. 369). This measurement was on a $\Delta F = 0$ hyperfine transition. A second paper (Bouchiat, Guena and Pottier 1984) reported a measurement on a $\Delta F = 1$ transition in cesium and found $\text{Im}(E_1^{PNC}/\beta) = (-1.78 \pm 0.26 \pm 0.12)$ mV/cm. "Within the quoted uncertainties, the two results clearly agree, so the two measurements successfully cross-check one another. It is then fair to combine them, which yields Im E_1^{PNC}/β = -1.56 ± 0.17 ± 0.12 mV/cm" (Bouchiat *et al.* 1984, p. 467). The theoretical value had changed slightly and was (-1.61 ± 0.07 ± 0.20), so theory and experiment were in agreement.

The latest experiment on cesium has been performed by Carl Wieman and his collaborators (Gilbert *et al.* 1985; Gilbert and Wieman 1986; Wieman, Gilbert and Noecker 1987). They found Im E_1^{PNC}/β = -1.65 ± 0.13 mV/cm, in good agreement with the previous measurement by M. Bouchiat *et al.* and with the theoretical prediction, discussed above. This was the first atomic parity-violation experiment to obtain an uncertainty of less than 10 percent. The experimental checks were extensive. They included four independent spatial reversals of experimental conditions to identify the parity-nonconserving signal when, in principle, only two are required to resolve the effect. This reduced the potential systematic error, because nearly all the factors that can affect the transition rate are correlated with, at most, one of these reversals. Other possible sources of systematic error were identified, and their possible effects measured in auxiliary experiments. The experimenters also introduced known nonreversing fields, misalignments, etc. The measured effect of these interventions agreed with their calculations of these effects and also indicated that these effects were small compared to the parity-violating signal. Their analysis of their data over time scales from minutes to days also indicated that the distribution of their measured values of Im E_1^{PNC}/β was completely statistical and that time dependent systematic effects were small.

It is fair to say that the current situation with respect to atomic parity-violation experiments and the Weinberg-Salam theory is that the preponderance of evidence favors the theory. The later experiments, which eliminated various sources of background and systematic uncertainty, are more credible than the earliest attempts to measure atomic-parity violation. No one knows with certainty why those early results were wrong. Nevertheless, since those early experiments physicists have found new sources of systematic error that were not dealt with in the early experiments. The redesign of the apparatus has, in many cases, precluded testing whether these effects were significant in the older apparatus. Although one cannot claim with certainty that these effects account for the earlier, presumably incorrect results, one does have reasonable grounds for believing that the later results are more accurate. The consistency of the later measurements, especially those done by different groups, enhances that belief.

The decision between the early atomic parity-violation results and the SLAC E122 result was determined by evidential weight based on epistemological criteria. The discord between the various atomic parity violation results was resolved by both the greater credibility of the later results, again based on epistemological criteria, and by a preponderance of evidence.

5. CONCLUSION

I have argued that in four separate episodes in modern physics the discord between experimental results was resolved by reasoned discussion based on epistemological and methodological criteria. This is an explicit challenge to social constructivists, who deny that this occurs. I believe that it is insufficient for them to claim that in these episodes the resolution *could* have been different. The fact that something may be possible does not imply that it is true. If constructivists want to demonstrate that physicists' choice between discordant results was not based on evidence and methodological criteria then I believe that they must argue either that the choice *was* based on different criteria, or that the decision *should* have been different.

In a recent paper, Alan Nelson (1994) has suggested that historical accounts, such as those I have given, are insufficient to establish the superiority of a rational or reasonable account over a constructivist one. He states, in discussing the atomic-parity violation episode that,

> Franklin does a lovely job of showing, once all the actual evidence was in, the Standard Model could have been regarded as more strongly supported than the hybrids. But, the constructivist should reply that this is yet another exercise in retrospective rationalism. *After* scientists make a choice in a case like this, they naturally go on to *construct* the kind of evidence that supports their choice. In a possible world where scientists preferred hybrid models, experiments would have been tuned differently, etc. so that the constructed evidence would have rationally supported a hybrid model. A Franklin counterpart in that possible world would be arguing that hybrid theories were chosen on rationalist grounds! (Nelson 1994, p. 546)

Nelson has placed the cart before the horse. Scientists decide what the valid experimental evidence is and then make their theory choice, not vice versa. Scientists have both an interest in producing scientific knowledge and a career interest in being correct, and such a procedure is far more likely to produce a correct choice. Without evidence as a guide how is the scientist supposed to make such a choice? He or she might just as well flip a coin. Constructivist accounts (see the work of Pickering and Collins) always seem to favor accepted theory, but there are numerous cases in which accepted theory has been overthrown by experimental evidence. Parity violation and CP (chrage-conjugation parity) violation are two such examples. (for details see Franklin 1986, chaps. 1 and 3). Agreement with accepted theory does not guarantee a correct theory choice.

I believe that Nelson also overestimates the plasticity of both Nature and of experimental practice. He is, of course, correct that experimenters do tend to modify their practice, both as they perform the experiment and as they analyze their data, in order to produce a result. Not all such possible procedures can, however, be justified. For example, a scientist who excluded all those experimental runs whose results didn't agree with his preferred theory would not be credible. If that fraud became known, the scientist would be ostracized. Not everything goes. In addition, I believe that Nelson overstates just how much one can change results by using legitimate procedures. It would, for example, require dramatic and unjustifiable modifications of apparatus and procedures to demonstrate that objects whose density is greater than that of air fall up when released. I do not believe that in our world any justifiable tuning could have produced evidence favoring hybrid models.[52] There may, of course, be possible worlds in which the hybrid models are correct. Valid experimental evidence in those worlds would demonstrate that.

To be fair, Collins and Pickering have offered constructivist accounts of two of the episodes I have discussed, atomic-parity violation and gravity waves. I believe their accounts are incorrect. In his most recent comments on the atomic-parity violation episode Pickering (1991) argues that my view that the decision of the physics community to accept the W-S theory on the basis of reasonable evidence fails because there are too many reasons. He presents four alternative scenarios, which he regards as equally reasonable. None of these alternatives actually occurred, leading Pickering to argue that because reason was unable to decide the issue, his account, based primarily on future research opportunities, is correct. Interestingly, constructivists never seem to consider alternative constructivist accounts. Three of the alternatives involve questions about the evaluation of experimental evidence; (1) the physics community might have decided that the atomic parity-violation results of Washington-Oxford were wrong and were therefore excluded,[53] (2) the physics community might have lumped the atomic parity-violation results together with those of SLAC E122 and concluded that they neutralized each other,[54] and (3) the community might have waited until E122 had been replicated before making a decision.[55] Pickering regards these alternatives as being as reasonable as accepting the SLAC E122 results and awaiting further work on atomic parity violation. He is, however, somewhat alone. None of these alternatives were pursued. He has presented no reasons why they should have been. He has merely asserted that they were equally reasonable. I have argued elsewhere that given the evidential context they were not equally reasonable (Franklin 1993b). (See also notes 53-55).

Pickering might also deny that the physics community engaged in an evaluation of the experimental evidence. (See, however, the quotations above from Dydak and Bouchiat, in which they do evaluate the evidence). In addition, the atomic parity violation experiments, including repetitions of the original Washington-Oxford experiments on bismuth, as well as others, have continued through the 1980s and into the 1990s, reaching agreement with the predictions of the W-S theory.[56] If the original Oxford-Washington results were simply regarded as wrong, there seems little reason for this to have happened. If, however, judgment was suspended concerning which of the discordant results was correct, then the subsequent experimental work certainly makes sense and even seems to be required.

Pickering also asks why a theorist might not have attempted to find a variant of electroweak gauge theory that might have reconciled the Washington-Oxford atomic parity results with the positive E122 result. (What such a theorist was supposed to do with the supportive atomic parity results of Berkeley and of Novosibirsk is never mentioned). "But though it is true that E122 analyzed their data in a way that displayed the improbability [the probability of the fit to the hybrid model was 6×10^{-4}] of a particular class of variant gauge theories, the so-called 'hybrid models', I do not believe that it would have been impossible to devise yet more variants" (Pickering 1991, p. 462). Pickering notes that open-ended recipes for constructing such variants had been written down as early as 1972 . I agree that it would certainly have been possible to do so, but one may ask whether or not a scientist might have wished to do so. If the scientist agreed with my view--that one had (a) reliable evidence (E122 and others) that supported the W-S theory and a set of conflicting and (b) uncertain results from atomic parity-violation experiments that gave an equivocal answer on support of the W-S theory--what reason would they have had to invent an alternative? Constructivists like to claim that they are only describing scientific practice and not making judgments. Pickering seems to ignore this dictum, as does Collins. They do, in fact, substitute their judgment for that of the scientific community. (See Pickering's discussion, given above, of the conflict between the early atomic-parity violation results and that of SLAC E122). In the case of gravity waves, Collins has stated, "Under these circumstances it is not obvious how the credibility of the high flux case [Weber's results] fell so low. In fact, it was not the single uncriticized experiment that was decisive;...Obviously the sheer weight of negative opinion was a factor, but given the tractability, as it were, of all the negative evidence, it did not *have* to add up so decisively. There was a way of assembling the evidence, noting the flaws in each grain, such that outright rejection of the high flux claim was not the necessary inference" (Collins 1985, p. 91). Collins also presents alternatives that were plausible to him but not to scientists working in the field. As I have shown, there were good reasons for rejecting Weber's results. One need not explain an incorrect result.

Nick Rasmussen (private communication) has suggested that I am holding constructivists to an impossibly high standard. He says that examination of the published record will never show scientists making a decision that goes against experimental evidence.[57] This is because scientists always give reasons for their decision that are meant to appeal to and persuade the scientific community. Why such reasons are persuasive to members of the scientific community is not discussed by constructivists. Rasmussen states that constructivists will never be able to show that the situation was different or that it should have been different, using such evidence.[58]

I disagree. Rasmussen's view requires that we believe that scientists do not give their "real" arguments, that they are presenting only those arguments that will persuade their fellow scientists. There is, however, no evidence that the public and private arguments are different. In one case where I have been able to examine both the private e-mail correspondence between the proposers of a Fifth Force and their published response to criticisms of the proposal, there was no such difference (Franklin 1993a, pp. 35-48).There are also other sources available to the historian of science. Notebooks, letters, e-mail, and the like could all show that the public and private reasons differ.

Collins has claimed, in his study of gravity waves, that the public and private reasons are different. Based on interviews with scientists, he concludes that the community need not have rejected Weber's results. Collins' claim disagrees with the published discussion (at the Seventh International Conference on General Relativity and Gravitation) mentioned earlier. Although individual scientists may find fault with particular bits of evidence, that doesn't mean that the overall decision, based on all of the evidence, is unreasonable.

I have argued that in four episodes from the history of modern physics the discord between experimental results was resolved by reasoned discussion based on epistemological and methodological criteria. Let us briefly review the resolution of these instances of discordant experimental results.

In the case of the Fifth Force tower gravity experiments, the measurements were correct. It was the comparison between theory and experiment that had led to the discord. It was shown that the results supporting the Fifth Force could be explained by inadequate theoretical models; either failure to account adequately for local terrain, or the failure to include plausible local density variations. In other words, the Sherlock Holmes strategy had been incorrectly applied. The experimenters had overlooked plausible alternative explanations of the results and possible sources of error.

There were, in addition, other credible negative tower gravity results that did not suffer from the same difficulties as did the positive results. There was also an overwhelming preponderance of evidence against the Fifth Force from other types of experiment. The decision as to which theory-experiment comparison was correct was not made solely on the basis of the experiments' and calculations themselves, although one could have justified this. Scientists examined all of the available evidence, and came to a reasoned decision about (a) which were the correct results and (b) that a Fifth Force did not exist.

In the composition-dependence experiments of Thieberger and of the Eöt-Wash group, the discordant results were an obvious problem for the physics community. Both experiments appeared to be carefully done, with all plausible and significant sources of possible error and background adequately accounted for. Yet the two experiments disagreed. These were attempts to both observe and measure the same quantity, a composition-dependent force, with very different apparatuses, a float experiment and a torsion pendulum. Was there in one of the experiments some unknown but crucial background that produced the wrong result? To this day, no one has found an error in Thieberger's experiment, but the consensus is that the Eöt-Wash group is correct and that Thieberger is wrong. There is no Fifth Force. How was the discord resolved? In this episode it was resolved by an overwhelming preponderance of evidence. The torsion-pendulum experiments were repeated by others including Fitch, Cowsik, Bennett, and Newman, and by Eöt-Wash themselves. These repetitions, in different locations and using different substances, gave consistently negative results. In addition, Bizzeti, using a float apparatus similar to that used by Thieberger, also obtained results showing no evidence of a Fifth Force. Bizzeti's result was quite important. Had he agreed with Thieberger then one might well have wondered whether, between torsion-balance experiments and float experiments, there was some systematic difference that gave rise to the conflicting results. This did not happen. There was an overwhelming preponderance of evidence against the composition-dependence of a Fifth Force.

In the early attempts to detect gravity waves, it clear that, according to the epistemological criteria discussed above, the critics' results were far more credible than Weber's. They had checked their results by independent confirmation, which included the sharing of data and analysis programs. They had also eliminated a plausible source of error--i.e. that of the pulses being longer than expected--by analyzing their results using the nonlinear algorithm and by looking for such long pulses. They had also calibrated their apparatuses by injecting known pulses of energy and observing the output.

Weber's reported result failed several tests suggested by these criteria. Weber had not eliminated the plausible error of a mistake in his computer program. It was, in fact, shown that this error could account for his result. It was also argued that Weber's analysis procedure, which varied the threshold accepted, could also have produced his result. Having increased the credibility of his result when he showed that it disappeared when the signal from one of the two detectors was delayed, he then undermined his result by obtaining a positive result when he thought two detectors were simultaneous, when, in fact, one of them had been delayed by 4 hours. As Garwin argued, Weber's result itself also argued against its credibility. The coincidence in the time-delay graph was too narrow to have been produced by Weber's apparatus. Weber's analysis procedure also failed to detect calibration pulses. The evidence against Weber's result was overwhelming, and used epistemological criteria.

It seems clear that the physics community's decision that there was no 17-keV neutrino was based on experimental evidence, discussion, and criticism--or, in other words, on epistemological criteria. It had been shown that the two most persuasive positive results had overlooked effects that mimicked the presence of a 17-keV neutrino. The Sherlock Holmes strategy had been incorrectly applied. The newer, negative results were persuasive not only because of their improved statistical accuracy, but also because they were able to demonstrate that their experimental apparatuses could detect a kink in the spectrum if one were present. This was a direct experimental check that there were no effects present that would mask the presence of a heavy neutrino. The Tokyo group had also shown that the shape correction factors used in their experiment did not mask any possible 17-keV neutrino effect This combination of almost overwhelming and persuasive evidence against the existence of a 17-keV neutrino, combined with the demonstrated and admitted problems with the positive results, decided the issue. There was no 17-keV neutrino.

It is fair to say that the current situation with respect to atomic parity-violation experiments and the Weinberg-Salam theory is that the preponderance of evidence favors the theory. The later experiments eliminated various sources of background and systematic uncertainty and are more credible than the earliest attempts to measure atomic-parity violation. No one knows with certainty why those early results were wrong. Nevertheless, since those early experiments physicists have found new sources of systematic error that were not dealt with in the early experiments. The redesign of the apparatus has, in many cases, precluded testing whether or not these effects were significant in the older apparatus. While one cannot claim with certainty that these effects account for the earlier, presumably incorrect results, one does have reasonable grounds for believing that the later results are more accurate. The consistency of the later measurements, especially those done by different groups, enhances that belief. The discord between the various atomic parity-violation results was resolved by both the greater credibility of the later results, again based on epistemological criteria, and by a preponderance of evidence. The decision between the early atomic parity violation results and the SLAC E122 result was, as discussed in detail earlier, determined by evidential weight based on epistemological criteria. I have argued that the discord between experimental results is resolved by reasoned discussion based on epistemological and methodological criteria. I have also argued elsewhere that there are good reasons for belief in experimental results. It follows then that we may reasonably use experimental evidence as the basis of scientific knowledge.

NOTES

[1] The time period needed in the cases I will discuss is of the order of years. Because the resolution of discord often involves the replication of experiments, this seems to be a reasonable time period.

[2] "For all its fallibility, science is the best institution for generating knowledge about the natural world that we have" (Collins 1985, p. 165).

[3] Noel Swerdlow had pointed out that it is a universal law and applies also to the Andromeda galaxy, which would seem to be outside the jurisdiction of the American Physical Society.

[4] An example of this is the measurement of the infrared absorption spectrum of organic molecules. Sometimes a pure sample of the substance could not be obtained, so that it was placed in an oil paste or a solution. The observed spectrum would then be a superposition of the spectra of both the substance and the paste or solute. Observation of the known spectrum of the paste would then provide grounds for believing that the apparatus is working properly.

[5] For details of this history see the work of Franklin (1993b).

[6] This gives rise to the gravitational force $F = G\, m_1 m_2 / r^2$.

[7] This type of calculation, known as upward or downward continuation was well-known. The results were quite sensitive to the surface gravity measurements and to the model of the earth used. This made knowledge of the local mass distribution and of the local terrain very important, a point we shall return to later.

[8] Typical values for G from mineshaft measurements were $G = (6.720 \pm 0.024) \times 10^{-11}$ m^3 kg^{-1} s^{-2} (Hilton mine) and $6.704^{+0.089}_{-0.025} \times 10^{-11}$ (Mount Isa mine) (from Stacey 1987). This should be compared to the best laboratory value at the time, i.e., $G = 6.6726(5) \times 10^{-11}$.

[9] The Moriond workshops were extremely important in the history of a Fifth Force. At these workshops many of those working in the field met, presented formal papers, and held informal discussions. It seems fair to say that, if you wanted to be up to date on what was going on in the field, you had to attend these workshops.

[10] Other experimental evidence was presented as early as January 1987. This will be discussed in the next section.

[11] Gravity measurements tend to be made on roads, rather than in ditches or surrounding fields. Roads are usually higher than their surroundings, giving rise to an elevation bias.

[12] Contrast this with the 16 points in the Greenland survey.

[13] Their final result was 60 ± 90 µGal.

[14] Parker and Zumberge could not do this for the Australian mine experiments because the data were proprietary.

[15] I will not discuss the positive results obtained by Boynton, which were subsequently superseded. This does not change anything essential in the story. For details see the work of Franklin (1993b).

[16] There were, at the time, theoretical explanations that allowed both results to be correct. These were eliminated by further experimental work.

[17] Some wag remarked that all that Thieberger's experiment showed was that any sensible float wanted to leave New Jersey.

[18] For a detailed discussion of this episode, along with a criticism of Collins' account of the same episode, see the work of Franklin (1994). In this discussion I will rely, primarily, on a panel discussion on gravitational waves that took place at the Seventh International Conference on General Relativity and Gravitation (GR7), Tel-Aviv University, 23-28 June 1974. The panel included Weber and three of his critics--Tyson, Kafka, and Drever--and included not only papers presented by the four scientists, but also included discussion, criticism, and questions. It includes almost all of the important and relevant arguments concerning the discordant results. The proceedings were published by Shaviv and Rosen (1975). Unless otherwise indicated all quotations in this section are from Shaviv and Rosen (1975). I shall give the author and page numbers in the text.

[19] Given any such threshold, there is a finite probability that a noise pulse will be larger than that threshold. The point is to show that there are pulses in excess of those expected statistically.

[20] One might worry that this cascading effect would give rise to spurious coincidences.

[21] Drever (pp. 287-88) looked explicitly for such longer pulses and found no evidence for them.

[22] Garwin's computer simulation used three different thresholds.

[23] For a detailed history of this episode see the work of Franklin (1995a).

[24] The units of mass are in keV/c^2, but physicists usually refer to the masses of particles in energy units such as keV. Physicists currently believe that the mass of the neutrino is zero--or very close to it.

[25] There was also suggestive, although not conclusive, evidence from a third type of experiment, that detecting internal bremsstrahlung in electron capture (IBEC), a form of β decay. Not all of the IBEC experiments gave positive results. In addition, as discussed below, one of the experiments that convinced the physics community that the 17-keV neutrino did not exist, that of Mortara et al. (1993), used the same type of solid-state detector that Simpson had used.

[26] For details of some early experiments see the work of see the work of Franklin (1990, chap. 1).

[27] Although, as we shall see later, there is good reason to doubt the existence of the 17-keV neutrino, I shall speak of it as if it existed.

[28] In a normal beta-decay spectrum the quantity $K = (N(E)/[f(Z,E) (E^2 - 1)^{1/2}E])^{1/2}$ is a linear function of E, the energy of the electron. A plot of that quantity as a function of E is called a Kurie plot.

[29] The maximum energy available in the decay of tritium is 18.6 keV.

[30] Simpson (1985, p. 1891) reported, "The decay of tritium has been followed with this detector over a period of four years and the halflife has been determined to be 12.35 ± 0.03 yr, in very good agreement with published values."

[31] Simpson was searching for a low mass neutrino with a mass of the order of 10s of eV.

[32] See discussion below on the issue of confirmation and the role of different experimental apparatuses. For a general discussion see (Franklin and Howson 1984).

[33] Simpson's work was based, in part, on theoretical work done by McKellar and by Schrock (McKellar 1980; Shrock 1980).

[34] These corrections were extremely important in analyzing the data (see below).

[35] For details see Franklin and Howson (1984).

[36] Kalbfleisch and Milton (1985) also argued that Simpson's analysis required an incorrect value for the endpoint energy of the tritium spectrum.

[37] Figure 18 shows the number of preprints on the subject of the 17-keV neutrino received at CERN, as a function of time. It is not an exact picture of work done in the field because not everyone sent preprints of their work to CERN. It does, however, give a good comparative view. As discussed below, work on the 17-keV neutrino continued during the period 1986-1991, albeit at a low level.

[38] One might speculate that the fact that these positive results were reported by someone other than Simpson was important.

[39] Bonvicini's work was very important. By showing that a smooth shape-correction factor might either mask or enhance a kink due to a 17-keV neutrino, he cast considerable doubt on the early negative results obtained with magnetic spectrometers. This work was influential in persuading scientists to perform the later, more stringent, experimental tests.

[40] In addition, Morrison (1992a) showed that Simpson's most persuasive reanalysis of Ohi's early negative result was dependent on a statistical fluctuation. Hetherington and others (1987) had also suggested that this might be a problem.

[41] These results were essentially the same as those reported by Kawakami et al.(1992). In his published paper Bonvicini (1993) agreed with this evaluation.

[42] These arguments provided good grounds for the belief that the 17-keV neutrino did not exist, but did not, of course, guarantee that it did not.

[43] Pickering (1984a) has also discussed this episode from a social constructivist view. Other discussions are contained in Ackermann (1991), Franklin (1990, chap. 8; 1993c), Lynch (1991), and Pickering (1991).

[44] I will discuss the uncertainty in the theoretical calculation later.

[45] The original experimental result of $(-8 \pm 3) \times 10^{-8}$ cited a 2-SD uncertainty, whereas the later result $(-0.7 \pm 3.2) \times 10^{-8}$ used a 1.5-SD uncertainty.

[46] Carl Wieman, whose work on atomic parity violation will be discussed below, informed me of this.

[47] As we shall see, Dydak's choice was justified by subsequent experimental and theoretical work.

[48] Recall that the theoretical predictions of the effect differed by a factor of two.

[49] The experimenters used several strategies to establish the validity of their result that I have discussed earlier as parts of an epistemology of experiment. The experimenters intervened and observed the predicted effects when they changed the angle of the calcite prism and when they varied the beam energy. They checked and calibrated their apparatus by using the unpolarized SLAC beam and observed no instrumental asymmetries and found that their apparatus could measure asymmetries of the expected size. They also used different counters--i.e., the lead glass shower counter and the gas Cerenkov counter--and obtained independent confirmation of the validity of their measurement.

[50] This result was also presented as an addendum to the proceedings of the Cargese Workshop.

[51] Recall that the Moscow group also saw such effects.

[52] I do not deny that we may be wrong in asserting the superiority of the Standard Model over the hybrids, but in that case our error will be shown by experimental evidence..

[53] Although I suspect that a majority of the physics community were skeptical of the Washington-Oxford results, there were no obvious reasons for believing these results were wrong. The systematic uncertainties that were cited by the Washington and Oxford groups did make the results uncertain.

[54] Although both the atomic parity-violation experiments and SLAC E122 used new techniques, they were, in fact, quite different apparatuses, subject to different backgrounds and sources of error and uncertainty. There were no good reasons to lump them together.

[55] As noted by Bouchiat, and discussed in detail by Franklin (1990, chap. 8), the SLAC E122 experiment was very carefully checked. There were good reasons to believe the result was valid without waiting for an expensive and time consuming replication. Contrary to Jacqueline Susann, once may be enough.

[56] The calculations of the effect predicted by the W-S theory have also changed. Recent calculations have reduced the size of the expected effect.

[57] I have never claimed that one must restrict oneself to published sources.

[58] This is a variant of the argument given by Nelson.

CHAPTER 8

CALIBRATION

Calibration, the use of a surrogate signal to standardize an instrument, is an important strategy for establishing the validity of experimental results. In this paper I present several examples, typical of physics experiments, which illustrate the adequacy of the surrogate. In addition, I discuss several episodes in which the question of calibration is both difficult to answer and of paramount importance. These include early attempts to detect gravity waves, the question of the existence of a 17-keV neutrino, and the existence of a fifth force in gravity. I argue that in these more complex cases the adequacy of calibration, in an extended sense, was both considered and established.

Calibration, the use of a surrogate signal to standardize an instrument, is an important strategy for the establishment of the validity of experimental results.[1] If an apparatus reproduces known phenomena, then we legitimately strengthen our belief that the apparatus is working properly and that the experimental results produced with that apparatus are reliable. If calibration fails, then we do not trust experimental results produced with that apparatus. Thus, if your spectrometer reproduces the known Balmer series in hydrogen, you have reason to believe that it is a reliable instrument. If it fails to do so, then it is not an adequate spectrometer.[2]

Harry Collins disagrees. In *Changing Order* (1985), he argues for what he calls the "experimenters' regress." This is the idea that a correct experimental result is one obtained with a good experimental apparatus, whereas a good experimental apparatus is one that produces the correct result. He argues that there is no way out of this regress; that scientists cannot provide independent grounds for the belief that an experimental apparatus is working properly. This obviously casts doubt on experimental results obtained with that apparatus. In particular, he argues that calibration cannot provide such grounds. "The use of calibration depends on the assumption of near identity of effect between the surrogate signal and the unknown signal that is to be measured (detected) with the instrument" (Collins 1985, p. 105). Collins further argues that the adequacy of the surrogate signal is not usually questioned by scientists and that calibration can only be performed provided that this assumption is not questioned too deeply.

I believe that Collins is wrong. The question of the adequacy of the surrogate signal is one that experimental physicists consider carefully, and they offer arguments for the adequacy of the surrogate. In many cases, as illustrated below, the adequacy of the calibration is clear and obvious.[3] There are, however, also instances, which involve discordant results or other controversies, in which the question of calibration may be both difficult to answer and of paramount importance. This is particularly true when a new type of experimental apparatus is used to search for a hitherto unobserved phenomenon. The episode that Collins uses to support his view of calibration, that of the early attempts to detect gravity waves, is just such an instance. I have argued elsewhere (Franklin 1994 and "How to Avoid the Experimenters' Regress") in detail that Collins' account of this episode is incorrect, and I will summarize the argument below. I will also present several other cases in which the adequacy of the calibration signal was crucial to the discussion.

Calibration of an experimental apparatus becomes a particularly interesting question when the experiment gives a null result--that is, when the phenomenon looked for is not found. One may legitimately ask whether the apparatus would have detected a signal had one been present. Here, the question of whether the surrogate signal is sufficiently similar to the signal one wishes to detect is central. Calibration is more than just the use of a surrogate signal to test whether an apparaus is operating properly. In an extended sense, calibration also includes aspects of the general issue of how one argues for the validity of an experimental result. This includes the question of whether there are background effects that might mimic or mask the phenomenon one wishes to observe, as well as that of the correctness of the analysis procedures used in the experiment.

One should distinguish here between experimental data and an experimental result. They are usually different. What I will refer to as "analysis procedures" are those processes that transform data into a result. These processes may involve computer analysis, making cuts on the data, and other procedures. I know of no general way of characterizing such procedures because they are experiment specific, but a few illustrations may help. In the K^+_{e2} experiment, discussed below, the experimental data were photographs of sparks produced in chambers in a magnetic field, along with various scaler readings. The analysis procedures transformed those sparks into particle trajectories and determined the momentum of the particle. The analysis also included cuts on the data to determine whether there actually were any events attributable to K^+_{e2} decay and to measure the branching ratio. (For details see Franklin 1990, chap. 6). In the gravity-wave experiments, which are also discussed below, the data consisted of electronic signals that are the output of piezo-electric crystals attached to the gravity-wave antenna. These signals were analyzed with computer programs to determine whether there was a significant change in the amplitude or phase of the signal. Such a pulse could then be compared to the output from a distant antenna. The experimental result was the number of coincidences between the signals from the two antennas as a function of time delay. Let us begin with a few illustrative cases in which the calibration of the experimental apparatus is unproblematic and in which the calibration of the apparatus is independent of the phenomenon one wishes to observe.

I. UNPROBLEMATIC CALIBRATION

Consider the problem I faced as an undergraduate assistant in a research laboratory. I was asked to determine the chemical composition of the gas in a discharge tube, and I was given an optical spectroscope. The procedure I followed was to use the spectroscope to measure the known spectral lines from various sources, such as hydrogen, sodium, and mercury. The fact that I could measure these known lines accurately showed that the apparatus was working properly. In addition to providing a check on whether I could measure spectral lines of optical wavelengths, this procedure also provided a numerical calibration of my apparatus. I determined small corrections to my results as a function of wavelength. I then proceeded to determine the composition of the gas in the discharge tube by measuring the spectral lines emitted, making the small corrections needed, and comparing them with known spectra. There was no doubt that the calibration procedure was adequate. The calibration lines measured spanned the same wavelength region as the ones I used to determine the composition.

A similar instance occurred in an experiment to measure the infrared spectrum of organic molecules (Randall et al. 1949). In this experiment, the experimenters knew in advance that the apparatus would produce an artifact in addition to the desired signal. In several cases, it was not possible to prepare a pure sample of the organic substance. The substance had to be placed in an oil paste or in solution. In such cases, one expects to observe the spectrum of the oil or of the solvent superimposed on the spectrum of the substance (Figure 1). The agreement of the spectral lines measured in the compound spectrum with the independently measured spectrum of the oil or of the solvent provides grounds for belief in the other spectral measurements and also provides a numerical calibration of the apparatus. As one can see in Figure 1, the range of the spectrum of the oil and that of the organic substance overlap.

Let us consider a more complex experimental apparatus, that used by the Princeton group (Christenson et al. 1964) in their experiment that observed the decay $K^o_L \rightarrow 2\pi$ and established the violation of the combined particle-antiparticle and space-reflection (CP) symmetry[4] (for details see Franklin 1986, chaps. 3, 6, 7). The Princeton group detected the decay by measuring the momenta of the two charged decay particles and reconstructing their invariant mass,[5] by assuming that the decay particles were pions, and by reconstructing the direction of the decaying particle relative to the beam. If it was a K^o_L decay into two pions, the reconstructed invariant mass of the decay particle would be the mass of the K^o_L and the angle between the reconstructed momentum of the decay particle and the beam would be zero. An excess of events was indeed found at the K^o_L mass and at zero angle to the beam (Figure 2). In order to demonstrate that the apparatus was functioning properly and that it could detect such decays, the Princeton group calibrated it by investigating the known phenomenon of the regeneration of K^o_S mesons, followed by those particles' decay into two pions. If the apparatus was operating properly, the distributions in mass and angle, in both the regenerated K^o_S decays and the proposed K^o_L decays, should have been identical. They were identical. The reconstructed mass and angle of the proposed K^o_L decay events were 499.1 ± 0.8 MeV/c^2 and 4.0 ± 0.7 milliradians, respectively, in good agreement with the values 498.1 ± 0.4 MeV/c^2 and 3.4 ± 0.3 milliradians obtained from the regenerated K^o_S events. Here, too, there was no doubt that the surrogate and the phenomenon were sufficiently similar. In both cases, one detected two particle decays of particles which had the K^o mass and that were travelling parallel to the beam.[6]

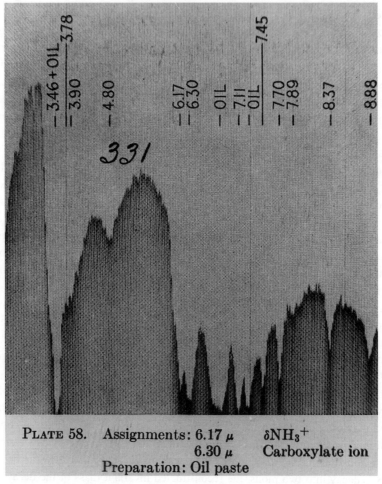

Figure 1. Infrared spectrum of an organic molecule prepared in an oil paste. The oil spectrum is clearly shown. From Randall et al. (1949)

A somewhat different example is provided by an experiment to measure the K^+_{e2} branching ratio (Bowen et al. 1967). In this case, the positron resulting from the decay was to be identified by its momentum, by its range in matter, and by its giving a signal in a Cerenkov counter set to detect positrons. The proper operation of the apparatus was shown, in part, by the experimental data themselves. Because the K^+_{e2} decay was very rare (approximately 10^{-5}) compared to other known K^+-decay modes, these other decays, in coincidence with noise in the Cerenkov counter, would also be detected. In particular, the muon from $K^+_{\mu2}$ decay, which has a known momentum of 236 MeV/c, was detected. A peak was observed at the predicted momentum, establishing that the apparatus could measure momentum accurately (Figure 3).

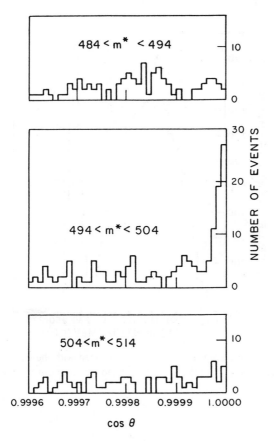

Figure 2. Angular distributions in three mass ranges for events with cosθ > 0.9995. From Christenson *et al.* (1964).

In addition, the width of the peak determined the experimental momentum resolution, a quantity needed for the analysis of the experiment. The Cerenkov counter was checked and its efficiency for positrons measured, by the comparison with a known positron detector in an independent experiment. The apparatus was also sensitive to K^+_{e3} decay. This decay produced high-energy positrons with a maximum momentum of 227 MeV/c, which was quite close to the 246 MeV/c momentum expected for K^+_{e2} decay. High-energy K^+_{e3} positrons were used to determine the range in matter expected for the K^+_{e2} positrons, and to demonstrate that the apparatus could indeed measure the range of positrons in that energy region.[7] In this case, too, there was no doubt as to the adequacy of this more complex calibration.

Another instance of calibration occurred in one of the experiments that demonstrated that parity, or left-right symmetry, was not conserved in the weak interactions (Wu *et al.* 1957). The experiment was designed to test parity conservation, by examining the counting rate of electrons from the β decay of polarized nuclei. If there was a difference

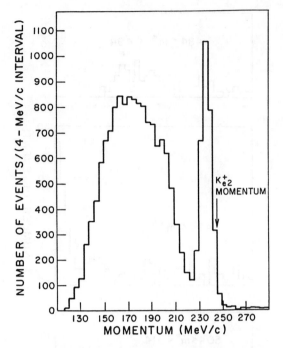

Figure 3. Momentum distribution of all K^+ decays obtained with the Cerenkov counter in the triggering logic. The momentum of the K^+_{e2} decay is shown. From Bowen et al. (1967).

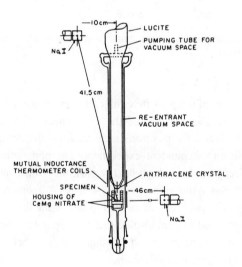

Figure 4. Schematic drawing of the experimental apparatus of Wu et al. (1957).

in the counting rates when the electrons were emitted either parallel to or antiparallel to the polarization, then parity would not be conserved. The experiment consisted of a layer of polarized ^{60}Co nuclei and a single fixed electron counter that was located in a direction either parallel or antiparallel to the orientation of the nuclei. The direction of polarization could be changed, and any difference in counting rate in the fixed counter could be observed. The experimental apparatus is shown in Figure 4. The experimental results, shown in Figure 5(b), show the presence of an asymmetry in counting rate and thus, parity nonconservation.

Let us examine the arguments offered by the experimenters to support their claim that their result was, in fact, due to the asymmetric β decay of polarized ^{60}Co nuclei. The experimenters calibrated the electron counter by observing the electrons from the known ^{137}Cs conversion line. This decay produced electrons with an energy of 513 keV, in comparison with the continuous energy range of 0 to 665 keV of electrons from ^{60}Co decay. The electron counter was also checked for stability against magnetic or temperature effects.

One might also ask whether the ^{60}Co nuclei were, in fact, polarized. It had been established in other experiments that ^{60}Co nuclei could indeed be polarized and that the degree of polarization determined by measuring the anisotropy of the γ rays emitted. The Wu *et al.* experiment included two sodium iodide counters (one in the equatorial plane and one near the polar direction [Figure 4]) to detect the γ rays, to measure the anisotropy, and, thus, to monitor the polarization. The measured difference in the counting rate in the two γ-ray counters, and thus in the polarization, is shown in Figure 5(a). The ability of the apparatus to reproduce this known effect provided grounds for belief that the ^{60}Co nuclei were polarized. Note that the polarization calibration measurement of γ rays was independent of the observed β-decay asymmetry, the phenomenon of interest. In addition, if the observed β-decay asymmetry was due to the decay of polarized ^{60}Co nuclei, the effect should disappear at the same time as the γ-ray anisotropy. As seen in Figure 5, it did.

In each of these five cases, the calibration of the apparatus did not depend on the outcome of the experiment in question. The proper operation of the experimental apparatus was demonstrated independently of the composition of the gas discharge, of what the spectrum of the organic substance was, of whether the K^0_L actually decays into two pions, of what the K^+_{e2} branching ratio was, or of whether parity was conserved. Clearly, five examples do not demonstrate that calibration is always unproblematic, because, as shown in the next section, it isn't. These experiments, however, are far more typical of the calibration procedures used in physics than is gravity-wave detection. I also believe that, in cases such as these, calibration is legitimately more decisive. Had any of these calibration procedures failed, then the results of the experiments would have been rejected. Scientists are quite good at employing the pragmatic epistemology of experiment.

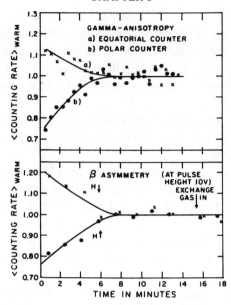

Figure 5. Gamma and β counting rates for polarizing field down (a) and up (b). The asymmetry is apparent. From Wu *et al.* (1957).

II. HITHERTO UNOBSERVED PHENOMENA, NEW EXPERIMENTAL APPARATUS, AND NULL EXPERIMENTS

A) EARLY ATTEMPTS TO DETECT GRAVITY WAVES[8]

The experiments that Collins uses as the prime example in his argument against the efficacy of calibration are the early attempts to detect gravity waves. Beginning in the late 1960s and continuing through the early 1970s, there were numerous attempts to detect gravitational radiation. Gravity waves were predicted by general relativity, but had not been detected at that time (Collins 1985; Franklin 1994).[9] Each of these early attempts used variants of a standard detector developed by Joseph Weber. Weber had used a massive aluminum-alloy bar, or antenna, (known as a "Weber bar"), which was supposed to oscillate when struck by gravitational radiation (Figure 6). The oscillation was to be detected by observing the amplified signal from piezoelectric crystals attached to the antenna. The expected signals were quite small (the gravitational force is quite weak in comparison to the electromagnetic force) and the bar had to be insulated from sources of noise such as electric, magnetic, thermal, acoustic, and seismic forces. Because the bar was at a temperature different from absolute zero, thermal noise could not be avoided, and to minimize its effect Weber set a threshold for pulse acceptance. Weber claimed to have observed above-threshold coincidences between two widely-separated detectors, in excess of those expected from thermal noise. Weber's claim was sufficiently plausible that others attempted to replicate his findings. By 1975, six other experimental results had been reported. None were in agreement with Weber's results.

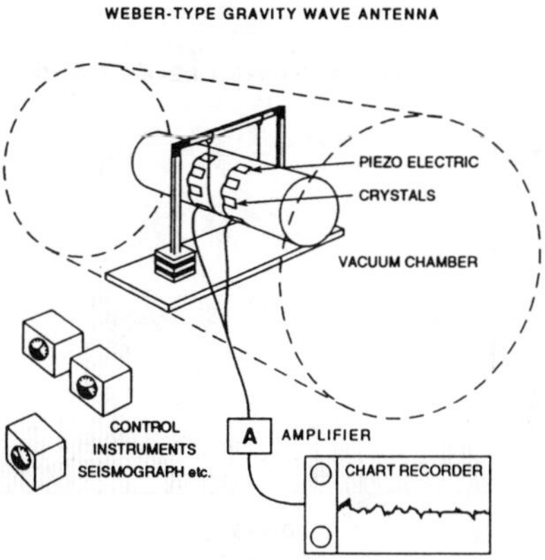

CALIBRATION

WEBER-TYPE GRAVITY WAVE ANTENNA

Figure 6. A Weber-type gravity wave detector. From Collins (1985).

The consensus was that Weber was wrong and that gravity waves had not been detected (Shaviv and Rosen 1975).

How was this agreement reached? This is a case in which the same apparatus or a slight variant of that apparatus was used to search for the same quantity. What complicated the search was that the phenomenon had not been previously observed. No one at the time knew how large a gravity-wave signal would be. Most physicists believed that the Weber bar was not sensitive enough to detect the size of the signal expected but,[10] because of Weber's positive result, they attempted to reproduce it. This is not to say that they believed that Weber's result was correct, but, rather, that they believed it was sufficiently credible to be worthy of further investigation.[11] The questions raised did not concern the adequacy of the detector but, rather, whether the detector was operating properly and whether the data were being analyzed correctly. This episode also illustrates the importance of including data analysis in discussions of experimental results. In this case there was no disagreement about what was good data. There was, as we shall see, considerable disagreement about what were the correct data analysis procedures, and thus, about what were the correct experimental results.

The first issue raised concerned the calibration of the apparatus. Scientists injected pulses of acoustic energy into the antenna and determined whether or not their apparatus could detect such pulses. Weber's apparatus failed to detect the pulses, whereas each of the six experiments performed by his critics detected them with high efficiency. This difference was due to a difference between the analysis procedures used by Weber and those used by his critics.

The question of determining whether or not there is a signal in a gravitational-wave detector, or whether two such detectors have fired simultaneously, is not easy to answer. There are several problems. One is that there are energy fluctuations in the bar, because of thermal, acoustic, electric, magnetic, and seismic noise, etc. When a gravity wave strikes the antenna its energy is added to the existing energy. This may change either the amplitude or the phase--or both--of the signal emerging from the bar. It is not

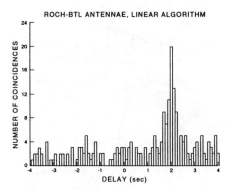

Figure 7. A plot showing the calibration pulses for the Rochester-Bell Laboratory collaboration. The peak due to the calibration pulses is clearly seen. From Shaviv and Rosen (1975).

just a simple case of observing a larger signal from the antenna after a gravitational wave strikes it. This difficulty informs the discussion of which was the best analysis procedure to use.

The nonlinear, or energy, algorithm preferred by Weber was sensitive only to changes in the amplitude of the signal. The linear algorithm, preferred by everyone else, was sensitive to changes in both the amplitude and the phase of the signal. Weber preferred the nonlinear procedure because it resulted in proliferation, several pulses exceeding threshold for each input pulse to his detector.[12] Weber admitted that the linear algorithm, preferred by his critics, was more efficient (by a factor of 20) at detecting calibration pulses. Tyson's results for calibration-pulse detection are shown in Figure 7, for the linear algorithm, and in Figure 8, for the nonlinear algorithm. There is a clear peak for the linear algorithm, whereas no such peak is apparent for the nonlinear procedure. (The calibration pulses were inserted periodically during data-taking runs. The peak was displaced by two seconds by the insertion of a time delay, so that the calibration pulses would not mask any possible real signal, which was expected at zero time delay).

Nevertheless, Weber preferred the nonlinear algorithm. His reason for this was that this procedure gave a more significant signal than did the linear one. This is illustrated in Figure 9, in which the data analyzed with the nonlinear algorithm is presented in the top graph and those analyzed with the linear procedure are presented in the bottom graph. Weber, in fact, was using the positive result to decide which was the better analysis procedure.

Weber's failure to successfully calibrate his apparatus was criticized by others. "Finally, Weber has not published any results in calibrating his system by the impulsive introduction of known amounts of mechanical energy into the bar, followed by the observation of the results either on the single detectors or in coincidence" (Levine and Garwin 1973, p. 177). His critics, however, analyzed their own data using both

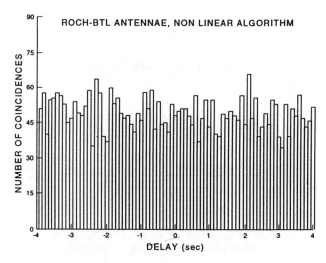

Figure 8. A time-delay plot for the Rochester-Bell Laboratory collaboration, using the non-linear algorithm. No sign of any zero-delay peak is seen. From Shaviv and Rosen (1975).

algorithms. They hypothesized that, unlike the calibration pulses for which the linear algorithm was superior, if using the linear algorithm either masked or failed to detect a real signal, then using the nonlinear algorithm on their own data should produce a clear signal. No signal appeared. Typical results are shown in Figures 8 and 10. Figure 8, which shows Tyson's data analyzed with the nonlinear algorithm, not only shows no calibration peak, but it does not show a signal peak at zero time delay. It is quite similar to the data analyzed with the linear algorithm, as shown in Figure 10, which also shows no signal (no calibration pulses were injected in this run).

Weber had an answer. He admitted that the linear algorithm was better for detecting calibration pulses, which were short. He claimed, however, that the real signal for gravitational waves was a pulse longer than most investigators thought it to be. He argued that the nonlinear algorithm was better for detecting these long pulses. If the gravity-wave signal was longer than expected, then one would have expected it to appear when the critics' data was processed with the nonlinear algorithm. It did not (see Figure 8).[13] Weber's experiment had failed the calibration test.

Let us summarize the evidential situation concerning gravity waves at the beginning of 1975. There were discordant results. Weber had reported positive results on gravitational radiation, whereas six other groups had reported no evidence for such radiation. The critics' results were not only more numerous, but had also been carefully cross-checked. The groups had exchanged both data and analysis programs and had confirmed their results. The critics also had investigated whether their analysis procedure, the use of a linear algorithm, could account for their failure to observe Weber's reported results. They had used Weber's preferred procedure, a nonlinear algorithm, to analyze their data, and still had found no sign of an effect. They also had

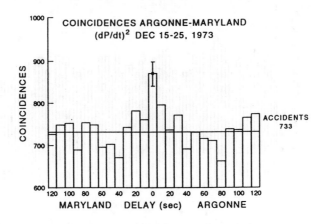

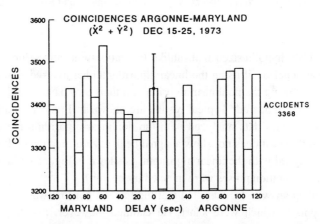

Figure 9. Weber's time-delay data for the Maryland-Argonne collaboration for the period Dec. 15-25, 1973. The top graph uses the non-linear algorithm, whereas the bottom uses the linear algorithm. The zero-delay peak is seen only with the non-linear algorithm. From Shaviv and Rosen (1975).

calibrated their experimental apparatuses by inserting electrostatic pulses of known energy and finding that they could detect a signal. Weber, on the other hand, as well as his critics using his analysis procedure, could not detect such calibration pulses. Under ordinary circumstances, Weber's calibration failure would have been decisive. It was because this episode is atypical--that is, one in which a new type of apparatus was used to search for a previously unobserved phenomenon--that it was not decisive. Other arguments were both needed and provided.

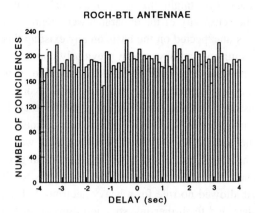

Figure 10. A time-delay plot for the Rochester-Bell Laboratory collaboration, using the linear algorithm. No sign of a zero-delay peak is seen. From Shaviv and Rosen (1975).

There were, in addition, other serious questions raised about Weber's analysis procedures. These included (1) an admitted programming error that generated spurious coincidences between Weber's two detectors; (2) Weber's report of coincidences between two detectors when the data had been taken four hours apart, and thus could not have produced real coincidences; (3) the question of selectivity in setting signal thresholds; and (4) whether Weber's experimental apparatus could produce the narrow coincidences claimed. The physics community's decision to reject Weber's result and accept those of his critics was based not only on his calibration failure, but also on the discovery of these other problems.

B) THE DISAPPEARING PARTICLE: THE CASE OF THE 17-KEV NEUTRINO

Experiments often give discordant results. This is nowhere better illustrated than in the recent history of experiments concerning the existence of a heavy, 17-keV neutrino (For details see Franklin 1995a or "The Appearance and Disappearance of the 17-keV Neutrino"). What makes this episode so intriguing is that both the original positive claim, as well as all subsequent positive claims, were obtained in experiments using one type of apparatus--namely those incorporating a solid-state detector--whereas the initial negative evidence resulted from experiments using another type of detector, a magnetic spectrometer. These were both seemingly reliable types of experimental apparatus. Solid-state detectors had been in wide use since the early 1960s, and their use was well understood. Magnetic spectrometers had been used in nuclear β-decay experiments since the 1930s, and both the problems and advantages of using this technique had been well studied. The discordant results were obtained using different types of experimental apparatus. One might worry--and the physics community did worry--that the discord was due to some crucial difference between the types of apparatus or to different sources of background that might mimic or mask the signal. Thus, the issues of calibration and of the sensitivity of the apparatus were central. In addition, Simpson's 1985 "discovery" was unexpected on the basis on any existing theory at the time. Such a particle, a heavy neutrino, had never been previously observed. As we shall see, the evidence against the existence of the 17-keV neutrino was provided by null experiments, experiments that did not observe the particle or its effects. It was only when it was demonstrated that the experimental apparatuses could detect the effect of such a particle--if it had been present--that the issue was decided.

To summarize the history briefly, the 17-keV neutrino was first "discovered" by Simpson in 1985. The initial replications of the experiment all gave negative results and suggestions were made that attempted to explain Simpson's result by use of accepted physics, physics that allowed no role for a heavy neutrino. Subsequent positive results by Simpson and others led to further investigation. Several of these later experiments found evidence supporting that claim, whereas others found no evidence for such a particle. The question of the existence of such a heavy neutrino remained unanswered for several years. Recently, doubt was cast on the two most convincing positive experimental results, and errors were found in those experiments. In addition, recent, extremely sensitive experiments have found no evidence for the 17-keV neutrino. The consensus is that it does not exist. Let us examine how this decision was reached.

The existence of the 17-keV neutrino was first reported in 1985 by Simpson (1985).[14] He had searched for a heavy neutrino by looking for a kink in the energy spectrum of electrons emitted in β decay or in the Kurie plot,[15] at an energy equal to the maximum-allowed decay energy minus the mass of the heavy neutrino, in energy units.[16] The fractional deviation in the Kurie plot value is $\Delta K/K \sim R[1 - M_2^2/(Q - E)^2]^{1/2}$, where M_2 is the mass of the heavy neutrino, R is the intensity of the second neutrino branch, Q is the total energy available for the transition, and E is the energy of the electron. Simpson's result is shown in Figure 11. A kink is clearly seen at an energy of 1.5 keV, corresponding to a 17 keV neutrino. The mixing probability for the 17-keV neutrino was 3%.

Simpson used a tritium β-decay source implanted in a solid-state detector. He devoted considerable effort to both the calibration of the apparatus and the details of data recording and analysis. Two of the key elements of the measurement were the energy calibration and the energy resolution. The energy was calibrated using x-rays of known energy from copper, molybdenum, and silver.[17] The calibration, as well as the stability of the entire recording apparatus, was monitored constantly. β-Decay spectrum data, as well as those data plus calibration data, were recorded with the use of a slotted wheel, or an x-ray chopper. This allowed x-rays from the copper-molybdenum calibration source to strike the detector when the slots were open. When the slots were closed, the calibration x-rays were excluded. The signal from the detector was routed to different halves of the same multichannel analyzer, depending on whether the slots were open. Thus, when the slots were closed, one should have observed only the β-decay spectrum and, when the slots were open, the β-decay spectrum with the x-ray calibration lines superimposed. This is seen in Figure 12.

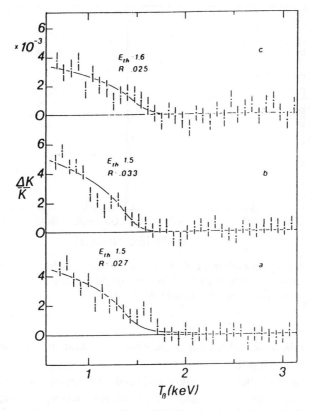

Figure 11. The data of three runs presented as ΔK/K (the fractional change in the Kurie plot) as a function of the kinetic energy of the β particles. E_{th} is the threshold energy, the difference between the endpoint energy and the mass of the heavy neutrino. A kink is clearly seen at E_{th} = 1.5 keV, or at a mass of 17.1 keV. Run a included active pileup rejection, whereas runs b and c did not. c was the same as b except that the detector was housed in a soundproof box. No difference is apparent. From Simpson (1985).

The energy resolution was determined at the same time using both copper and molybdenum x-rays, and in separate experiments using x-rays from iron and silver. Energy calibration and energy resolution were crucial to this experiment. If the detector did not have a linear energy response, one would not know what energy spectrum one expected so that one could search for the kink due to a heavy neutrino. If the energy resolution was too broad it might have masked the kink.

In the heavy-neutrino search, one had to worry about these factors at low energy, approximately 1.5 keV. "Because of the difficulty of energy calibrating an x-ray detector below about 6 keV the calibration was established in the following way. The x-rays from Cu and Br, and the Mo K_α were used to determine a linear calibration (with a typical rms deviation of 6 eV). The precision pulser was then used to measure the pulse-height response over the whole ADC [analog to digital converter] range. This was

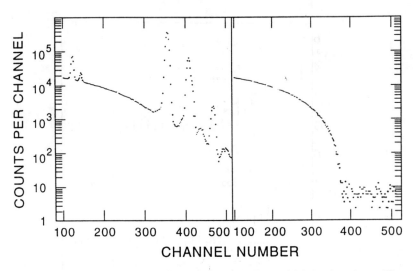

Figure 12. Logarithmic display of a typical spectrum in the multichannel analyzer. The x-rays shown on the left side are those of Cu, Mo, and Ag. The ability of the chopper system to eliminate the x-rays is clear. From Simpson (1981a).

combined with the x-ray calibration to determine a calibration over the whole energy range" (Simpson 1985, p. 1891).

Although Simpson's energy calibration and his determination of the energy resolution increased the credibility of his results, they did not play an important role in the subsequent discussion of whether or not the 17-keV neutrino existed. Although all subsequent experiments calibrated their apparatuses in similar ways, the discussion was centered on calibration as it is more broadly construed. This included the question of whether the experiment could detect the presence of a 17-keV neutrino--if it existed-- and whether the experiment had sufficient statistical accuracy to detect a small kink in the energy spectrum. Questions were also raised concerning the analysis procedures used in calculating the experimental result and in the theory-experiment comparison (discussed in detail below). These questions were raised about both Simpson's original result and about those of his critics. Once again, as was the case for gravity waves, the issue of analysis procedures was of central importance.

Simpson's original result was published in April, 1985. By the end of that year, there had been five attempts to replicate his result. All attempts were negative. A typical result, that of Ohi et al. (1985), is shown in Figure 13. These experiments were not Heraclitean, or exact, replications of Simpson's experiment. Whereas Simpson had used a tritium source implanted in a solid-state detector, all of these experiments used a ^{35}S source. Three of them, (Altzitzoglou et al. 1985; Apalikov et al. 1985; Markey and Boehm 1985), used magnetic spectrometers and two, (Datar et al. 1985; Ohi et al. 1985), used solid-state detectors, but with an external, rather than an implanted source.[18]

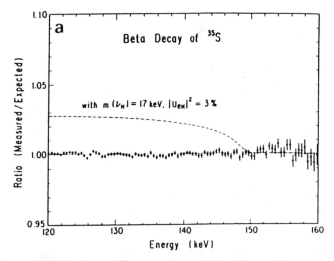

Figure 13. The ratio of the measured ^{35}S beta-ray spectrum to the theoretical spectrum. A three percent mixing of a 17-keV neutrino should distort the spectrum as indicated by the dashed curve. From Ohi et al. (1985).

The use of two different sources raised the first of the questions concerning the analysis procedures used in these experiments. In order to demonstrate that a kink existed in the β-decay spectrum of tritium, Simpson had to compare his measured spectrum with that predicted theoretically. This involved a rather complex calculation, which included various atomic- physics effects--in particular screening by atomic electrons. It was Simpson's calculation of these effects that was questioned. In particular, Lindhard and Hansen (1986) argued that with a different, and presumably better, calculation of the atomic-physics effects little of Simpson's claimed effect remained. The atomic-physics effects were far smaller for the higher-energy, ^{35}S decay.

Simpson took this criticism quite seriously and redid his calculation using the screening potential suggested by his critics (Simpson 1986b). He found that the effect was reduced by approximately 20% but was still clearly present. Simpson also offered several criticisms of the negative results. He noted that these experiments fitted the β-decay spectrum over a broad energy range, whereas the effect due to a heavy neutrino was concentrated in a narrow energy region near threshold. This criticism was related to another criticism of magnetic spectrometer experiments, which was offered somewhat later. In order to fit the observed β-decay spectra in such experiments, the experimenters had to use a shape-correction factor. One might worry that this slowly varying factor might mask the small effect due to the presence of a heavy neutrino. (This factor was typically $1 + \alpha E$, where α was a fitted parameter and E was the electron energy). Simpson, in fact, presented a reanalysis of Ohi et al.'s negative result,

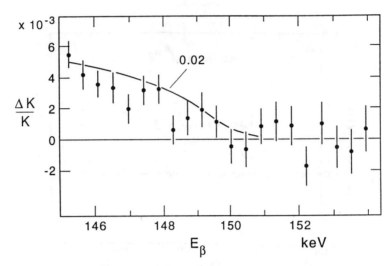

Figure 14. ΔK/K for the ^{35}S spectra of (Ohi et al. 1985) as recalculated by Simpson. From Simpson (1986a).

using a narrow energy range (Simpson 1986a). This reanalysis gave a positive result (Figure 14).

These early negative results had a chilling effect on work in the field. For the next few years negative evidence continued to accumulate but at a slow rate. In 1987 a group at Chalk River, using a magnetic spectrometer and the spectrum of ^{63}Ni, reported an upper limit of 0.3% for the presence of the 17-keV neutrino, which was in contrast to Simpson's original 3% effect (Hetherington et al. 1987). The experimenters analyzed the data using both a broad and a narrow energy range and obtained negative results for both. They also cautioned that, "concentration on a too narrow region can lead to misinterpretation of a local statistical anomaly as a more general trend" (p. 1512). This experiment was generally regarded as the most complete and persuasive magnetic spectrometer experiment done to that point (Bonvicini 1993, p. 98).

The tide began to turn in favor of the existence of the 17-keV neutrino in 1989, with the publication of two new experimental results by Simpson and Hime (Hime and Simpson 1989; Simpson and Hime 1989). The new experiments were on tritium, the element originally used by Simpson, and on ^{35}S, the element used in the five original negative experiments. The ^{35}S experiment used a solid-state detector, but with an external source. One problem with such experiments is that backscattering of electrons was both significant (approximately 30%) and could be a problem if it were not energy independent, a point that Simpson and Hime had checked. In these experiments, the mixing probability for the 17-keV neutrino was reduced to 1%. For the tritium experiment this was due, in large part, to the use of a different atomic- screening potential. This 1% result was confirmed in 1991 in the two most persuasive positive experiments, those of Hime and Jelley (1991) and of the Berkeley group headed by

Norman (Sur et al. 1991). In mid-1992 the evidential situation with respect to the existence of the 17-keV neutrino was quite uncertain. There was persuasive experimental evidence on both sides of the issue (see Table 1).

Bonvicini's work (issued first as a 1992 CERN report [CERN-PPE/92-54] and later published under his name as Bonvicini 1993)[19] made the situation more uncertain and also set the stage for further developments. In this work, Bonvicini discussed the question of whether, in the energy spectrum, a kink in due to an admixture of a 17-keV neutrino could be masked by the presence of unknown distortions, such as the shape correction factors used in magnetic spectrometer experiments. He performed a detailed analysis and Monte Carlo simulation of what were then generally regarded as the best experiments on either side of the 17-keV neutrino issue--the positive result from ^{35}S by Hime and Jelley (1991), and the negative result from ^{63}Ni by Hetherington et al. (1987). He also analyzed the early negative experiments that used magnetic spectrometers.[20] He concluded that "My analysis ... shows that [given the limited statistics of the experiments] *large continuous distortions in the spectrum can indeed mask or fake a discontinuous kink*" (Bonvicini 1993, p. 97; emphasis added).

Bonvicini also concluded that the positive Hime and Jelley result was statistically sound. He cautioned, however, that the electron response function (ERF), the efficiency for detection of electrons, in this experiment had been only partially measured and that this might be a possible problem. The ERF had been measured at only a single energy, whereas the energy spectrum was continuous. There was a possibility that the ERF could be energy dependent and the calibration inadequate. Bonvicini suggested that the ERF should be measured at several energies spanning the fitted energy spectrum. His analysis of the experiment of Hetherington et al. concluded that although their use of a 2.5 percent shape correction factor was certainly acceptable when searching for a 3 percent kink (Simpson's original result), when one looked for a 0.8 percent kink (the more recent value) more work was needed. His summary of the overall situation was as follows. "A look at the published data seems to indicate that the statistical criteria listed above would eliminate all the negative experiments considered here, but it is left to the authors to look at their data" (Bonvicini 1993, p. 114). Bonvicini's work argued quite strongly that the negative results of the previous magnetic-spectrometer experiments were inconclusive and suggested the design of experiments that either used no shape-correction factor or had such overwhelming statistical accuracy that a kink would always be visible. His work led to the design, construction, and performance of further experiments that would avoid the experimental difficulties.

The most detailed summary of the evidence at the time,as well as a moderate position, was provided by Hime in early 1992 (Hime 1992). Although Hime was an active participant in the controversy and one of those who provided persuasive evidence in favor of the 17-keV neutrino, his summary seems quite fair and judicious. He provided a reasonably complete history of the experiments and their results and devoted considerable attention to possible experimental problems or difficulties.

He considered the issue of the atomic-physics corrections to the tritium results and noted that taking account of the criticism had reduced the size of Simpson's original result from 3% to ~1%. He observed that part of the difficulty with these calculations was that the experiments did not use free tritium, but, rather, used tritium bound in a crystal lattice. Hime also discussed Simpson's reanalysis of the early negative results in ^{35}S, and remarked that they were based on a reanalysis of the data over only a narrow band of energy. "The difficulty remains, however, that an analysis using such a narrow region could mistake statistical fluctuations as a physical effect. The claim of positive effects in these cases [by Simpson] should be taken lightly without a more rigorous treatment of the data" (Hime 1992, p. 1303).[21]

He also discussed the issue of the uniformly negative results provided by magnetic-spectrometer experiments, citing Bonvicini's work. Hime observed that "given the obvious disagreement between magnetic spectrometer searches on the one hand and the positive results with solid state detectors on the other it is now generally agreed that insight into the discrepancy could be made if the sensitivity of a magnetic spectrometer to uncover a heavy neutrino signal could be experimentally demonstrated. Proposals include measurements with a mixed source (such as 99% ^{35}S + 1% ^{14}C), or artificially invoking energy loss in part of the spectrum at some predetermined level. This latter approach was suggested by the Caltech group and has been implemented in their program" (Hime 1992, p. 1310). He concluded that the existence of the 17-keV neutrino was still an open question (see Table 1).

The new Caltech result Hime referred to appeared in the work of Radcliffe et al. (1992). This experiment also looked at the ^{35}S spectrum with a magnetic spectrometer. Radcliffe et al. took data in two different runs--a wide energy range, 130-167 keV and a narrow scan of 10 keV around the kink expected at 150 keV for the 17-keV neutrino. Both runs were consistent with no heavy neutrino and excluded a 17-keV neutrino, with a 0.85% mixing probability at the 99.3% confidence level and the 99.9% confidence level, for the wide- and narrow-scan runs, respectively. A novel feature of this experiment was the attempt to simulate a kink in the spectrum. All of the previous searches for a heavy neutrino, with magnetic spectrometers, had been negative, and a question had been raised as to whether this type of apparatus, in fact, was capable of detecting such a kink. The experimenters shielded 10% of their detector with a 17-micron aluminum foil. The electrons would lose energy in passing through the foil and they expected this energy loss to produce in the spectrum a kink that would simulate a heavy neutrino with a 1% admixture. Their results with the foil in place are shown in Figure 15. The small kink that is visible gave a best fit for a mass of 15.6 keV with a mixing factor of 2.5%. One might legitimately wonder whether the apparatus was sensitive enough to detect a heavy neutrino with 1% mixing.[22] In addition, the shape of the spectrum distortion produced by the energy loss was different than that expected for a heavy neutrino. Although this was a reasonable attempt at calibration, to show that magnetic spectrometer experiments were sensitive to the presence of a heavy neutrino, it was not successful. Because of the problems with the shape and size of the distortion produced, it did not persuade the physics community that magnetic spectrometers were sensitive to the presence of a heavy neutrino.

Table 1: Summary of Results of Experiments to Detect the 17-keV Neutrino as Described by Hime (1992). [a] INS = Institute of Nuclear Studies; LBL = Lawrence Berkeley Laboratory; IBEC = Internal Bremsstrahlung; ITEP = Institute for Theoretical and Experimental Physics $(\sin^2\theta) \times 100$

Experiment and location	Isotope	(Confidence limit)	M_{2b} (keV)	Reference
Solid-state detector:				
Guelph	^{3}H in Si(Li)	2-3	17.1	Simpson (1985)
INS Tokyo	^{35}S	<0.15 (90% CL)	17	Ohi et al.(1985)
Bombay	^{35}S	<0.60 (90% CL)	17	Datar et al.(1985)
Guelph	^{3}H in Si(Li)	1.10 ± 0.30	17.07 ± 0.09	Hime (1989)
	^{3}H in HPGe	1.11 ± 0.14	16.93 ± 0.07	Hime (1989)
	^{35}S	0.73 ± 0.11	16.9 ± 0.4	Simpson (1989)
Oxford	^{35}S	0.78 ± 0.09	16.95 ± 0.35	Hime (1991)
	^{63}Ni	0.99 ± 0.22	16.75 ± 0.36	Oxford Report
LBL	^{14}C in HPGe	1.2 ± 0.3	17.1 ± 0.6	Sur et al.(1991)
IBEC Studies:				
CERN/Isolde	^{125}I	<2.0 (98% CL)	17	Borge et al.(1986)
Zagreb	^{55}Fe	<1.6 (95% CL)	15-45	Zlimen et al. (1988, 1990)
	^{71}Ge	1.6 ± 0.8	17.1 ± 1.3	Zlimen et al. (1991)
LBL	^{55}Fe	0.85 ± 0.45	21 ± 2	Norman et al.(1991)
Buenos Aires	^{71}Ge	0.80 ± 0.25	13.8 ± 1.8	TANDAR Preprint
Magnetic spectrometer:				
Princeton	^{35}S	<0.40 (99% CL)	17	Altzitzoglou et al. (1985)
ITEP	^{35}S	<0.17 (90% CL)	17	Apalikov et al. (1985)
Caltech	^{35}S	<0.25 (90% CL)	17	Markey and Boehm (1985)
	^{63}Ni	<0.25 (90% CL)	17	Wark and Boehm (1986)
Chalk River	^{63}Ni	<0.28 (90% CL)	17	Hetherington et al.(1987)
Caltech	^{35}S	<0.60 (90% CL)	17	Becker et al. (1991)
Munich	^{177}Lu	<0.80 (83% CL)	17	Conference Report

Morrison also cast doubt on the support that Simpson's reanalysis of Ohi *et al.*'s data had provided for the the 17-keV neutrino:

> The question then is, How could the apparently negative evidence of Figure 1a [13] become the positive evidence of Figure 1b [14]? The explanation is given in Figure 1c [16], where a part of the spectrum near 150 keV is enlarged. Dr. Simpson only considered the region 150 keV ± 4 keV (or more exactly + 4.1 and -4.9 keV). The procedure was to fit a

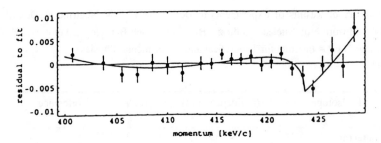

Figure 15. Synthetic kink induced in the beta spectrum of ^{35}S by a 17 μm aluminum foil. The solid curve is the spectrum expected with a 2.5% admixture of a 15.6-keV neutrino. Radcliffe *et al.* (1992).

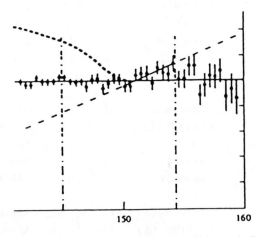

Figure 16. Morrison's reanalysis of Simpson's reanalysis of Ohi's result. From Morrison (1992).

straight line, shown solid, through the points in the 4 keV interval above 150 keV, and then to make this the base-line by rotating it down through about 20° to make it horizontal. This had the effect of making the points in the interval 4 keV below 150 keV appear above the extrapolated dotted line.

This, however, creates some problems, as it appears that a small statistical fluctuation between 151 and 154 keV is being used: the neighboring points between 154 and 167, and below 145 keV, are being neglected although they are many standard deviations away from the fitted line. Furthermore, it is important, when analyzing any data, to make sure that the fitted curve passes through the end-point of about 167 keV, which it clearly does not" (Morrison 1992a, p. 600).

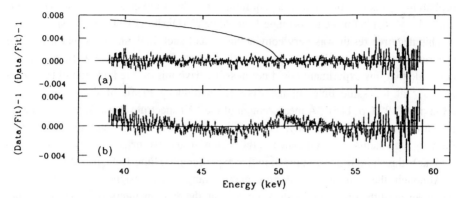

Figure 17. Deviations from the best global fit with $|U|^2$ free (a) and fixed to 1% (b). The curve in (a) indicates the size of a 1% mixing effect of the 17-keV neutrino. From Ohshima (1993).

The effect seen by Simpson was quite sensitive to the energy interval chosen. In general, an experimental result should be robust against such changes. Recall also the earlier comments of Hetherington et al. (1987) and those of Hime (1992), concerning the danger of mistaking a statistical fluctuation for a physical effect.

Further argument against the existence of the 17-keV neutrino was provided by the theoretical reanalysis of the data of Hime and Jelley by Piilonen and Abashian (1992). They performed a detailed simulation of the Hime-Jelley experiment, using a more complete electron response function than had been used originally and concluded, "We agree with Hime and Jelley that there is a serious distortion in their ^{35}S data, though we cannot pinpoint any definite cause for it. We believe that if the original data is reanalyzed by Hime and Jelley with a more realistic electron response function such as we have derived in our simulation, then the consistency of this distortion with a two-component neutrino hypothesis (with M_2 = 17 keV) will disappear" (p. 233).

Support for the existence of the 17-keV neutrino began to erode shortly after the publication of Hime's review. A very high-statistics magnetic-spectrometer experiment on ^{63}Ni was reported by the Tokyo group (Kawakami et al. 1992; Ohshima 1993; Ohshima et al. 1993). The experimenters noted some of the problems of experiments that used wide energy regions and commented that, "We have concentrated on performing a measurement of high statistical accuracy, in a narrow energy region, using very fine energy steps. Such a restricted energy scan ... also reduced the degree of energy-dependent corrections and other related systematic uncertainties" (Kawakami et al. 1992, p. 45). The narrow energy range minimized any effect of the shape correction factor. The results of their experiment are shown in Figure 17.

No effect is seen. Note the enormous statistical accuracy of this experiment in comparison with previous experiments (Figures 11 and 13). Their best value for the mixing probability of a 17-keV neutrino was -0.011% ± 0.033% (statistical) ± 0.030% (systematic), with an upper limit of 0.073% at the 95% confidence level for the mixing probability. This was the most stringent limit yet. "The result clearly excludes neutrinos with $|U|^2 \geq 0.1\%$ for the mass range 11 to 24 keV" (Ohshima 1993, p. 1128).[23]

This negative result was very convincing. It had such high statistics that it met the criteria for a good experiment set previously by both Bonvicini and by Hime. "Thus I conclude that this experiment could not possibly have missed the kink and obtain[ed] a good χ^2 at the same time, in the case of an unlucky misfit of the shape factor" (Bonvicini 1993, p. 115). "A measurement of the ^{63}Ni spectrum [Kawakami et al. 1992] has circumvented this difficulty. The sufficiently narrow energy interval studied, and the very high statistics accumulated in the region of interest, makes it very unlikely that a 17-keV threshold has been missed in this experiment" (Hime 1993, p. 165).

Although the experiment's narrow energy range was designed to minimize the dependence of the result on the shape correction, the experimenters also checked on the sensitivity of their result to that correction. They normalized their data in the three energy regions using the counts in the overlapping regions, and divided their data into two parts: (a) below 50 keV, which would be sensitive to the presence of a 17-keV neutrino, and (b) above 50 keV, which would not. They then fit their data within region (b) and extrapolated the fit to region (a). The resulting fit was far better than one that included a 1% mixture of the 17-keV neutrino, which demonstrated that the shape correction was not masking a possible effect of a heavy neutrino.

The 17-keV neutrino received another severe blow when Hime, following the suggestion of Piilonen and Abashian, extended his calculation of the electron response function of his detector to include electron scattering effects that had not been included previously and found that he could fit the positive results of Hime and Jelley without the need for a 17-keV neutrino (Hime 1993). This seemed to remove one of the most persuasive pieces of evidence for the heavy neutrino. "It will be shown that scattering effects are sufficient to describe the Oxford β-decay measurements and that the model can be verified using existing calibration data. Surprisingly, the β spectra are very sensitive to the small corrections considered. Consequently, any reinterpretation of the data is reliable only if the scattering amplitudes can be computed or measured accurately, and *independent* of the β-decay measurements" (Hime 1993, p. 166).

There remained a possibility that Hime's new calculation was incorrect. Hime was able to independently confirm his model by measuring the electron response function using monoenergetic internal-conversion electron sources occupying the same geometry as the β-decay sources used in the original experiments. The comparison between the measurements and the calculation is shown in Figure 18. "The solid curve drawn through these residuals is taken directly from the calculations presented above, including the effects of baffle-scattering, aperture penetration, and back-diffusion from the source substrate. The data reveal a structure that agrees well with the model, both in overall shape and intensity" (Hime 1993, p. 170).[24]

The Argonne group provided the evidence that sounded the death knell for the 17-keV neutrino (Mortara et al. 1993). This experiment used a solid-state, Si(Li) detector, an external ^{35}S source, and a solenoidal magnetic field to focus the decay electrons. The field also had the effect of reducing the backscattering of the decay electrons, a possible

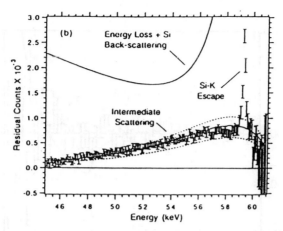

Figure 18. ^{109}Cd spectrum accumulated in Oxford geometry. Residuals extracted from the 61-keV K=IC tail when intermediate scattering efffects are neglected. The solid curve shows the effect calculated for intermediate scattering. From Hime (1993).

problem. Their final result, shown in Figure 19, was $\sin^2\theta = -0.0004 \pm 0.0008$ (statistical) ± 0.0008 (systematic), for the mixing probability of the 17-keV neutrino.

What made this result so convincing was that the experimenters were able to demonstrate the sensitivity of their apparatus to a possible 17-keV neutrino.

> To assess the reliability of our procedure, we introduced a known distortion into the ^{35}S beta spectrum and attempted to detect it. A drop of ^{14}C-doped valine ($E_o - m_e \sim 156$ keV) was deposited on a carbon foil and a much stronger ^{35}S source was deposited over it. The data from the composite source were fitted using the ^{35}S theory, ignoring the ^{14}C contaminant. The residuals are shown in Figure 5 [20]. The distribution is not flat; the solid curve shows the expected deviations from the single component spectrum with the measured amount of ^{14}C. The fraction of decays from ^{14}C determined from the fit to the beta spectrum is $(1.4 \pm 0.1)\%$. This agrees with the value of 1.34% inferred from measuring the total decay rate of the ^{14}C alone while the source was being prepared. This exercise demonstrates that our method is sensitive to a distortion at the level of the positive experiments. Indeed, the smoother distortion with the composite source is more difficult to detect than the discontinuity expected from the massive neutrino.
>
> In conclusion, we have performed a solid-state counter search for a 17 keV neutrino with an apparatus with demonstrated sensitivity. We find no evidence for a heavy neutrino, in serious conflict with some previous experiments. (p. 396)

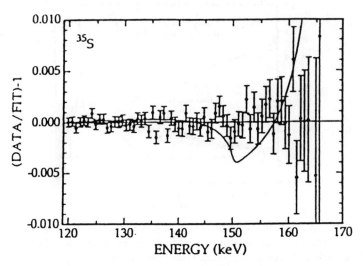

Figure 19. Residuals from a fit to the pile-up corrected ^{35}S data assuming no massive neutrino; the reduced χ^2 for the fit is 0.88. The solid curve represents the residuals expected for decay with a 17-keV neutrino and $\sin^2\theta = 0.85\%$; the reduced χ^2 of the data is 2.82. From Mortara et al. (1993).

This experiment had met the criteria that Hime had specified in his review--namely high statistics and a demonstrated ability to detect a kink in the spectrum--and had found no trace of the 17-keV neutrino.

The Berkeley group also withdrew their positive result on the 17-keV neutrino. They had found a problem with their experimental apparatus. The cause of the artifact, found in 1993, was quite subtle. The way in which the center detector was separated from the guard ring was by cutting a groove in the detector. "The n$^+$ is divided by a 1-mm-wide circular groove into a 'center region' 3.2 cm in diameter, and an outer 'guard ring.' By operating the guard ring in anticoincidence mode, one can reject events occurring near the boundary which are not fully contained within the center region" (Sur et al. 1991, p. 444). Such events would not give a full energy signal and would thus distort the observed spectrum.

What the Berkeley group found was that ^{14}C decays occurring under the groove shared the energy between both regions without necessarily giving a veto signal, and thus gave an incorrect event energy, which distorted the spectrum. They also found that, although their earlier tests had indicated that the ^{14}C was uniformly distributed in the detector, their new tests showed that between one third and one half of the ^{14}C was localized in grains. They also found that approximately 1% of the grains were located under the groove. Thus, the localization of the ^{14}C combined with the energy sharing gave rise to a spectrum distortion that simulated that expected from a 17-keV neutrino (E.B. Norman, private communication and Wietfeldt et al. [1993b]).

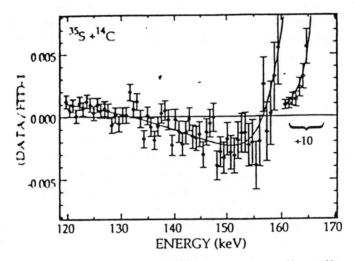

Figure 20. Residuals from fitting the beta spectrum of a mixed source of ^{14}C and ^{35}S with a pure ^{35}S shape; the reduced χ^2 of the data is 3.59. The solid curve indicates residuals expected from the known ^{14}C contamination. The best fit yields a mixing of $(1.4 \pm 0.1)\%$ and reduced χ^2 of 1.06. From Mortara et al. (1993).

There was virtually no evidence left that supported the existence of the 17-keV neutrino. The positive results of Hime and Jelley and of the Berkeley group were gone. These withdrawals, combined with the convincing negative results of the Tokyo and Argonne groups, had decided the issue. There was no 17-keV neutrino.

Calibration in the extended sense was both discussed and questioned throughout this episode and played a crucial role in its resolution. The question of whether the experimental apparatus, including the analysis procedure used, could detect the phenomenon of interest was central to the discussion. There was no doubt that the experimental apparatuses could detect electrons. What was questioned was whether the experimental apparatus itself or the analysis procedure either masked or mimicked the presence of the 17-keV neutrino. At various times, as we have seen, they did both.

Ultimately, it was shown that the two most persuasive experiments supporting the existence of the 17-keV neutrino had experimental problems that mimicked the presence of the particle. For the Hime-Jelley experiment it was electron scattering in the apparatus, and for the Berkeley result it was energy sharing in the detector. The early negative results, despite their apparent strength were, in fact, uncertain. As Bonvicini showed, given their limited statistics, the shape-correction factor used in the comparison of the theoretical prediction and the experimental result could either mask or mimic the presence of the particle. It was only when experiments could overcome this problem that the negative results could be convincing. In the case of the Tokyo results the overwhelming statistics of their result answered Bonvicini's criticisms. The Argonne result was convincing because the group demonstrated that the experimental apparatus and analysis procedure could indeed detect the presence of a 1% kink in their energy spectrum, which was exactly the effect expected if the 17-keV neutrino existed. This was a null experiment in which it was shown that if there had been an effect it would have been observed. This was calibration in the extended sense, and it was decisive.

C) THE FIFTH FORCE

The "Fifth Force" was a proposed modification of Newton's Law of Universal Gravitation. Based on a reanalysis of the original Eötvös experiment[25] Fischbach et al. (1986) suggested modifying the gravitational potential between two masses from $V = -Gm_1m_2/r$ to $V = -Gm_1m_2/r\,[1 + \alpha e^{-r/\lambda}]$, where the second term gives the Fifth Force with strength α and range λ. The reanalysis also suggested that α was approximately 0.01 and λ was approximately 100m. In addition, in contrast to the ordinary gravitational force, the Fifth Force was composition dependent. The Fifth Force between a copper mass and an aluminum mass would differ from that between a copper mass and a lead mass. (For details of this episode see Franklin 1993a).

In this episode, we also have a hitherto unobserved phenomenon as well as discordant experimental results. The first two experiments gave contradictory answers. One experiment supported the existence of the Fifth Force, whereas the other found no evidence for it. Here, too, we must consider calibration in an extended sense. As we shall see, there was no problem in detecting a force, but there were questions as to whether there were background effects that might simulate or mask the presence of the Fifth Force. In both experiments, the experimenters examined the plausible sources of such backgrounds. They magnified the size of these possible backgrounds to sizes larger than those found in their experiments and looked for measurable effects. When none were found they concluded that the backgrounds were negligible.[26]

The first experiment, that of Thieberger, looked for a composition-dependent force using a new type of experimental apparatus, which measured the differential acceleration between copper and water (Thieberger 1987a). The experiment was conducted near the edge of the Palisades cliff in New Jersey to enhance the effect of an intermediate-range force. The experimental apparatus is shown in Figure 21. The horizontal acceleration of the copper sphere relative to the water can be determined by measuring the steady-state velocity of the sphere and applying Stokes' law for motion in a resistive medium. Thieberger's results are shown in Figure 22. The sphere clearly has a velocity, indicating the presence of a force. He found a 4.7 ± 0.2 mm/h velocity in the y-direction (perpendicular to the cliff, as predicted) and 0.6 ± 0.2 mm/h in the x-direction. Thieberger concluded, "The present results are compatible with the existence of a medium-range, substance-dependent force" (p. 1068).

The ability of the apparatus to respond to a force, a calibration of the apparatus, was demonstrated with the use of magnetic positioning coils, which produced a known, non-uniform, direct-current magnetic field. The field gradient interacting with the different diamagnetic constants of water and copper produced a known force, which moved the

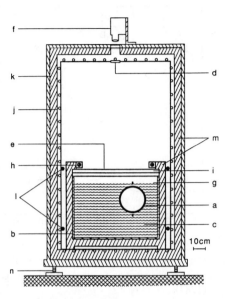

Figure 21. Schematic diagram of the differential accelerometer used in Thieberger's experiment. A precisely balanced hollow copper sphere (a) floats in a copper-lined tank (b) filled with distilled water (c). The sphere can be viewed through windows (d) and (e) by means of a television camera (f). The multiple-pane window (e) is provided with a transparent x-y coordinate grid for position determination on top with a fine copper mesh (g) on the bottom. The sphere is illuminated for one second per hour by four lamps (h) provided with infrared filters (i). Constant temperature is maintained by means of a thermostatically controlled copper shield (j) surrounded by a wooden box lined withStyrofoam insulation (m). The Mumetal shield (k) reduces possible effects due to magnetic field gradients and four circular coils (l) are used for positioning the sphere through forces due to ac-produced eddy currents, and for dc tests. From Thieberger (1987).

sphere. The test produced a measured velocity of 14 ± 1 mm/h when the field was turned on, which was in good agreement with the predicted value of 15 ± 2 mm/h. The velocity produced by the calibration was also similar (within a factor of three) to the experimental result. After the test was performed, the field was turned off and the entire experimental apparatus surrounded by a Mumetal shield, which reduced any external magnetic fields. The test motion required a field gradient of 10 G/m, whereas the measured field inside the shield was less than 0.1 G, suggesting that magnetic force effects on the experimental result were negligible.

A determined critic might object, as Collins did, that the apparatus was calibrated with a magnetic force and not with a gravitational force. This criticism is unjustified. There is considerable evidence that the acceleration of an object is independent of the source of the force acting on it.[27]

Other possible sources of a spurious signal were temperature gradients, leveling error, or instrumental asymmetries.

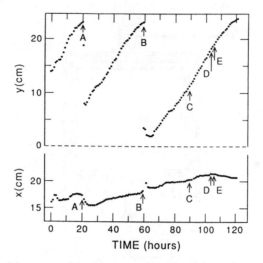

Figure 22. Position of the center of the sphere as a function of time. The y axis points away from the cliff. The position of the sphere was reset at points A and B by engaging the coils shown in Figure 21. From Thieberger (1987).

For fourteen hours, between points C and D on Fig. 2 [22], the temperature of the external west wall of the box was elevated by an average of 6°C above the east-wall temperature to test for possible sensitivity to external temperature gradients. The difference is over twice as high as the maximum difference ever observed between these two walls and over ten times higher than the average difference. No appreciable effect on the slope is observed. To estimate possible effects of leveling errors the east side of the instrument was dropped by 4.6mm at the point labelled E. This variation is over ten times larger than the maximum error estimated for the rest of the experiment. Again no effect can be seen on the y motion, but a small unexplained effect on the x motion seems to have occurred. (Thieberger 1987a, p. 1067-8)

Thieberger also tested for possible instrumental asymmetries. He rotated the apparatus by 90° and obtained similar results--4.5 ± 0.5 mm/h normal to the cliff. This indicated the absence of any large instrumental asymmetries. He also performed the experiment in the absence of the cliff but under otherwise similar conditions. This was intended to verify that the effect seen was, in fact, due to the presence of the cliff. He found x and y velocity components of -0.9 ± 0.2 and -1.2 ± 0.2 mm/h, respectively. These observations were smaller by a factor of four than the observed effect at the Palisades. Nevertheless, such a positive result might lead one to question the validity of Thieberger's result at the Palisades. With no cliff present, one expects zero velocity, and a positive result might indicate the presence of unaccounted for systematic effects, a point we shall return to later.

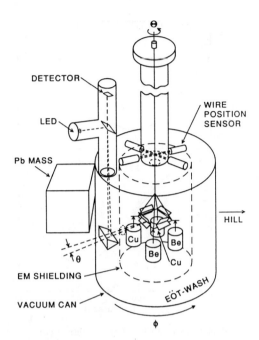

Figure 23. Schematic view of the University of Washington torsion pendulum experiment. The Helmholtz coils are not shown. From Stubbs *et al.* (1987).

The second experiment, by the whimsically named Eöt-Wash group, was also designed to look for a substance-dependent, intermediate range force (Raab 1987; Stubbs *et al.* 1987). It was located on a hillside on the University of Washington campus, in Seattle. The experimental apparatus is shown in Figure 23. If the hill attracted the copper and beryllium bodies differently, then the torsion pendulum would experience a net torque. This torque could be observed by measuring shifts in the equilibrium angle of the torsion pendulum as the pendulum was moved relative to a fixed geophysical point. Their experimental results are shown in Figure 24. The theoretical curves were calculated with the assumed values of 0.01 and 100m, for the Fifth Force parameters α and λ, respectively. These were the best values for the parameters at the time. There is clearly no evidence for a Fifth Force.

The question of background effects was explicitly addressed by the Eöt-Wash group. "We paid particular attention to systematic effects that could either produce a false signal or possibly cancel a true signal. The most important sources of such errors are (1) departures from fourfold rotational symmetry in the torsion pendulum ...; (2) deviation of the can rotation axis from true vertical ['tilt']; and (3) thermal gradients across the apparatus" (Stubbs *et al.*, p. 1071).

The test bodies were machined to be identical, and were coated with a thin gold film to minimize electrostatic forces. The experimenters minimized the magnetic forces by surrounding the rotating magnetic shield with stationary Helmholtz coils, which

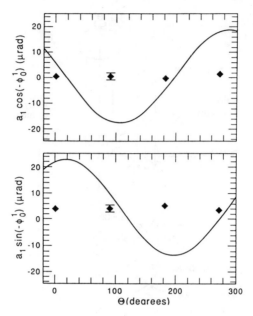

Figure 24. Deflection signal as a function of θ. The theoretical curves correspond to the signal expected for $\alpha = 0.01$ and $\lambda = 100$m. From Raab (1987).

reduced the field at the outer surface of the shield to approximately 10 mG. "Magnetic perturbations of the pendulum were negligible; reversing the currents in the Helmholtz coils caused a_1 [the Fifth Force signal] to change by only 3.8 ± 2.3 μrad. By scaling this result to our normal operating conditions we inferred that magnetic effects contributed to systematic errors at the 0.1 μrad level" (Stubbs et al., p. 1071-2).

The "tilt" of the apparatus was constantly monitored, because the apparatus was quite sensitive to tilt. The experimenters deliberately induced a tilt of 250 μrad and found a spurious a_1 signal of 20 μrad. They included in their final data sample only those runs for which the measured tilt was < 25 μrad, which gave rise to a correction of < 0.71 μrad.

The temperature of the apparatus was kept constant with a measured temperature gradient of 0.01K across the apparatus. The experimenters enhanced the effect of thermal influence by heating one side of the apparatus with a heat gun. They found a positive effect that was consistent with a temperature effect on the detector and the electronics but that did not simulate an external torque on the pendulum. They established an upper limit of 0.11 ± 0.17 μrad for thermal effects. All of these effects-- thermal, magnetic, and tilt--were well below the 20 μrad signal expected for the Fifth Force.

Ultimately, the discord between Thieberger's result and that of the Eöt-Wash group was resolved by an overwhelming preponderance of evidence in favor of the Eöt-Wash result (Franklin 1995b).[28] To this day, however, no one knows with certainty what was wrong with Thieberger's experiment. If calibration provides grounds for reasonable belief in an experimental result then one might ask what went wrong with the procedure. Thieberger examined all the sources of background that seemed plausible to him. There is, of course, no guarantee that he examined all the sources of background. The presence of a positive signal in the absence of the cliff, or other local mass asymmetry might very well have indicated that other such sources were present. Thieberger was quite aware of the novelty of his experimental apparatus and of the subsequent negative results. As he himself noted in response to a suggestion of a possible source of background, "The observed motion could indeed have been due to ordinary forces. Unanticipated spurious effects can easily appear when a new method is used for the first time to detect a weak signal .... Even though the sites and the substances vary, effects of the magnitude expected have not been observed. Therefore, although convection of the type proposed by Keyser does not seem to be the explanation, it now seems likely that some other spurious effect may have caused the motion observed at the Palisades cliff" (Thieberger 1989), p. 810).

In this episode, we have also seen that calibration in an extended sense, although difficult, is not impossible. Faced with a hitherto unobserved phenomenon, and in one case using a new type of experimental apparatus, physicists were able to demonstrate that if such effects existed they would have been detected by the apparatus, and also that there were no plausible sources of background that might mimic or mask the effect.

III. DISCUSSION

These cases not only illustrate successful calibration but also suggest that calibration is more than just using a surrogate signal to standardize an instrument. In its extended sense, calibration includes examination of sources of background that might mask or simulate the desired signal. This process often involves magnifying plausible backgrounds to see if they produce a significant effect. We have also seen that one must include the analysis procedures as an essential element in producing a credible experimental result. In both the case of gravity waves and that of the 17-keV neutrino, problems with experimental results were traceable to difficulties with the analysis procedures.

The more complex cases considered -- gravity waves, the 17-keV neutrino, and the Fifth Force--show that the adequacy of the calibration, that is the "near-enough" identity of the surrogate signal with the desired signal-- in fact was discussed critically. It was both essential and difficult to demonstrate the adequacy of the calibration procedures. These were experiments in which one looked for hitherto unobserved phenomena. In two of the cases--gravity waves and Thieberger's float experiment on the Fifth Force--new types of experimental apparatuses were used. In addition, in all three episodes one had not only discordant but also null, experimental results. In those cases, it was essential to show that the experimental apparatus could detect a signal if one were present. In each of these cases, the discord between the results was resolved on the basis of epistemological and methodological criteria (For more extended discussion see Franklin 1995b).

None of the unproblematic cases considered offer support for Collins' view that the assumption of surrogate adequacy is not examined critically. In such cases, it is clear to anyone working in the field that the calibration is sufficient. One need not belabor the obvious. I note, however, that in four cases (I exclude my own experiment, which was never published) the calibration procedure is described in the published paper and is thereby available for critical scrutiny.

It is clear, however, that the enumeration and discussion of cases, no matter how numerous, can never establish that all experimental calibration procedures are adequate. I do not believe that they are. Recall Weber's calibration failure. This type of failure often manifests itself in discordant experimental results. This is why critical discussion of calibration is not only essential in cases of discord but, as we have seen, plays a major role in the resolution of such discord. I believe that these cases, spanning a variety of subfields within physics, support my view that calibration is critically discussed and can provide grounds for belief in experimental results.

A skeptical reader may well wonder how, if calibration is so pervasive and successful, experimental results can be wrong. As illustrated above, not all calibration procedures are adequate. Weber's experiment failed the calibration test and was shown by both critical discussion and by an other evidence to be incorrect. In Thieberger's experiment, although the cause of his error is unknown, the overwhelming preponderance of evidence has convinced both the physics community and Thieberger himself that a source of background that simulates the presence of the Fifth Force has been overlooked. Calibration does not guarantee a correct result; but its successful performance does argue for the validity[29] of the result.

NOTES

[1] By valid, I mean that the experimental result has been argued for in the correct way, using epistemological strategies such as those discussed by Franklin (1986, chap, 6; 1990, chap. 6). These strategies include, in addition to experimental checks and calibration: (1) the reproducion of artifacts that are known in advance to be present; (2) intervention, in which the experimenter manipulates the object under observation; (3) independent confirmation using independent experiments; (4) elimination of plausible sources of error and alternative explanations of the result (the Sherlock Holmes strategy); (5) the use of the results themselves to argue for their validity; (6) the use of an independently well-corroborated theory of the phenomena to explain the results; (7) the use of an apparatus based on a well-corroborated theory; and 8) the use of statistical arguments.

[2] Calibration is distinguished from measurement in that the result expected is known in advance; in a measurement, the result is presumably not known with certainty. Calibration not only argues for the validity of an experimental result, but also, as discussed below, can also provide a numerical scale for the measurements or for a numerical correction.

[3] In some experimental papers, in which the reliability of a standard piece of experimental apparatus has been established previously, the discussion of calibration may be omitted.

[4] CP conservation required that the K^o_L decay into three pions. CP conservation forbade K^o_L decay into two pions.

[5] The invariant mass of a particle $m_o = [(\Sigma E_i)^2 - (\Sigma p_i)^2 c^2]^{1/2}$, where E_i and p_i are, respectively, the energies and momenta of the particles emitted in the decay. The invariant mass is the same for all observers.

[6] The very small mass difference between the K^o_L meson and the K^o_S meson is insignificant in this context. It is far smaller than the experimental uncertainty of the reconstructed mass.

[7] The approximately 10 percent difference in momentum was considered small enough, given the known behavior of positrons in this energy region.

[8] For a detailed discussion of this episode see the article by Franklin (1994) or "How to Avoid the Experimenters' Regress." In this discussion I will rely, primarily, on a panel discussion on gravitational waves that took place at the Seventh International Conference on General Relativity and Gravitation (GR7), Tel-Aviv University, June 23-28, 1974. The panel included Weber and three of his critics--Tyson, Kafka, and Drever--and included not only papers presented by the four scientists but also discussion, criticism, and questions. It included almost all the important and relevant arguments concerning the discordant results. The proceedings were published as Shaviv and Rosen (1975). Unless otherwise indicated, all references in this section are from the report edited by Shaviv and Rosen (1975). For other citations, I shall give the author and the page numbers in the text.

[9] Gravity waves have been observed by measuring the change in period of a binary pulsar. The change in period is $(2.427 \pm 0.026) \times 10^{-12}$ seconds/second (measured) (Taylor and Weisberg 1989) and $(2.402576 \pm 0.000069) \times 10^{-12}$ seconds/second (theory) (Damour and Taylor 1991). If General Relativity is correct, Weber should not have observed a positive result.

[10] In retrospect, they were correct.

[11] This is the distinction between pursuit and justification (Franklin 1993b). "Pursuit" is the further investigation of a hypothesis or experimetal result. "Justification" is the process by which a hypotheis or result comes to be accepted as scientific knowledge.

[12] One might worry that this cascading effect would give rise to spurious coincidences.

[13] Drever (Shaviv and Rosen 1975, pp. 287-88) looked explicitly for such longer pulses and found no evidence for them.

[14] Although there is good reason to doubt the existence of the 17-keV neutrino, I shall speak of it as if it existed.

[15] In a normal β-decay spectrum the quantity $K = (N(E)/[f(Z,E) (E^2 - 1)^{1/2} E])^{1/2}$ is a linear function of E, the energy of the decay electron. A plot of that quantity as a function of E, the energy of the decay electron, is called a Kurie plot.

[16] Simpson used tritium as his β-decay source.

[17] It had already been established that electrons and x-rays behaved similarly in these detectors, so the x-rays could be used as a calibration signal for an electron experiment.

[18] It is true that two different experiments provide more support for an experimental result than do two repetitions of the same experiment. However, this is true only if they give the same result. If they differ, one wonders which of the two is correct and, also about whether there is some

crucial difference in the experimental apparatuses or in the analyses that has produced the discord

[19] The major difference between the CERN report and the published paper is a detailed discussion of the negative Tokyo experiment (Ohshima *et al.* 1993), discussed below. Quotations are from the 1993 published paper.

[20] Bonvicini ignored experiments on tritium on the grounds that the Coulomb correction factor in such experiments is quite large for low-energy electrons (where the kink due to the 17-keV neutrino would be seen) and is difficult to calculate precisely. He also suggested that experiments on tritium be avoided in future work.

[21] This agreed with the earlier cautionary note by Hetherington *et al.* (1987).

[22] The shape of the spectrum distortion produced was also different from that expected for a heavy neutrino.

[23] The published value in the report of Kawakami *et al.* (1992) was (0.018% ± 0.033% ± 0.033%), with an upper limit of 0.095%. $|U|^2$ is the mixing probability.

[24] Hime attributed the small peak at the high-energy end to electron ionization of the silicon K-shell with the subsequent escape of silicon K x-rays. It does not cast any doubt on the confirmation of the model.

[25] The original Eötvös experiment was designed to measure the ratio of the gravitational mass to the inertial mass of different substances. Eötvös found that these two masses were equal to approximately 1 part/1 million. Fischbach *et al.* reanalyzed Eötvös' data and found a composition-dependent effect, which they interpreted as evidence for the Fifth Force.

[26] There is a possibility that there are saturation effects present, that the effect observed at large backgrounds might be the same as that at small backgrounds. In these cases, this was not an issue, because the size of the effect seen at large background was negligible when compared to the signal. I am grateful to my colleague David Bartlett for pointing this out.

[27] As Mr. Ed might say, "A force is a force, of course, of course."

[28] The issue was actually more complex. There were also discordant results on the distance dependence of the Fifth Force. For details see Franklin (1995b) or "The Resolution of Discordant Results."

[29] As noted earlier, the validity of a result is characterized by the application of correct epistemological and methodological procedures.

CHAPTER 9

LAWS AND EXPERIMENT

Traditionally, the role of experiment has been limited to the testing of laws or theories.[1] During the past decade we have recognized that the relation between laws and experiment is more complex, and that experiment may play varied roles. In addition to the confirmation or refutation of theory, experimental results may also call for a new theory, give us hints as to the structure of that theory, help to decide the mathematical form of the theory, or provide evidence for some future theory to explain. In addition, experiment may also have a life of its own, independent of theory.[2] In this essay I will discuss some of the problems associated with the role of experiment in the evaluation of laws.

1. ENABLING LAWS

I will begin, however, with a less-discussed interaction of experiment and theory. This is the use of laws in the design of experiment, what Peter Galison has called an "enabling theory." He illustrates this in his discussion of the experiments to measure the gyromagnetic ratio of the electron. The experiments were begun by Maxwell and failed because his apparatus was not sensitive enough to detect the effect. Maxwell had no model of the phenomenon which might have given him an estimate of the size of the effect he was looking for and indicated to him that his experiment would not work. It was only when Einstein and de Haas used the model of an electron orbiting an atomic nucleus that anyone had an idea of the size of the effect expected and experiments could be designed to measure them. (For details see Galison 1987, Ch.2).

I would like to further illustrate this use of laws in experimental design by considering an experiment to measure the K^+_{e2} branching ratio, the fraction of all K^+ mesons that decay, as the subscript indicates, into two particles, namely a positron, e^+, and a neutrino, v_e. The $K^+_{\mu 2}$ branching ratio signifies the decay of K^+ mesons into a positive muon, μ^+, and a muon neutrino, v_μ. (Bowen et al. 1967. See also Franklin 1990, Ch.6 for a discussion of this experiment). In addition, the experiment was a test of the V-A theory, the accepted theory of weak interactions at the time the experiment was done in the 1960s. The V-A theory predicted that the ratio of the decay rates for $(K^+ \rightarrow e^+ + v_e)/(K^+ \rightarrow \mu^+ + v_\mu)$, that is the ratio $(K^+_{e2}/K^+_{\mu 2})$ would be 2.6×10^{-5} for a pure axial vector interaction, A, (without radiative corrections), whereras a pure pseudoscalar interaction, P, gave a value of 1.02. Thus, a measurement of the K^+_{e2} branching ratio provides a stringent test of the theory, and on the presence of the P interaction.

Because $K^+_{\mu 2}$ decays had a branching ratio of approximately 63 percent, this gave a predicted K^+_{e2} branching ratio of 1.44×10^{-5}. (This includes the effect of radiative corrections). This set the scale for the experiment. If one wanted to detect say 10 K^+_{e2} decays, one would need to have a sample of approximately 700,000 K^+ decays.[3]

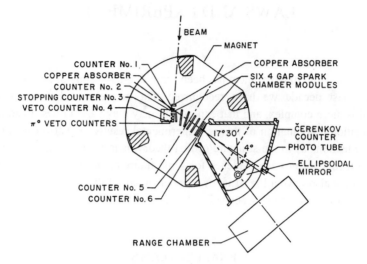

Figure 1. Details of the experimental apparatus for the K^+_{e2} experiment of Bowen *et al.* (1967).

The experimental apparatus is shown in Figure 1. The kaons were separated from pions and protons in the beam and identified by their range in matter and by time of flight. They were then stopped in counter C_3.

Particles from decays at approximately 90° to the incident beam passed through a set of thin plate spark chambers, the momentum chambers, located in a magnetic field. This permitted the measurement of the particle momentum with an experimentally measured resolution of 1.9%. The decay particles were detected by two scintillation counters, C_5 and C_6, and then passed through a gas Cerenkov counter, which was set to detect high energy positrons. This Cerenkov counter had been independently checked, and had a measured efficiency for 250 MeV positrons of between 95% and 99%, depending on the trajectory of the positron through the counter. For particles other than electrons, the measured efficiency was approximately 0.38%. Thus, the counter served to identify decay positrons.

A thick plate range spark chamber was placed behind the Cerenkov counter. Its total thickness of 80 g/cm² was enough to stop all particles resulting from the decay of the K^+ meson, so that one could use the range in matter, along with momentum and detection by the Cerenkov counter, to help identify decay particles.

Let us consider what signal the experimenters expected to find and what the possible sources of background, processes that might simulate K^+_{e2} decay, were. This sort of calculation is often required in order to estimate whether an experiment is feasible. If the backgrounds are too large the experiment cannot be done.

The positron from K^+_{e2} decay has a momentum of 246.9 MeV/c in the K^+ center of mass. This is higher than the momentum of any other direct decay product of K^+ decay. The closest competitor is the muon from $K^+_{\mu2}$ decay, which has a momentum of 235.6

MeV/c. The principal sources of high-momentum positrons that might mimic a real K^+_{e2} decay are

1) $K^+ \rightarrow e^+ + \pi^\circ + \nu_e$, K^+_{e3} decay, (where π° is an additional particle, namely the uncharged π meson) with a maximum positron momentum of 228 MeV/c and a branching ratio of about 5%.

2) $K^+ \rightarrow \mu^+ + \nu_\mu$, followed by $\mu^+ \rightarrow e^+ + \nu_e + \nu_\mu$, (where ν_μ is the muon antineutrino), with a maximum momentum of 246.9 MeV/c and a branching ratio of approximately 1.2×10^{-4} per foot of muon path. I note that this decay rate per foot is about a factor of 5 larger than the total expected K^+_{e2} decay rate. If this source of background could not be eliminated or greatly reduced the experiment could not be done.

Using the measured momentum resolution of 1.9%, the accuracy with which the momentum of a particle could be determined in the apparatus, and the known K^+_{e3} decay rate and momentum spectrum one could calculate that the number of K^+_{e3} events expected in the K^+_{e2} decay region, 242 MeV/c - 252 MeV/c, was less than 5% of the expected K^+_{e2} rate. If the K^+ decayed in flight then the momentum of the positron from K^+_{e3} decay might be higher than 228 MeV/c. This possible source of background was completely eliminated when "prompt" events were removed from the event sample.

The background due to K^+ decay into a muon, followed by muon decay into a positron was also calculated. This involved a detailed calculation which included the decay rate, the momentum and angular distribution of the decay positrons relative to the muons, the momentum and angular resolution in the thin plate chambers, and the ability to extrapolate the decay particle trajectory from the thin plate chambers into the range chamber. The result of this calculation was an expected background of about 15% of the expected K^+_{e2} decay rate. The experimenters expected a total background of approximately 20% of the expected K^+_{e2} rate. There were other sources of background, due to the operation of the experimental apparatus, but these were not calculable in advance. Thus, calculations based on the law the experimenters were proposing to test were vital in determining the feasibility of the experiment. Without such calculations the experimenters would not have known the relative sizes of the predicted effect and the sources of background.

An interesting point is that a theory does not have to be correct in order to serve as a useful enabling theory. In the case of the Fifth Force hypothesis, to be discussed later, the law is now generally considered to be incorrect, yet it was extremely important in the design of the experiments that ultimately refuted it. The suggestion was of a small addition to the normal gravitational force, which had a strength of approximately one percent that of the normal gravitational coupling and a range of about 100 meters. The strength allowed experimenters to estimate the size of the effects expected, whereas the range suggested improvements in the experimental sensitivity by placing the apparatus in an environment with a large matter asymmetry, such as a cliff or a hillside, or near a large laboratory mass.[4] I also note that the model of an orbiting electron to explain magnetism is incorrect. The current explanation is that magnetism is due to electron spin, a quantity unknown to either Einstein or de Haas. The order of magnitude of the effect is, however, the same.

2. DO EXPERIMENTS TEST LAWS?

A) THE THEORY-LADENNESS OF OBSERVATION

One might well ask whether the importance of the theory of the phenomenon in the experimental design precludes using the experiment as a test of that theory. Recall that the K^+_{e2} experiment was designed to be a rather stringent test of the V-A theory. In this case it seems clear that there is no real problem. The theory was involved only in determining the feasibility of the experiment, and did not, in any way, influence the results of the experiment, nor did it guarantee results in agreement with that theory. The experiment would have produced valid data and results even if the theory was wrong. The experimental apparatus and analysis procedure were shown to operate properly, independent of the theory. If the actual K^+_{e2} branching ratio had been smaller than that predicted, then the experimenters would not have seen any excess events. If it had been larger, the observed number of events would have been far larger than the estimated background. In the actual case the experimenters found $6^{+5.2}_{-3.7}$ events. This gave a branching ratio of $(2.1^{+1.8}_{-1.3}) \times 10^{-5}$ in good agreement with the theoretical prediction of 1.44×10^{-5}.

This discussion is related to the more general problem of the theory-ladenness of observation.[5] This is the fact that terms used in observation statements or measurement reports are laden with a particular theory. For example, although terms like "force" and "mass" appear in observation statements, they take their meanings only within the context of Newtonian dynamics. The question then arises whether or not such theory-laden terms can be used in experiments that test the theories. Kuhn (1970), Feyerabend (1975), and Barnes (1982), have all argued that there can be no comparison of competing paradigms or theories based solely on experimental evidence. "There is no appropriate scale available with which to weigh the merits of alternative paradigms: they are incommensurable" (Barnes 1982. p. 65).[6]

Briefly stated the argument is as follows. There can be no theory-neutral observation language. All observation terms are theory-laden, and thus we cannot compare experimental results, because in different paradigms, terms describing experimental results have different meanings, even when the words used are the same. An example of this would be the term "mass," which is a constant in Newtonian mechanics, whereas in special relativity it is a function of velocity.

Let us consider the following thought experiment. I shall demonstrate that an experiment described in procedural, theory-neutral (between the two competing theories or paradigms) terms, gives different results when interpreted within the two alternative paradigms. Thus, a measurement of the quantity derived will unambiguously distinguish between the two. I shall take as an example of Kuhn's own exemplars: the difference between Newtonian and Einsteinian mechanics. The case in point can be loosely described as the scattering of equal-"mass" objects.

The experimental procedure is as follows: Consider a class of objects, let us say billiard balls. The objects are examined pairwise by placing a compressed spring between them. The spring is allowed to expand freely, and the velocities of the two objects measured. Because we restrict ourselves to a single frame of reference in the laboratory, the measurement is theory-neutral. We then select two balls whose velocities were equal. A Newtonian would say that they have the same mass, M_N, whereas an Einsteinian would say that the relativistic masses $M_R = M_{OR} (1 - v^2/c^2)^{-1/2}$ are equal and thus that the rest masses M_{OR} are equal. This is agreed, but the point is that the procedure itself is theory-neutral. One of the objects is placed at rest in the laboratory and the other is given a velocity V_1 (again, theory-neutral), and the objects are allowed to scatter off each other. Care is taken to make the collision elastic (no energy given off). The angle between the two objects after the collision is measured. Although the assignments of energies and momenta will be different, the measurement of the angle does not depend on these assignments. As is well known, the predicted value of this angle differs in Newtonian and relativistic mechanics (for details see Franklin 1986 pp. 109-13). For a Newtonian $\theta_N = 90°$, whereas for an Einsteinian, $\theta_R < 90°$. A measurement of the angle between the velocities of the two outgoing objects will clearly distinguish between the two paradigms. They are commensurable.

This is not, of course, a general proof that such experiments can always be constructed. It should be pointed out, too, that the experiments of Lummer and Pringsheim and of Rubens and Kurlbaum on the spectrum of black-body radiation, which led to Planck's suggestion of quantization, also seem to be neutral with respect to classical and quantum physics. If this type of theory-neutral experiment can be found for what are two of the most revolutionary changes in physics, then I suspect that they can be found for almost all cases. However plausible the notion of the theory-ladenness of observation is, it has not been demonstrated that it prohibits theory testing.

B) THE DUHEM-QUINE PROBLEM

One might ask, however, whether one should trust the results of any particular experiment. I have previously suggested an epistemology of experiment, a set of strategies that provides grounds for reasonable belief in the validity of results. These include: 1) experimental checks and calibration, in which the experimental apparatus reproduces known phenomena; 2) reproducing artifacts that are known in advance to be present; 3) intervention, in which the experimenter manipulates the object under observation; 4) independent confirmation using different experiments; 5) elimination of plausible sources of error and alternative explanations of the result (the Sherlock Holmes strategy); 6) using the results themselves to argue for their validity; 7) using an independently well-corroborated theory of the phenomena to explain the results; 8) using an apparatus based on a well-corroborated theory; and 9) using statistical arguments. (See Franklin 1986, Ch. 6 and 1990, Ch. 6 for details). I believe that such results may then be legitimately used in testing theory.

Philosophers of science have raised another question concerning the use of experiment in the testing of theory. This is known as the Duhem-Quine problem [See Harding, 1976]. In the usual *modus tollens* if a hypothesis h entails an experimental result e then $\neg e$ (not e) entails $\neg h$. As Duhem and Quine pointed out it is not just h that entails e but rather h and b, where b includes background knowledge and auxiliary hypotheses. Thus, $\neg e$ entails $\neg h$ or $\neg B$ and we don't know where to place the blame. As Quine put it, "Any statement can be held true come what may, if we make drastic enough adjustments elsewhere in the system" (Quine, 1953, p. 43). One might, of course, also question the experimental result $\neg e$. This is where one applies the epistemology of experiment, to justify our belief in the experimental result.

In practice, however, scientists never confront all the logically possible explanations of a given result. There are usually only a reasonable number of plausible or physically interesting alternatives on offer. Scientists evaluate the cost of accepting one of these alternatives in the light of all existing evidence. One of these alternatives may be far better supported by the evidence than any of the others. The alternatives themselves may also be tested and we may be left with only one explanation.

Are there any hints that one might offer as to how one goes about solving the problem of where to place the blame? Noretta Koertge (1978) has made some very useful suggestions.[7] She suggests two strategies.

1) Check the most accessible source of trouble first. In other words, check those alternative explanations that are most easily tested.

2) Check on the most probable source of trouble early on.[8]

These are, in fact, the strategies scientists tend to use. She goes on to discuss the appraisals that go into a solution, X, of the Duhem-Quine problem.

1) "How interesting or informative or explanatory would X be if it were true.

2) What is the probability that X is true?" (Koertge 1978, p. 263).

This will, of course, involve appraisals of the plausibility or probability of the alternatives. It will also involve estimates of the scientific interest of the alternatives. As Koertge points out, it is here that the most serious differences of opinion will occur within the scientific community. Even if one had agreed upon and clear measures of content, simplicity, depth, heuristic power, etc. different scientists might well give these criteria different weights. These criteria may also be applied in different ways in different situations.

Should the fact that there is no prescriptive algorithm for the solution of the problem worry us excessively. I don't think so. It is precisely because of these differing judgments that more of the alternatives are likely to be explored. Sometimes, even a very implausible hypothesis turns out to be correct. The history of science shows us that in practice the Duhem-Quine problem is solved in the way Koertge and I suggest, using valid experimental evidence. (See Franklin 1990, Ch. 7 for details).

C) DISCORDANT RESULTS

Although the epistemology of experiment, discussed earlier, provides us with reasonable grounds for belief in experimental results, it does not guarantee that those results are correct, or even that two different experiments will give the same result. If we are to use experimental results to evaluate theory then we must solve this problem. How does the scientific community deal with this question of the fallibility of results or decide between two conflicting results? Fallibility can be dealt with. One uses the best available evidence, with the knowledge that it is fallible and that decisions made may turn out to be incorrect. Reasonable belief does not guarantee certainty.[9]

Discordant results are not unusual occurrences in the history of physics, particularly when a new experimental technology is used to investigate previously unobserved phenomena. Thus, in the early days of experimental investigation of the Weinberg-Salam unified theory of electroweak interactions, several groups reported such discordant results. In 1976 and 1977 groups at the University of Washington and Oxford University (Baird *et al.* 1976, 1977 Lewis *et al.* 1977) reported results on atomic parity violation that disagreed with the predictions of W-S theory. In 1978 and 1979 groups at Berkeley (Conti *et al.* 1979) and Novosibirsk (Barkov and Zolotorev 1978a,b, 1979) reported results in agreement with the theory. (I note here that only two of these experiments were replications in the sense that they measured the same quantity. Oxford and Novosibirsk measured parity violation in a particular transition in atomic bismuth, whereas Washington and Berkeley looked for similar effects in a different transition in bismuth and in thallium, respectively. All of the experiments did, however, investigate parity violation in atomic systems.)[10] How was the decision made between the conflicting atomic-parity violation results? The early Washington and Oxford results were internally inconsistent and also contained systematic uncertainties that were approximately as large as the predicted effects. In addition, both experiments used a new type of experimental apparatus. Both the Washington and Oxford groups continued their investigations into the 1980s, making improvements in their apparatus. Other atomic-parity violation experiments were also performed. All of the subsequent experiments gave results in agreement with the theoretical predictions.[11] Repetition of the experiments and improvements in the apparatuses provided convincing support for one set of early results, those that agreed with W-S theory. In this case there were also theoretical attempts to reconcile the discordant early results. This, so-called, hybrid model was experimentally refuted.

A similar case of discordant results occurred recently in the investigation of the "Fifth Force," a proposed modification of the law of gravity.[12] The two initial experimental results, those of Thieberger (1987) and of the Eöt-Wash group, a group at the University of Washington,[13] (Stubbs *et al.* 1987), gave conflicting results, one favoring the existence of the Fifth Force and one opposed. These were followed shortly thereafter by Boynton's (1987) "marginally observed" (his words) positive result.

The subsequent history seems to be an illustration of one way in which the scientific community deals with conflicting experimental evidence. Rather than making an immediate decision as to which were the valid results, this seemed extremely difficult to do on methodological or epistemological grounds, the community chose to await further measurements and analysis before coming to any conclusion about the evidence. The torsion-balance experiments of Eöt-Wash and Boynton were repeated by others including Fitch *et al.* (1988), Cowsik *et al.* (1988, 1990), Bennett (1989), and Newman *et al.* (1989), and Nelson *et al.* (1990) and by Eöt-Wash (Adelberger et al. 1989; Stubbs *et al.* 1989) and Boynton (1990) themselves. These repetitions, in different locations and using different substances, gave consistently negative results. In addition, Bizzeti *et al.*(1989a,b), using a float apparatus similar to that of Thieberger, also obtained results showing no evidence of a Fifth Force. There is, in fact, no explanation of either Thieberger's or of Boynton's original, presumably incorrect, results. The scientific community has chosen, I believe quite reasonably, to regard the preponderance of negative results as conclusive.[14] In this case too, there were theoretical attempts to reconcile the discordant results. Once the decision as made that the results arguing against the existence of the Fifth Force were correct those theoretical attempts became academic. There was nothing left to explain.

3. CONCLUSION

In this essay I have discussed several aspects of the complex interaction between laws and experiment. We have seen that laws may often be important in determining the feasibility of experiment and in experimental design. I have also argued that this important role of laws does not preclude the testing of those same laws by the experiments they were used to design. We have seen a case, that of the Fifth Force, where the hypothesis used in the experimental design was, in fact, refuted by those experiments. We have also examined several other problems concerning the role of experimental evidence in the evaluation of laws. These include the theory-ladenness of observation, the Duhem-Quine problem, the fallibility of experimental results, and the question of discordant experimental results. I have argued that all of these problems can be reasonably solved, so that we may use experimental evidence to evaluate laws. There is an evidential basis for scientific laws.

NOTES

[1] In this paper I will use the terms "laws" and "theory" interchangably. In general, I think of laws as generalizations of empirical data, (I expect that other contributors to this volume will discuss the nature of laws) such as Kepler's Laws of Planetary Motion: 1) the planets move in ellipses with the sun at one focus; 2) the line joining the planet to the sun sweeps out equal areas in equal times; and 3) the ratio of the square of the period to the cube of the semi-major axis of its elliptical orbit is the same for all planets. I tend to think of theories as being somewhat deeper

and having explanatory power. Nevertheless some laws clearly go beyond experimental generalizations. We may justly regard the conjunction of Newton's three Laws of Motion with his Law of Universal Gravitation as a theory. They certainly entail Kepler's laws, and indeed Kepler's laws are regarded as a confirmation of Newtonian mechanics.

[2] For a discussion of these roles see Franklin (1990, Ch. 7).

[3] This is an oversimplification. The detector had a small solid angle, and thus detected only a small fraction of the decays of the stopped K^+ mesons. In addition, the solid angle was a function of the decay particle momentum. The result of this was that one needed a considerably larger number of stopped kaons to observe the K^+_{e2} decay. For the entire experimental run a total of approximately 1.3×10^9 kaons were stopped and a total of 6 K^+_{e2} events observed.

[4] For details see Franklin (1993a).

[5] Hacking (1983, pp. 171-2) and Hanson (1969).

[6] In fairness to Kuhn, it should be noted that he has modified his views somewhat in later work (Kuhn 1983). He argues there for "local holism," that individual terms can be learned only as part of a system of terms, and for "local incommensurability," which implies that vocabularies of two rival successor theories preserve meanings across theory change, and that only "core terms" remain incommensurable. Here, incommensurable means simply untranslatable. Although Kuhn has not specified clearly what he means by core terms, I believe that under any reasonable view, mass must be such a term. Thus, my argument still holds. I am grateful to Friedel Weinert for suggesting this point.

[7] Koertge's discussion is influenced, in part, by *Zen and the Art of Motorcycle Maintenance* (Pirsig 1974). In case of ignition failure test the spark plugs before dismantling the carburetor.

[8] If your carburetor has a history of trouble, check it early on.

[9] See discussions of the fallibility of experimental results, of theoretical calculation, and of the comparison between theory and experiment in Franklin (1990).

[10] The interesting and important question of whether the evidence supported the W-S theory was made more difficult when another experiment on the scattering of polarized electrons from deuterium, the SLAC E122 experiment, confirmed W-S theory (Prescott *et al.* 1978, 1979). This decision was based on the greater evidential weight of the carefully done and checked E122 experiment in comparison to the uncertain atomic parity results. For details of this see Franklin (1990, Ch. 8). For differing views of the episode see Ackermann (1991), Lynch (1991), and Pickering (1991).

[11] By this time the theoretical calculations had also changed.

[12] For details see Franklin (1993a).

[13] The Eöt-Wash group took its nickname from the fact that some of the original evidence for the Fifth Force came from a reanalysis of an experiment done early in the twentieth century by Eötvös and his collaborators.

[14] It is a fact of experimental life that experiments rarely work when they are initially turned on and that experimental results can be wrong, even if there is no apparent error. It is not necessary to know the exact source of an error in order to discount or to distrust a particular experimental result. Its disagreement with numerous other results can, I believe, be sufficient.

POSTSCRIPT

These essays, along with my previous books--*The Neglect of Experiment; Experiment, Right or Wrong;* and *The Rise and Fall of the Fifth Force*--are my efforts over the past twenty years to present science as it is actually done. I believe that these studies show that scientific decisions are based on epistemological and methodological criteria and that it is constructed from valid experimental evidence and from reasoned and critical discussion. In short, it is "good science." It is not the clumsy at best, evil at worst, science portrayed in *The Golem*. Does this mean that scientists are infallible, rationality machines who do not have ordinary human motivations such as desires for fame, career enhancement, or economic gain. Of course not! What these essays show is that "justification," the process by which an experimental result or a hypothesis or theory is accepted as scientific knowledge, is a reasonable, dare one say rational, process.

I believe that these essays also show that the actual history of science is far messier than that given in introductory textbooks or in popularizations. The episodes discussed in this book--gravity waves, the 17-keV neutrino, the Fifth Force, K-meson decay, and atomic-parity violation--each involved discordant experimental results. But, the discord was resolved by the application of methodological and epistemological criteria. (For somewhat less messy cases such as the discovery of parity violation in the weak interactions see my previous work).

The reader may wonder why I have concentrated on such messy episodes, those which involve discord and controversy. The reason is that the methodologies of science show up more clearly in such cases than they do in cases in which the experimental results all agree and in which these results agree with theoretical predictions. In the latter everything seems inevitable. The methodology while present, is in the background.

What are the methods of science? Surprisingly, I agree that science is a social construction. It is, after all, constructed by the community of scientists. But it is constructed on the basis of experimental evidence and reasoned and critical discussion.

REFERENCES

Ackermann, R. 1989. The New Experimentalism. *British Journal for the Philosophy of Science* 40: 185-190.

Ackermann, R. 1991. Allan Franklin, Right or Wrong. *PSA 1990, Volume 2.* A. Fine, M. Forbes and L. Wessels. East Lansing, MI, Philosophy of Science Association: 451-457.

Adelberger, E.G. 1988. Constraints on Composition-dependent Interactions from the Eöt-Wash Experiment. *5th Force Neutrino Physics: Eighth Moriond Workshop.* O. Fackler and J. Tran Thanh Van. Gif sur Yvette: Editions Frontieres: 445-456.

Adelberger, E.G. 1989. High-Sensitivity Hillside Results from the Eöt-Wash Experiment. *Tests of Fundamental Laws in Physics: Ninth Moriond Workshop.* O. Fackler and J. Tran Thanh Van. Les Arcs, France, Editions Frontieres: 485-499.

Adelberger, E.G., C.W. Stubbs, W.F. Rogers, et al. 1987. New Constraints on Composition- Dependent Interactions Weaker than Gravity. *Physical Review Letters* 59: 849-852.

Altherr, T., P. Chardonnet and P. Salati 1991. The 17 keV Neutrino in the Light of Astrophysics and Cosmology. *Physics Letters* 265B: 251-257.

Altzitzoglou, T., F. Calaprice, M. Dewey, et al. 1985. Experimental Search for a Heavy Neutrino in the Beta Spectrum of ^{35}S. *Physical Review Letters* 55: 799-802.

Ander, M., M.A. Zumberge, T. Lautzenhiser, et al. 1989. Test of Newton's Inverse-Square Law in the Greenland Ice Cap. *Physical Review Letters* 62: 985-988.

Apalikov, A.M., S.D. Boris, A.I. Golutvin, et al. 1985. Search for Heavy Neutrinos in β Decay. *JETP Letters* 42: 289-293.

Aronson, S.H., G.J. Bock, H.Y. Cheng, et al. 1982. Determination of the Fundamental Parameters of the $K^°$-$\bar{K}^°$ System in the Energy Range 30-110 GeV. *Physical Review Letters* 48: 1306-1309.

Aronson, S.H., G.J. Bock, H.Y. Cheng, et al. 1983a. Energy Dependence of the Fundamental Parameters of the $K^°$-$\bar{K}^°$ System. I. Experimental Analysis. *Physical Review D* 28: 476-494.

Aronson, S.H., G.J. Bock, H.Y. Cheng, et al. 1983b. Energy Dependence of the Fundamental Parameters of the $K^°$-$\bar{K}^°$ System. II. Theoretical Formalism. *Physical Review D* 28: 494-523.

Aronson, S.H., H.Y. Cheng, E. Fischbach, et al. 1986. Experimental Signals for Hyperphotons. *Physical Review Letters* 56: 1342-1345.

Astone, P., M. Bassan, P. Bonifazi, et al. 1993. Long-term Operation of the Rome 'Explorer' Cryogenic Gravity Wave Detector. *Physical Review D* 47: 362-375.

Auerbach, L.B., J.M. Dobbs, A.K. Mann, et al. 1967. Measurement of the Branching Ratios of $K^+_{\mu 2}$, $K^+_{\pi 2}$, K^+_{e3}, and $K^+_{\mu 3}$. *Physical Review* 155: 1505-1515.

Bahran, M. and G. Kalbfleisch 1991. Search for Heavy Neutrino in Tritium Beta Decay. *Joint International Lepton-Photon Symposium and Europhysics Conference on High Energy Physics*, Geneva, Switzerland, World Scientific: 606-608.

Bahran, M. and G.R. Kalbfleisch 1992. Limit on Heavy Neutrino in Tritium Beta Decay. *Physics Letters* 291B: 336-340.

Baird, P.E.G., M.W.S. Brimicombe, R.G. Hunt, *et al.* 1977. Search for Parity-Nonconserving Optical Rotation in Atomic Bismuth. *Physical Review Letters* 39: 798-801.

Baird, P.E.G., M.W.S. Brimicombe, G.J. Roberts, *et al.* 1976. Search for Parity Non-Conserving Optical Rotation in Atomic Bismuth. *Nature* 264: 528-529.

Barkov, L.M. and M.S. Zolotorev 1978a. Observations of Parity Nonconservation in Atomic Transitions. *JETP Letters* 27: 357-361.

Barkov, L.M. and M.S. Zolotorev 1978b. Measurement of Optical Activity of Bismuth Vapor. *JETP Letters* 28: 503-506.

Barkov, L.M. and M.S. Zolotorev 1979a. Parity Violation in Atomic Bismuth. *Physics Letters* 85B: 308-313.

Barkov, L.M. and M.S. Zolotorev 1979b. Parity Violation in Bismuth: Experiment. *International Workshop on Neutral Current Interactions in Atoms in Cargese*. W. L. Williams. Washington, National Science Foundation: 52-76.

Barnes, B. 1982. *T.S. Kuhn and Social Science*. New York: Macmillan.

Barnes, B. 1991. How Not to Do the Sociology of Knowledge. *Rethinking Objectivity, part1. Special Issue of Annals of Scholarship* 8(3-4): 321-335.

Bars, I. and M. Visser 1986. Feeble Intermediate-Range Forces from Higher Dimensions. *Physical Review Letters* 57: 25-28.

Bartlett, D.F. and W.L. Tew 1989a. The Fifth Force: Terrain and Pseudoterrain. *Tests of Fundamental Laws in Physics: Ninth Moriond Workshop*, Les Arcs, France, Editions Frontieres: 543-545.

Bartlett, D.F. and W.L. Tew 1989b. Possible Effect of the Local Terrain on the Australian Fifth- Force Measurement. *Physical Review D* 40: 673-675.

Bartlett, D.F. and W.L. Tew 1990. Terrain and geology near the WTVD Tower in North Carolina: Implications for Non-Newtonian Gravity. *Journal of Geophysical Research* 95: 363-369.

Becker, H.W., D. Imel, H. Hendrikson, *et al.* 1991. Experimental Studies of the ^{35}S Beta- Spectrum Anomalies and Heavy Neutrino Admixture? Massive Neutrinos, Tests of Fundamental Symmetries: Proceedings of the XXVIth Rencontre de Moriond, Les Arcs, France, Editions Frontieres: 159-164.

Bell, J.S. and J. Perring 1964. 2π Decay of the K^{o}_{2} Meson. *Physical Review Letters* 13: 348-349.

Bellotti, E., E. Fiorini and A. Pullia 1966. An Experimental Investigation of the K_{e3} Decay. *Physics Letters* 20: 690-692.

Bennett, W.R. 1989. Modulated-Source Eötvös Experiment at Little Goose Lock. *Physical Review Letters* 62: 365-368.

Benvenuti, A., D. Cline, F. Messing, et al. 1976. Evidence for Parity Nonconservation in the Weak Neutral Current. *Physical Review Letters* 37: 1039-1042.

Bernstein, J., N. Cabibbo and T.D. Lee 1964. CP Invariance and the 2π Decay of the $K^o{}_2$. *Physics Letters* 12: 146-148.

Beyerchen, A. 1977. *Scientists Under Hitler: Politics and the Physics Community in the Third Reich.* New Haven: Yale University Press.

Birich, G.N., Y.V. Bogdanov, S.I. Kanorskii, et al. 1984. Nonconservation of Parity in Atomic Bismuth. *JETP* 60: 442-449.

Bizzeti, P.G. 1986. Significance of the Eötvös Method for the Investigation of Intermediate Range Forces. *Il Nuovo Cimento* 94B: 80-86.

Bizzeti, P.G., A.M. Bizzeti-Sona, T. Fazzini, 1988. New Search for the 'Fifth Force' with the Floating Body Method: Status of the Vallambrosa Experiment. *Fifth Force Neutrino Physics: Eighth Moriond Workshop*. O. Fackler and J. Tran Tahn Van. Gif sur Yvette, Editions Frontieres: 501-513.

Bizzeti, P.G., A.M. Bizzeti-Sona, T. Fazzini, et al. 1989a. Search for a Composition Dependent Fifth Force: Results of the Vallambrosa Experiment. *Tran Thanh Van, J.* O. Fackler. Gif sur Yvette, Editions Frontieres: 511-524.

Bizzeti, P.G., A.M. Bizzeti-Sona, T. Fazzini, et al. 1989b. Search for a Composition-dependent Fifth Force. *Physical Review Letters* 62: 2901-2904.

Bock, G.J., S.H. Aronson, K. Freudenreich, et al. 1979. Coherent K_S Regeneration by Protons from 30 to 130 GeV/c. *Physical Review Letters* 42: 350-353.

Boehm, F. and P. Vogel 1984. Low-Energy Neutrino Physics and Neutrino Mass. *Annual Reviews of Nuclear and Particle Science* 34: 125-153.

Boehm, F. and P. Vogel 1987. *Physics of Massive Neutrinos.* Cambridge: Cambridge University Press.

Bogdanov, Y.V., I.I. Sobel'man, V.N. Sorokin, et al. 1980a. Investigation of Optical Activity of Bi Vapors. *JETP Letters* 31: 214-219.

Bogdanov, Y.V., I.I. Sobel'man, V.N. Sorokin, et al. 1980b. Parity Nonconservation in Atomic Bismuth. *JETP Letters* 31: 522-526.

Bogen, J., and J. Woodward. 1988. Saving the Phenomena. *The Philosophical Review* 97:303- 352.

Bonvicini, G. 1993. Statistical Issues in the 17-keV Neutrino Experiments. *Zeitschrift fur Physik A* 345: 97-117.

Borge, M.J.G., A. De Rujula, P.G. Hansen, et al. 1986. Limits on Neutrino-Mixing from the Internal Bremsstrahlung Spectrum of ^{125}I. *Physica Scripta* 34: 591-596.

Borreani, G., G. Rinaudo and A.E. Werbrouck 1964. Positron Spectrum in $K^+ \to e^+ + \pi^o + \nu$. *Physics Letters* 12: 123-126.

Bouchiat, C. 1980. Neutral Current Interactions in Atoms. *Proceedings, International Workshop on Neutral Current Interactions in Atoms*. W. L. Williams. Washington, D.C., National Science Foundation: 357-369.

Bouchiat, M.A. and C. Bouchiat 1974. Weak Neutral Currents in Atomic Physics. *Physics Letters* 48B: 111-114.

Bouchiat, M.A., J. Guena, L. Hunter, *et al.* 1982. Observation of Parity Violation in Cesium. *Physics Letters* 117B: 358-364.

Bouchiat, M.A., J. Guena and L. Pottier 1984. New Observation of a Parity Violation in Cesium. *Physics Letters* 134B: 463-468.

Bouchiat, M.A. and L. Pottier 1984. Atomic Parity Violation Experiments. *Atomic Physics 9*. R. Van Dyck and E. Fortson. Singapore, World Scientific: 246-271.

Bowen, D.R., A.K. Mann, W.K. McFarlane, *et al.* 1967. Measurement of the K^+_{e2} Branching Ratio. *Physical Review* 154:1314-1322.

Boynton, P. 1990. New Limits on the Detection of a Composition-dependent Macroscopic Force. *New and Exotic Phenomena '90: Tenth Moriond Workshop*. O. Fackler and J. Tran Thanh Van. Gif sur Yvette, Editions Frontieres: 207-224.

Boynton, P., D. Crosby, P. Ekstrom, *et al.* 1987. Search for an Intermediate-Range Composition- dependent Force. *Physical Review Letters* 59: 1385-1389.

Boynton, P. and P. Peters 1989. Torsion Pendulums, Fluid Flows and the Coriolis Force. *Tests of Fundamental Laws in Physics: Ninth Moriond Workshop*. O. Fackler and J. Tran Tahnh Van. Gif sur Yvette, Editions Frontieres: 501-510.

Braginskii, V.B. and V.I. Panov 1972. Verification of the Equivalence of Inertial and Gravitational Mass. *Zhurnal Experimental'noi i Teoreticheskoi Fiziki* (JETP) 34: 463-466.

Brans, C. and R.H. Dicke 1961. Mach's Principle and a Relativistic Theory of Gravitation. *Physical Review* 124: 925-935.

Brimicombe, M.W.S., C.E. Loving and P.G.H. Sandars 1976. Calculation of Parity Nonconserving Optical Rotation in Atomic Bismuth. *Journal of Physics B*: L237-L240.

Brown, J.L., J.A. Kadyk, G.H. Trilling, *et al.* 1961. Experimental Study of the K^+_{e3} Decay Interaction.*Physical Review Letters* 7: 423-426.

Bucksbaum, P., E. Commins and L. Hunter 1981a. New Observation of Parity Nonconservation in Atomic Thallium. *Physical Review Letters* 46: 640-643.

Bucksbaum, P., E. Commins and L. Hunter 1981b. Observation of Parity Nonconservation in Atomic Thallium. *Physcial Review* 24D: 1134-1138.

Callahan, A.C., U. Camerini, R.D. Hantman, *et al.* 1966. Measurement of the $K^+_{\mu3}$ Decay Parameters. *Physical Review* 150: 1153-1164.

Cartwright, N. 1983. *How the Laws of Physics Lie*. Oxford: Oxford University Press

Cester, R., P.T. Eschstruth, O'Neill, G.K., *et al.* 1966. Positron Momentum Spectrum and Branching Ratio of K^+_{e3}. *Physics Letters* 21: 343-347.

Christenson, J.H., J.W. Cronin, V.L. Fitch, *et al.* 1964. Evidence for the 2π Decay of the K^o_2 Meson. *Physical Review Letters* 13: 138-140.

Chu, S.Y. and R.H. Dicke 1986. New Force or Thermal Gradient in the Eötvös Experiment. *Physical Review Letters* 57: 1823-1824.

Churchland, P. and C. Hooker, Eds. 1985. *Images of Science*. Chicago: University of Chicago Press.

Chwolson, O.D. 1927. *Die Physik*. Braunschweig: F. Vieweg and Sohn.

Close, F.E. 1976. Parity Violation in Atoms? *Nature* 264: 505-506.

Colella, R., A.W. Overhauser and S.A. Werner 1975. Observations of Gravitationally Induced Quantum Interference. *Physical Review Letters* 34: 1472-1474.

Collins, H. 1985. *Changing Order: Replication and Induction in Scientific Practice.* London: Sage Publications.

Collins, H. 1994. A Strong Confirmation of the Experimenters' Regress. *Studies in History and Philosophy of Modern Physics* 25(3): 493-503.

Collins, H. and T. Pinch 1993. *The Golem: What Everyone Should Know About Science.* Cambridge: Cambridge University Press.

Commins, E.D. and P. Kusch 1958. Upper Limit to the Magnetic Moment of He^6. *Physical Review Letters* 1: 208-209.

Conti, R., P. Bucksbaum, S. Chu, *et al.* 1979. Preliminary Observation of Parity Nonconservation in Atomic Thallium. *Physical Review Letters* 42: 343-346.

Conway, D. and W. Johnston 1959. Determination of the Low-Energy Region of the Tritium Beta Spectrum. *Physical Review* 116: 1544-1547.

Coupal, D.P., R.H. Bernstein, G.J. Bock, *et al.* 1985. Measurement of the Ratio $\Gamma(K_L \to \pi^+\pi^-)/\Gamma(K_L \to \pi l \nu)$ for K_L with 65 GeV/c Laboratory Momentum. *Physical Review Letters* 55: 566-569.

Cowsik, R., N. Krishnan, S.N. Tandor, *et al.* 1988. Limit on the Strength of Intermediate-Range Forces Coupling to Isospin. *Physical Review Letters* 61: 2179-2181.

Cowsik, R., N. Krishnan, S.N. Tandor, *et al.* 1990. Strength of Intermediate-Range Forces Coupling to Isospin. *Physical Review Letters* 64: 336-339.

Damour, T. and J.H. Taylor 1991. On the Orbital Period Change of the Binary Pulsar PSR 1913 + 16. *The Astrophysical Journal* 366: 501-511.

Datar, V.M., C. Baba, S.K. Bhattacherjee, *et al.* 1985. Search for a heavy neutrino in the β-decay of ^{35}S. *Nature* 318: 547-548.

Dawkins, R. 1995. *River out of Eden.* London: Weidenfeld and Nicolson.

De Bouard, X., D. Dekkers, B. Jordan, *et al.* 1965. Two Pion Decay of the K^0_2 at 10 GeV/c. *Physics Letters* 15: 58-61.

De Rujula, A. 1981. A New Way to Measure Neutrino Masses. *Nuclear Physics* B188: 414-458.

De Rujula, A. 1986a. Are There More Than Four? *Nature* 323: 760-761.

De Rujula, A. 1986b. On Weaker Forces than Gravity. *Physics Letters* 180B: 213-220.

Douglass, D.H., R.Q. Gram, J.A. Tyson, *et al.* 1975. Two-Detector-Coincidence Search for Bursts of Gravitational Radiation. *Physical Review Letters* 35: 480-483.

Dehmelt, H. 1990. Experiments on the Structure of an Individual Elementary Particle. *Science* 247: 539-545.

Drukarev, E.G. and M.I. Strikman 1986. Final-state Interaction of β Electrons and Related Phenomena. *JETP* 64: 686-692.

Dydak, F. 1979. Neutral Currents. *Proceeding of the Conference on High Energy Physics.* Geneva, CERN: 25-49.

Eckhardt, D.H., C. Jekeli, A.R. Lazarewicz, *et al.* 1988a. Results of a Tower Gravity Experiment. *Fifth Force Neutrino Physics: Eighth Moriond Workshop.* O. Fackler and J. Tran Thanh Van. Gif sur Yvette, Editions Frontieres: 577-583.

Eckhardt, D.H., C. Jekeli, A.R. Lazarewicz, *et al.* 1988b. Tower Gravity Experiment: Evidence for Non-Newtonian Gravity. *Physical Review Letters* 60: 2567-2570.

Eckhardt, D.H., C. Jekeli, A.J. Romaides, *et al.* 1990. The North Carolina Tower Gravity Experiment: A Null Result. *New and Exotic Phenomena '90: Tenth Moriond Workshop.* O. Fackler and J. Tran Thanh Van. Gif sur Yvette, Editions Frontieres: 237-244.

Ehrenhaft, F. (1914). "Die Quanten der Elektrizitat." *Annalen der Physik* 44: 657-700.

Elizalde, E. 1986. About the Eötvös Experiment and the Hypercharge Theory. *Physics Letters* 116A: 162-166.

Elliot, J., E. Dunham and R. Millis 1977. Discovering the Rings of Uranus. *Sky and Telescope* 53: 412-416.

Eman, B. and D. Tadic 1986. Distortion in the β-decay Spectrum for Low Electron Kinetic Energies. *Physical Review C* 33: 2128-2131.

Emmons, T.P., J.M. Reeve and E.N. Fortson 1983. Parity-Nonconserving Optical Rotation in Atomic Lead. *Physical Review Letters* 53: 2089-2092.

Eötvös, R., D. Pekar and E. Fekete 1922. Beitrage zum Gesetze der Proportionalitat von Tragheit und Gravitat. *Annalen der Physik (Leipzig)* 68: 11-66.

Eschstruth, P.T., A.D. Franklin, E.B. Hughes, *et al.* 1968. Positron Momentum Spectrum and Branching Ratio of K^+_{e3} Decay. *Physical Review* 165: 1487-1490.

Fairbank, W.M. 1988. Summary Talk on Fifth Force Papers. *5th Force Neutrino Physics: Eighth Moriond Workshop*, Les Arcs, France, Editions Frontieres: 629-644.

Feyerabend, P. 1975. *Against Method.* London: Humanities Press.

Fischbach, E. 1980. Tests of General Relativity at the Quantum Level. *Cosmology and Gravitation.* P. Bergmann and V. De Sabbata. New York, Plenum: 359-373.

Fischbach, E. 1988. The Fifth Force: An Introduction to Current Research. *Fifth Force Neutrino Physics: Eighth Moriond Workshop.* O. Fackler and J. Tran Thanh Van. Gif sur Yvette, Editions Frontieres: 369-382.

Fischbach, E., S. Aronson, C. Talmadge, *et al.* 1986. Reanalysis of the Eötvös Experiment. *Physical Review Letters* 56: 3-6.

Fischbach, E., H.Y. Cheng, S.H. Aronson, *et al.* 1982. Interaction of the K^0-$\bar{K}^0$ System with External Fields. *Physics Letters* 116B: 73-76.

Fischbach, E. and B. Freeman 1979. Testing General Relativity at the Quantum Level. *General Relativity and Gravitation* 11: 377-381.

Fischbach, E., M.P. Haugan, D. Tadic, *et al.* 1985. Lorentz Invariance and the Eötvös Experiments. *Physical Review D* 32: 154-162.

Fitch, V.L., M.V. Isaila and M.A. Palmer 1988. Limits on the Existence of a Material-dependent Intermediate-Range Force. *Physical Review Letters* 60: 1801-1804.

Fortson, E.N. and L.L. Lewis 1984. Atomic Parity Nonconservation. *Physics Reports* 113: 289-344.

Franklin, A. 1986. *The Neglect of Experiment*. Cambridge: Cambridge University Press.

Franklin, A. 1990. *Experiment, Right or Wrong*. Cambridge: Cambridge University Press.

Franklin, A. 1993a. *The Rise and Fall of the Fifth Force: Discovery, Pursuit, and Justification in Modern Physics*. New York: American Institute of Physics.

Franklin, A. 1993b. Discovery, Pursuit, and Justification. *Perspectives on Science* 1: 252-284.

Franklin, A. 1993c. Experimental Questions. *Perspectives on Science* 1: 127-146.

Franklin, A. 1994. How to Avoid the Experimenters' Regress. *Studies in the History and Philosophy of Science* 25: 97-121.

Franklin, A. 1995a. The Appearance and Disappearance of the 17-keV Neutrino. *Reviews of Modern Physics* 67: 457-490.

Franklin, A. 1995b. The Resolution of Discordant Results. *Perspectives on Science* 3: 346-420.

Franklin, A., M. Anderson, D. Brock, et al. 1989. Can a Theory-Laden Observation Test the Theory? *British Journal for the Philosophy of Science* 40: 229-231.

Franklin, A. and C. Howson 1984. Why Do Scientists Prefer to Vary Their Experiments? *Studies in History and Philosophy of Science* 15: 51-62.

Franklin, A. and C. Howson 1988. It Probably is a Valid Experimental Result: A Bayesian Approach to the Epistemology of Experiment. *Studies in the History and Philosophy of Science* 19: 419-427.

Fujii, Y. 1971. Dilatonal Possible Non-Newtonian Gravity. *Nature* 234: 5-7.

Fujii, Y. 1972. Scale Invariance and Gravity of Hadrons. *Annals of Physics (N.Y.)* 69: 494-521.

Fujii, Y. 1974. Scalar-Tensor Theory of Gravitation and Spontaneous Breakdown of Scale Invariance. *Physical Review D* 9: 874-876.

Galbraith, W., G. Manning, A.E. Taylor, et al. 1965. Two-pion Decay of the K^0_2 Meson. *Physical Review Letters* 14: 383-386.

Galileo 1954. *Dialogues Concerning Two New Sciences*. H.Crew and A. DeSalvio (trans.).New York: Dover Publishing.

Galison, P. 1987. *How Experiments End*. Chicago: University of Chicago Press.

Galison, P. 1997. *Image and Logic*. Chicago: University of Chicago Press.

Garwin, R.L. 1974. Detection of Gravity Waves Challenged. *Physics Today* 27(12): 9-11.

Geison, G. 1995. *The Private Science of Louis Pasteur*. Princeton, N.J.: Princeton University Press.

Gibbons, G.W. and B.F. Whiting 1981. Newtonian Gravity Measurements Impose Constraints on Unification Theories. *Nature* 291: 636-638.

Gilbert, S.L., M.C. Noecker, R.N. Watts, et al. 1985. Measurement of Parity Nonconservation in Atomic Cesium. *Physical Review Letters* 55: 2680-2683.

Gilbert, S.L. and C.E. Wieman 1986. Atomic Beam Measurement of Parity Nonconservation in Cesium. *Physical Review* 34A: 792-803.

Glashow, S.L. 1991. A Novel Neutrino Mass Hierarchy. *Physics Letters* 256B: 255-257.

Gross, P.R. and N. Levitt 1994. *Higher Superstition: The Acadenuc Left and Its Quarrels with Science*. Baltimore: The Johns Hopkins University Press.

Grossman, N., K. Heller, C. James, et al. 1987. Measurement of the Lifetime of K_S^0 Mesons in the Momentum Range 100 - 350 GeV/c. *Physical Review Letters* 59: 18-21.

Gullstrand, A. 1965. Presentation Speech. *Nobel Lectures in Physics 1922-1941*. Amsterdam: Elsevier.

Hanson, N.R. 1969. *Patterns of Discovery*. Cambridge: Cambridge University Press.

Hacking, I. 1983. *Representing and Intervening*. Cambridge: Cambridge University Press.

Harding, S. 1986. *The Science Question in Feminism*. Ithaca: Cornell University Press.

Harding, S., Ed. 1976. *Can Theories Be Refuted*. Dordrecht: Reidel.

Harding, S. 1996. Thinking Science, Thinking Society. *Social Text* 46-47: 15-26.

Haxton, W.C. 1985. Atomic Effects and Heavy Neutrino Emission in Beta Decay. *Physical Review Letters* 55: 807-809.

Heckel, B.R., E.G. Adelberger, C.W. Stubbs, et al. 1989. Experimental Bounds on Interactions Mediated by Ultralow-Mass Bosons. *Physical Review Letters* 63: 2705-2708.

Henley, E.M. and L. Wilets 1976. Parity Nonconservation in Tl and Bi Atoms. *Physical Review A* 14: 1411-1417.

Hetherington, D.W., R.L. Graham, M.A. Lone, et al. 1986. Search for Evidence of a 17-keV Neutrino in the Beta Spectrum of ^{63}Ni. *Nuclear Beta Decays and Neutrino: Proceedings of the International Symposium, Osaka, Japan, June 1986*. T. Kotani, H. Ejiri and E. Takasugi. Singapore, World Scientific: 387-390.

Hetherington, D.W., R.L. Graham, M.A. Lone, et al. 1987. Upper Limits on the Mixing of Heavy Neutrinos in the Beta Decay of ^{63}Ni. *Physical Review C* 36: 1504-1513.

Hime, A. 1992. Pursuing the 17 keV Neutrino. *Modern Physics Letters A* 7: 1301-1314.

Hime, A. 1993. Do Scattering Effects Resolve the 17-keV Conundrum? *Physics Letters* 299B: 165- 173.

Hime, A. and N.A. Jelley 1991. New Evidence for the 17 keV Neutrino. *Physics Letters* 257B: 441-449.

Hime, A. and J.J. Simpson 1989. Evidence of the 17-keV Neutrino in the β Spectrum of ^{3}H. *Physical Review D* 39: 1837-1850.

Holding, S.C., F.D. Stacey and G.J. Tuck 1986. Gravity in Mines--an Investigation of Newton's Law. *Physical Review D* 33: 3487-3494.

Holding, S.C. and G.J. Tuck 1984. A New Mine Determination of the Newtonian Gravitational Constant. *Nature* 307: 714-716.

Hollister, J.H., G.R. Apperson, L.L. Lewis, et al. 1981. Measurement of Parity Nonconservation in Atomic Bismuth. *Physical Review Letters* 46: 643-646.

Holton, G. 1978. Subelectrons, Presuppositions, and the Millikan-Ehrenhaft Debate. *Historical Studies in the Physical Sciences* 9: 166-224.

Hulse, R.A. and J.H. Taylor 1975. A Deep Sample of New Pulsars and Their Spatial Extent in the Galaxy. *The Astrophysical Journal* 201: L55-L59.

Imlay, R.L., P.T. Eschstruth, A.D. Franklin, *et al.* 1967. Energy Dependence of the Form Factor in K^+_{e3} Decay. *Physical Review* 160: 1203-1211.

Jekeli, C., D.H. Eckhardt and A.J. Romaides 1990. Tower Gravity Experiment: No Evidence for Non-Newtonian Gravity. *Physical Review Letters* 64: 1204-1206.

Jensen, G.L., F.S. Shaklee, B.P. Roe, *et al.* 1964. Study of the Three-Body Leptonic Decay Modes of the K^+ Meson. *Physical Review* 136B: 1431-1438.

Joravsky, D. 1970. *The Lysenko Affair*. Cambridge, MA: Harvard University Press.

Kalbfleisch, G.R. and K.A. Milton 1985. Heavy-Neutrino Emission. *Physical Review Letters* 55: 2225.

Kammeraad, J., P. Kasameyer, O. Fackler, *et al.* 1990. New Results from Nevada: A Test of Newton's Law Using the BREN Tower and a High Density Gravity Survey. *New and Exotic Phenomena '90: Tenth Moriond Workshop*, Les Arcs, France, Editions Frontieres, 245-254.

Kasameyer, P., J. Thomas, O. Fackler, *et al.* 1989. A Test of Newton's Law of Gravity Using the BREN Tower, Nevada. *Tests of Fundamental Laws in Physics: Ninth Moriond Workshop*, Les Arcs, France, Editions Frontieres, 529-542.

Kawakami, H., S. Kato, T. Ohshima, *et al.* 1992. High Sensitivity Search for a 17 keV Neutrino. Negative Indication with an Upper Limit of 0.095%. *Physics Letters* 287B: 45-50.

Keyser, P.T., T. Niebauer and J.E. Faller 1986. Comment on "Renalysis of the Eötvös Experiment". *Physical Review Letters* 56: 2425.

Kim, Y.E. 1986. The Local Baryon Gauge Invariance and the Eötvös Experiment. *Physics Letters* 177B: 255-259.

Koertge, N. 1978. Towards a New Theory of Scientific Inquiry. *Progress and Rationality in Science*. G. Radnitzky and G. Andersson. Dordrecht, Reidel.

Koertge, N., Ed. 1997. *A House Built on Sand: Flaws in Postmodernist Accounts of Science*. Oxford: Oxford University Press, forthcoming.

Konstantinowsky, D. 1915. Elektrische Ladungen und Brownsche Bewegung sehr kleiner Metallteilchen im Gase. *Annalen der Physik* 46: 261-297.

Koonin, S. 1991. Environmental Fine Structure in Low-Energy β-particle Spectra. *Nature* 354: 468- 470.

Kuhn, T.S. (1962) 1970. *The Structure of Scientific Revolutions*. Chicago: The University of Chicago Press.

Kuhn, T.S. 1983. Commensurability, Comparability, and Communicability. *PSA 1982*. P. D. Asquith and T. Nickles. East Lansing, 669-688.

Kuroda, K. and N. Mio 1989a. Galilean Test for Composition-dependent Force. *Proceedings of the Fifth Marcel Grossman Conference on General Relativity*. D. G. Blair and M. J. Buckingham. Singapore, World Scientific: 1569-1572.

Kuroda, K. and N. Mio 1989b. Test of a Composition-Dependent Force by a Free-Fall Interferometer. *Physical Review Letters* 62: 1941-1944.

Kuroda, K. and N. Mio 1990. Limits on a Possible Composition-dependent Force by a Galilean Experiment. *Physical Review D* 42: 3903-3907.

Laudan, L. 1980. Why Was the Logic of Discovery Abandoned? *Scientific Discovery, Logic and Rationality*. T. Nickles. Boston, Reidel: 173-183.

Leplin, J., Ed. 1984. *Scientific Realism*. Berkeley: University of California Press.

Levine, G. 1996a. Letter. *New York Review of Books*. 43: 54.

Levine, G. 1996b. What Is Science Studies for and Who Cares? *Social Text* 46-47: 113-127.

Levine, J.L. and R.L. Garwin 1973. Single Gravity-Wave Detector Results Contrasted with Previous Coincidence Detections. *Physical Review Letters* 31: 176-180.

Levine, J.L. and R.L. Garwin 1974. New Negative Result for Gravitational Wave Detection, and Comparison with Reported Detection. *Physical Review Letters* 33: 794-797.

Lewis, L.L., J.H. Hollister, D.C. Soreide, *et al.* 1977. Upper Limit on Parity-Nonconserving Optical Rotation in Atomic Bismuth. *Physical Review Letters* 39: 795-798.

Lindhard, J. and P.G. Hansen 1986. Atomic Effects in Low-Energy Beta Decay: The Case of Tritium. *Physical Review Letters* 57: 965-967.

Long, D.R. 1974. Why Do We Believe Newtonian Gravitation at Laboratory Dimensions? *Physical Review D* 9: 50-52.

Long, D.R. 1976. Experimental Examination of the Gravitational Inverse Square Law. *Nature* 260: 417-418.

Long, D.R. 1981. Current Measurements of the Gravitational "Constant" as a Function of Mass Separation. *Il Nuovo Cimento* 62B: 130-138.

Lubimov, V.A., E.G. Novikov, V.Z. Nozik, *et al.* 1980. An Estimate of the v_e Mass from the β- Spectrum of Tritium in the Valine Molecule. *Physics Letters* 94B: 266-268.

Lusignoli, M. and A. Pugliese 1986. Hyperphotons and K-Meson Decays. *Physics Letters* 171B: 468-470.

Lynch, M. 1991. Allan Franklin's Transcendental Physics. *PSA 1990, Volume 2*. A. Fine, M. Forbes and L. Wessels. East Lansing: MI, Philosophy of Science Association: 471-485.

MacKenzie, D. 1989. From Kwajelein to Armageddon? Testing and the Social Construction of Missile Accuracy. *The Uses of Experiment*. D. Gooding, T. Pinch and S. Shaffer. Cambridge, Cambridge University Press: 409-435.

Maddox, J. 1986. Newtonian Gravity Corrected. *Nature* 319: 173.

Madsen, J. 1992. Bose Condensates, Big Bang Nucleosynthesis, and Cosmological Decay of a 17 keV Neutrino. *Physical Review Letters* 69: 571-574.

Markey, H. and F. Boehm 1985. Search for Admixture of Heavy Neutrinos with Masses between 5 and 55 keV. *Physical Review C* 32: 2215-2216.

Martensson, A.M., E.N. Henley and L. Wilets 1981. Calculation of Parity-Nonconserving Optical Rotation in Atomic Bismuth. *Physical Review* 24A: 308-317.

McKellar, B.H.J. 1980. The Influence of Mixing of Finite Mass Neutrinos on Beta Decay Spectra. *Physics Letters* 97B: 93-94.

Mikkelsen, D.R. and M.J. Newman 1977. Constraints on the Gravitational Constant at Large Distances. *Physical Review D* 16: 919-926.

Milgrom, M. 1986. On the Use of Eötvös-Type Experiments to Dectect Medium-Range Forces. *Nuclear Physics* 227B: 509-512.

Miller, D.J. 1977. Elementary Particles - A Rich Harvest. *Nature* 269: 286-288.

Millikan, R.A. 1911. The Isolation of an Ion, A Precision Measurement of Its Charge, and the Correction of Stokes's Law. *Physical Review* 32: 349-397.

Millikan, R.A. 1913. On the Elementary Electrical Charge and the Avogadro Constant. *Physical Review* 2: 109-143.

Millikan, R.A. 1916. The Existence of a Subelectron? *Physical Review* 8: 595-625.

Moore, G.I., W. Zurn, K. Lindner, *et al.* 1988. Determination of the Gravitational Constant at an Effective Mass Separation of 22 m. *Physical Review D* 38: 1023-1029.

Morrison, D. 1992a. Review of 17 keV Neutrino Experiments. *Joint International Lepton-Photon Symposium and Europhysics Conference on High Energy Physics*. S. Hegarty, K. Potter and E. Quercigh. Geneva, Switzerland, World Scientific. **1**: 599-605.

Morrison, D. 1992b. Updated Review of 17 keV Neutrino Experiments. *Progress in Atomic Physics, Neutrinos and Gravitation: Proceeding of the XXVIIth Rencontre de Moriond*. G. Chardin, O. Fackler and J. Trab Thanh Van. Les Arcs, France, Editions Frontieres: 207-215.

Morrison, D.R.O. 1993. The Rise and Fall of the 17-keV Neutrino. *Nature* 366: 29-32.

Morrison, M. 1990. Theory, Intervention, and Realism. *Synthese* 82: 1-22.

Mortara, J.L., I. Ahmad, K.P. Coulter, *et al.* 1993. Evidence Against a 17 keV Neutrino from ^{35}S Beta Decay. *Physical Review Letters* 70: 394-397.

Muller, G., F.D. Stacey, G.J. Tuck, *et al.* 1989. Determination of the Gravitational Constant by an Experiment at a Pumped-Storage Reservoir. *Physical Review Letters* 63: 2621-2624.

Nebeker, F. 1994. Experimental Style in High-Energy Physics. *Historical Studies in the Physical and Biological Sciences* 24: 137-164.

Nelson, A. 1994. How Could Scientific Facts be Socially Constructed? *Studies in History and Philosophy of Science* 25(4): 535-547.

Nelson, P.G., D.M. Graham and R.D. Newman 1990. Search for an Intermediate-Range Composition-dependent Force Coupling to N-Z. *Physical Review D* 42: 963-976.

Neufeld, D.A. 1986. Upper Limit on Any Intermediate-Range Force Associated with Baryon Number. *Physical Review Letters* 56: 2344-2346.

Newman, R., D. Graham and P. Nelson 1989. A "Fifth Force" Search for Differential Accleration of Lead and Copper toward Lead. *Tests of Fundamental Laws in Physics: Ninth Moriond Workshop*. O. Fackler and J. Tran Thanh Van. Gif sur Yvette, Editions Frontieres: 459-472.

Nickles, T. 1980. Introductory Essay: Scientific Discovery and the Future of Philosophy of Science. *Scientific Discovery, Logic and Rationality*. T. Nickles. Boston, Reidel: 1-59.

Niebauer, T.M., M.P. McHugh and J.E. Faller 1987. Galilean Test for the Fifth Force. *Physical Review Letters* 59: 609-612.

Norman, E.B., 1994. private communication.

Norman, E.B., Y. Chan, M.T.F. Da Cruz, *et al.* 1992. A Massive Neutrino in Nuclear Beta Decay? *XXVI International Conference on High Energy Physics*, Dallas, American Institute of Physics: 1123-1127.

Norman, E.B., B. Sur, K.T. Lesko, *et al.* 1991. Evidence for the Emission of a Massive Neutrino in Nuclear Beta Decay. *Journal of Physics G* 17: S291-S299.

Novikov, V.N., V.P. Sushkov and I.B. Khriplovich 1976. Optical Activity of Heavy Metal Vapors - A Manifestation of the Weak Interaction of Electrons and Nucleons. *JETP* 44: 872-880.

O'Hanlon, J. 1972. Intermediate-Range Gravity: A Generally Covariant Model. *Physical Review Letters* 29: 137-138.

Ohi, T., M. Nakajima, H. Tamura, *et al.* 1985. Search for Heavy Neutrinos in the Beta Decay of ^{35}S. Evidence Against the 17 keV Heavy Neutrino. *Physics Letters* 160B: 322-324.

Ohshima, T. 1993. 0.073% (95% CL) Upper Limit on 17 keV Neutrino Admixture. *XXVI International Conference on High Energy Physics*. J. R. Sanford. Dallas, American Institute of Physics. **1:** 1128-1135.

Ohshima, T., H. Sakamoto, T. Sato, *et al.* 1993. No 17 keV Neutrino: Admixture < 0.073% (95% C.L.). *Physical Review D* 47: 4840-4856.

Paik, H.J. 1987. Terrestrial Experiments to Test Theories of Gravitation. *General Relativity and Gravitation*. M. A. H. MacCallum. New York, Cambridge University Press: 388-396.

Panov, V.I. and V.N. Frontov 1979. The Cavendish Experiment at Large Distances. *JETP* 50: 852-856.

Parker, R.L. and M.A. Zumberge 1989. An Analysis of Geophysical Experiments to Test Newton's Law of Gravity. *Nature* 342: 29-32.

Perutz, M. (1995). The Pioneer Defended. *New York Review of Books*. XLII: 54-58.

Pickering, A. 1981. The Hunting of the Quark. *Isis* 72: 216-236.

Pickering, A. 1984a. *Constructing Quarks*. Chicago: University of Chicago Press.

Pickering, A. 1984b. Against Putting the Phenomena First: The Discovery of the Weak Neutral Current. *Studies in the History and Philosophy of Science* 15: 85-117.

Pickering, A. 1987. Against Correspondence: A Constructivist View of Experiment and the Real. *PSA 1986*. A. Fine and P. Machamer. Pittsburgh, Philsophy of Science Association. 2: 196-206.

Pickering, A. 1991. Reason Enough? More on Parity Violation Experiments and Electroweak Gauge Theory. *PSA 1990, Volume 2*. A. Fine, M. Forbes and L. Wessels. East Lansing, MI, Philosophy of Science Association: 459-469.

Pickering, A. 1995. *The Mangle of Practice*. Chicago: University of Chicago Press.

Piilonen, L. and A. Abashian 1992. On the Strength of the Evidence for the 17 keV Neutrino. *Progress in Atomic Physics, Neutrinos and Gravitation: Proceedings of the XXVIIth Rencontre de Moriond*, Les Arcs, France, Editions Frontieres: 225-234.
Pinch, T. 1986. *Confronting Nature*. Dordrecht: Reidel.
Pirsig, R.M. 1974. *Zen and the Art of Motorcycle Maintenance*. New York: William Morrow.
Popper, K. 1959. *The Logic of Scientific Discovery*. New York: Basic Books.
Prescott, C.Y., W.B. Atwood, R.L.A. Cottrell, et al. 1978. Parity Non-Conservation in Inelastic Electron Scattering. *Physics Letters* 77B: 347-352.
Prescott, C.Y., W.B. Atwood, R.L.A. Cottrell, et al. 1979. Further Measurements of Parity Non- Conservation in Inelastic Electron Scattering. *Physics Letters* 84B: 524-528.
Quine, W. 1953. *From a Logical Point of View*. Cambridge: Harvard University Press.
Raab, F.J. 1987. Search for an Intermediate-Range Interaction: Results of the Eöt-Wash I Experiment. *New and Exotic Phenomena: Seventh Moriond Workshop*. O. Fackler and J. Tran Thanh Van. Les Arcs, France, Editions Frontieres: 567-577.
Radcliffe, T., M. Chen, D. Imel, et al. 1992. New Limits on the 17 keV Neutrino. *Progress in Atomic Physics, Neutrinos and Gravitation: Proceedings of the XXVIIth Rencontre de Moriond*, Les Arcs, France, Editions Frontieres: 217-224.
Randall, H.M., R.G. Fowler, N. Fuson, et al. 1949. *Infrared Determination of Organic Structures*. New York: Van Nostrand.
Rapp, R.H. 1974. Current Estimate of Mean Earth Ellipsoid Parameters. *Geophysics Research Letters* 1: 35-38.
Rapp, R.H. 1977. Determination of Potential Coefficients to Degree 52 by 5° Mean Gravity Anomalies. *Bulletin Geodesique* 51: 301-323.
Rasmussen, N. 1993. Facts, Artifacts, and Mesosomes: Practicing Epistemology with the Electron Microscope. *Studies in History and Philosophy of Science* 24: 227-265.
Reichenbach, H. 1938. *Experience and Predeiction*. Chicago: University of Chicago Press.
Riisager, K. 1986. Limits for the Electron Neutrino Mass from Internal Bremsstrahlung. *'86 Massive Neutrinos in Astrophysics and in Particle Physics:Proceeding of the Sixth Moriond Workshop*. O. Fackler and J. Tran Thanh Van. Gif sur Yvette, Editions Frontieres: 557-563.
Roehrig, J., A. Gsponer, W.R. Molzon, et al. 1977. Coherent Regeneration of K_s's by Carbon as a Test of Regge-Pole-Exchange Theory. *Physical Review Letters* 38: 1116-1119.
Roll, P.G., R. Krotkov and R.H. Dicke 1964. The Equivalence of Inertial and Passive Gravitational Mass. *Annals of Physics (N.Y.)* 26: 442-517.
Ross, A. 1991. *Strange Weather: Culture, Science, and Technology in the Age of Limits*. London: Verso Press.
Sandars, P.G.H. 1980. Many Body Aspects of Parity Nonconservation in Heavy Atoms. *Physica Scripta* 21: 284-292.
Scherk, J. 1979. Antigravity: A Crazy Idea. *Physics Letters* 88B: 265-267.

Schreckenbach, K., G. Colvin and F. von Feilitzsch 1983. Search for Mixing of Heavy Neutrinos in the β^+ and β^- Spectra of the ^{64}Cu Decay. *Physics Letters* 129B: 265-268.

Schwarzschild, B. 1986. Reanalysis of Old Eotvos Data Suggests 5th Force...to Some. *Physics Today* 39(10): 17-20.

Schwarzschild, B. 1991. Four of Five New Experiments Claim Evidence for 17-keV Neutrinos. *Physics Today* 44(5): 17-19.

Schwarzschild, B. 1993. In Old and New Experiments, the 17-keV Neutrino Goes Away. *Physics Today* 46(4): 17-18.

Sellars, W. 1962. *Science, Perception, and Reality*. New York: Humanities Press.

Shapere, D. 1982. The Concept of Observation in Science and Philosophy. *Philosophy of Science* 49: 482-525.

Shaviv, G. and J. Rosen, Eds. 1975. *General Relativity and Gravitation: Proceedings of the Seventh International Conference (GR7), Tel-Aviv University, June 23-28, 1974*. New York: John Wiley.

Shrock, R.E. 1980. New Tests For and Bounds on Neutrino Masses and Lepton Mixing. *Physics Letters* 96B: 159-164.

Simpson, J.J. 1981a. Measurement of the β-energy Spectrum of ^{3}H to Determine the Antineutrino Mass. *Physical Review D* 23: 649-662.

Simpson, J.J. 1981b. Limits on the Emission of Heavy Neutrinos in ^{3}H Decay. *Physical Review D* 24: 2971-2972.

Simpson, J.J. 1985. Evidence of Heavy-Neutrino Emission in Beta Decay. *Physical Review Letters* 54: 1891-1893.

Simpson, J.J. 1986a. Is There Evidence for a 17 keV neutrino in the ^{35}S β Spectrum? The Case of Ohi et al,. *Physics Letters* 174B: 113-114.

Simpson, J.J. 1986b. Evidence for a 17-keV Neutrino in ^{3}H and ^{35}S β Spectra. *'86 Massive Neutrinos in Astrophysics and in Particle Physics: Proceedings of the Sixth Moriond Workshop*. O. Fackler and J. Tran Thanh Van. Gif sur Yvette, Editions Frontieres: 565-577.

Simpson, J. 1991. The 17-keV Neutrino. *Joint International Lepton-Photon Symposium and Europhysics Conference on High Energy Physics*, Geneva, Switzerland, World Scientific: 596-598.

Simpson, J.J. 1993 private communication.

Simpson, J.J. and A. Hime 1989. Evidence of the 17-keV Neutrino in the β Spectrum of ^{35}S. *Physical Review D* 39: 1825-1836.

Snow, C.P. 1959. *The Two Cultures and the Scientific Revolution*. New York: Cambridge University Press.

Sokal, A. 1996a. Transgressing the Boundaries: Toward a Transformative Hermeneutics of Quantum Gravity. *Social Text* 46-47: 217-252.

Sokal, A. 1996b. A Physicist Experiments With Cultural Studies. *Lingua Franca* 6(4): 62-64.

Speake, C.C. and T.J. Quinn 1988. Search for a Short-Range Isospin-coupling of the Fifth Force with Use of a Beam Balance. *Physical Review Letters* 61: 1340-1343.

Speake, C.C., T.M. Niebauer, M.P. McHugh, et al. 1990. Test of the Inverse-Square Law of Gravitation Using the 300-m Tower at Erie Colorado. *Physical Review Letters* 65: 1967- 1971

Spero, R., J.K. Hoskins, R. Newman, et al. 1980. Tests of the Gravitational Inverse-Square Law at Laboratory Distances. *Physical Review Letters* 44: 1645-1648.

Stacey, F.D. 1978. Possibility of a Geophysical Determination of the Newtonian Gravitational Constant. *Geophysics Research Letters* 5: 377-378.

Stacey, F.D. and G.J. Tuck 1981. Geophysical Evidence for Non-Newtonian Gravity. *Nature* 292: 230-232.

Stacey, F.D., G.J. Tuck, S.C. Holding, et al. 1981. Constraint on the Planetary Scale Value of the Newtonian Gravitational Constant from the Gravity Profile with a Mine. *Physical Review D* 23: 1683-1692.

Stacey, F.D., G.J. Tuck, G.I. Moore, et al. 1987a. Geophysical Tests of the Inverse Square Law of Gravity. *New and Exotic Phenomena: Seventh Moriond Workshop*. O. Fackler and J. Tran Thanh Van. Gif sur Yvette, Editions Frontieres: 557-565.

Stacey, F.D., G.J. Tuck, G.I. Moore, et al. 1987b. Geophysics and the Law of Gravity. *Reviews of Modern Physics* 59: 157-174.

Stubbs, C.W. 1989. Eöt-Wash Constraints on Multiple Yukawa Interactions and on a Coupling to "Isospin". *Tests of Fundamental Laws in Physics: Ninth Moriond Workshop*. O. Fackler and J. Tran Thanh Van. Gif sur Yvette, Editions Frontieres: 473-484.

Stubbs, C.W. 1990. Seeking New Interactions: An Assessment and Overview. *New and Exotic Phenomena '90: Tenth Moriond Workshp*, Les Arcs, France, Editions Frontieres: 175-185.

Stubbs, C.W., E.G. Adelberger, B.R. Heckel, et al. 1989. Limits on Composition-dependent Interactions using a Laboratory Source: Is There a "Fifth Force?". *Physical Review Letters* 62: 609-612.

Stubbs, C.W., E.G. Adelberger, F.J. Raab, et al. 1987. Search for an Intermediate-Range Interaction. *Physical Review Letters* 58: 1070-1073.

Sur, B., E.B. Norman, K.T. Lesko, et al. 1991. Evidence for the Emission of a 17-keV Neutrino in the β Decay of ^{14}C. *Physical Review Letters* 66: 2444-2447.

Taylor, J.D., P.E.G. Baird, R.G. Hunt, et al. (1987). Parity Non-Conservation in Bismuth, Oxford University.

Taylor, J.H. and J.M. Weisberg 1989. Further Experimental Tests of Relativistic Gravity Using the Binary Pulsar PSR 1913 + 16. *The Astrophysical Journal* 345: 434-450.

Thieberger, P. 1986. Hypercharge Fields and Eötvös-Type Experiments. *Physical Review Letters* 56: 2347-2349.

Thieberger, P. 1987a. Search for a Substance-Dependent Force with a New Differential Accelerometer. *Physical Review Letters* 58: 1066-1069.

Thieberger, P. 1987b. Search for a New Force. *New and Exotic Phenomena: Seventh Moriond Workshop*. O. Fackler and J. Tran Thanh Van. Gif sur Yvette, Editions Frontieres: 579-589.

Thieberger, P. 1989. Thieberger Replies. *Physical Review Letters* 62: 810.

Thodberg, H.H. 1986. Comment on the Sign in the Reanalysis of the Eötvös Experiment. *Physical Review Letters* 56: 2423.

Thomas, J., P. Kasameyer, O. Fackler, *et al.* 1989. Testing the Inverse-Square Law of Gravity of a 465m Tower. *Physical Review Letters* 63: 1902-1905.

Thomas, J., P. Vogel and P. Kasameyer 1988. Gravity Anomalies at the Nevada Test Site. *5th Force, Neutrino Physics: Eight Moriond Workshop*, Les Arcs, France, Editions Frontieres: 585-592.

Tuck, G.J. 1989. Gravity Gradients at Mount Isa and Hilton Mines. *Abstracts of Contributed Papers, Twelfth International Conference on General Relativity and Gravitation*, Boulder, CO.

Van Fraassen, B.C. 1980. *The Scientific Image*. Oxford: Clarendon Press.

Wagoner, R.V. 1970. Scalar-Tensor Teory and Gravitational Waves. *Physical Review D* 1: 3209- 3216.

Wark, D. and F. Boehm 1986. A Search for 17-keV Neutrinos in the β–Spectrum of ^{63}Ni. *Nuclear Beta Decays and Neutrino: Proceedings of the International Symposium*, Osaka, Japan, June 1986. Singapore, World Scientific: 319-393.

Weber, J. 1975. Weber Responds. *Physics Today* 28(11): 13.

Weber, J., M. Lee, D.J. Gretz, *et al.* 1973. New Gravitational Radiation Experiments. *Physical Review Letters* 31: 779-783.

Weinberg, S. 1964. Do Hyperphotons Exist? *Physical Review Letters* 13: 495-497.

Weinberg, S., 1993. Private communication.

Weinberg, S. (1996). Letter. *New York Review of Books*. 43: 55-56.

Weisnagel, S. and J. Law 1989. Corrections to the Tritium β Decay Spectrum Arising from Radiative and Atomic Effects and Their Relationship to Neutrino Mass Experiments. *Canadian Journal of Physics* 67: 904-911.

Westfall, R.S. 1980. *Never at Rest*. Cambridge: Cambridge University Press.

Wieman, C.E., S.L. Gilbert and M.C. Noecker 1987. A New Measurement of Parity Nonconservation in Atomic Cesium. *Atomic Physics 10*. H. Narumi and I. Shimamura. Amsterdam, Elsevier Scientific Publishers: 65-76.

Wietfeldt, F.E., Y.D. Chan, M.T.F. da Cruz, *et al.* 1993a. Search for a 17 keV Neutrino in the Electron-Capture Decay of ^{55}Fe. *Physical Review Letters* 70: 1759-1762.

Wietfeldt, F.E., Y.D. Chan, M.T.F. DaCruz, *et al.* 1993b. Further Studies of a ^{14}C- Doped Germanium Detector. *Bulletin of the American Physical Society* 38: 1855-1856.

Wietfeldt, F.E., E.B. Norman, Y.D. Chan, *et al.* (1994). Search for a 17-keV Neutrino Using a ^{14}C- Doped Germanium Detector. Berkeley, CA, Lawrence Berkeley Laboratory.

Will, C. 1981. *Theory and Experiment in Gravitational Physics*. Cambridge: Cambridge University Press.

Will, C. 1984. *Was Einstein Right?* New York: Basic Books.

Wolpert, L. 1992. *The Unnatural Nature of Science*. Cambridge: Harvard University Press.

Weisberg, J.M. and J.L. Taylor 1984. Observations of Post-Newtonian Timing Effects in the Binary Pulsar PSR 1913 + 16. *Physical Review Letters* 52: 1348-1350.

Wu, C.S., E. Ambler, R.W. Hayward, *et al.* 1957. Experimental Test of Parity Nonconservation in Beta Decay. *Physical Review* 105: 1413-1415.

Zee, A. 1979. Broken-Symmetric Theory of Gravity. *Physical Review Letters* 42: 417-421.

Zee, A. 1980. Horizon Problem and the Broken Symmetric Theory of Gravity. *Physical Review Letters* 44: 703-706.

Zerner, F. 1915. Zur Kritik des Elementarquantums for Elektrizität. *Physikalische Zeitschrift* 16: 10- 13.

Zlimen, I., S. Kaucic and A. Ljubicic 1988. Search for Neutrinos with Masses in the Range of 16.4 → 17.4 keV. *Physica Scripta* 38: 539-542.

Zlimen, I., A. Ljubicic, S. Kaucic, *et al.* 1990. Search for Neutrinos with Masses in the Range 15 to 45 keV. *Fizika* 22: 423-426.

Zlimen, I., A. Ljubicic, S. Kaucic, *et al.* 1991. Evidence for a 17-keV Neutrino. *Physical Review Letters* 67: 560-563.

Zumberge, M., M.E. Ander, T.V. Lautzenhiser, *et al.* 1988. Results from the 1987 Greenland G Experiment. *Eos* 69: 1946.

INDEX

Abashian, A., 68, 75, 86, 209, 258
Ackermann, R., 169, 170--171, 172
ADC. *See* Analog to digital converter
Adelberger, Eric, 193
Algorithms, 18, 27
Altzitzoglou, T., 44, 45(table), 49, 50, 65(table)
Aluminum, 139, 196--197
  baffle, 75
  foil, 68, 69(fig.)
Analog to digital converter (ADC) range, 42, 251
Annealing, 53, 54
Antineutrino, 40, 50, 215
Antiscatter baffles, 51, 58(fig.), 209
Apalikov, A. M., 44, 45(table), 65(table)
A priori position, and empirical evidence, 5
Argonne National Laboratory, 75, 78--79, 81(figs.), 87, 209--211, 260. *See also* Maryland-Argonne collaboration
Aronson, Sam, 134, 135, 138, 142, 173--174
Astrophysics, 87--88
Atomic-parity violation, 2
  constructivist view, 6, 166
  and experimental evidence, 6, 84, 165--172, 212, 214--217, 221--226, 241--243
  and Weinberg-Salam unified theory, 6, 156, 164, 165, 166, 178, 212--214
Atomic theory, 157, 158
Australia, 177

Backscatter, 55, 64
  reduction, 78
Bahran, M., 64--65
Barkov, L. M., 216, 225(table)
Barnes, B., 276
Bars, I., 175
Bartlett, D. F., 177, 189, 190
Baryon, 139, 143, 144, 174, 176
Beam energy, 219, 220(fig.)
Beam particles, 100, 101(fig.)
Becker, H. W., 45(table), 58, 62, 65(table)

BEFS. *See* Beta environment fine structure
Bell, J. S., 134
Bennett, W.R., 195
Berenyi, D., 52
Berkeley group, 45(table), 59--61, 63, 65(table),71, 72, 78, 79, 80--82, 86, 91, 166, 167, 211, 262
Bernstein, J., 134
Beryllium block, 119, 120
ß (beta) decays, 42, 49, 62
  detections, 84, 260
  and kinks, 44, 65--66, 79
  and scattering effects, 74
  spectra, 55--56, 85--86. *See also* Tritium beta spectrum
ß electron, 46
Beta environment fine structure (BEFS), 63--64
Bismuth, 167, 214, 217, 215(table), 222--225
Bizzeti, P. G., 177, 195, 230
Black-body radiation spectrum, 277
Bock, G. J., 135
Boehm, F., 43, 44, 49, 50
Bogen, J., 20, 155
Bohr, Nils, 157
Bombay experiments, 45(table)
Bonvicini, G., 65--67, 74, 75, 79, 88, 209, 254--255, 271(n20)
Boreholes, gravity measurement in, 189
Borge, M.J.G., 45(table), 49, 50, 52, 65(table)
Bosons, 156
Bouchiat, C., 166, 170, 217, 220, 225
Bouchiat, M. A., 225, 226
Boynton, P., 176, 177, 280
Braginskii, V. B., 142
Branching ratio measurement, 98, 103, 104, 105--118
Brans, C., 139, 173
Brans-Dicke theory, 134, 139, 172
  modifications, 139, 173
Bratislava conference on nuclear physics (1900), 58, 61

INDEX

Bremsstrahlung (electromagnetic radiation), 104, 107
Bubble chambers, 98, 119
Buenos Aires experiments, 45(table)

^{14}C, 59, 60, 61, 63(table), 65(table), 71, 73, 78, 80, 82, 209, 210, 211(fig.), 262
Cabibbo, N., 134
Calcite prism, 218
Calibration
 defined, 14, 237
 and Fifth Force, 263--268, 269
 on gravity wave detector, 18, 19, 20, 21, 23, 198--200, 203, 244--248
 of magnetic spectrometer, 41, 252, 255
 and measurement, 270(n2)
 and meson decay, 240--243
 and 17-keV neutrino, 248--263
 unproblematic, 238--243, 269
 validity, 83, 169, 186, 263, 268--269
Callahan, A. C., 110, 125
Caltech experiments, 45(table), 64, 65(table), 67, 256
Cartwright, N., 149, 155
^{109}Cd, 79
^{139}Ce, 78
Cerenkov counter, 33, 98, 102, 104, 108--109, 111--112, 113--114, 119, 126--127, 128(n7), 172, 218, 219(fig.), 240, 274
CERN-ISOLDE experiments, 45(table), 65(table), 87(fig.), 207(fig.)
Cesium, 78, 226
Chalk River experiments, 45(table), 65(table), 254
Changing Order: Replication and Induction in Scientific Practice (Collins), 13, 237
Charge quantization, 158--159
Charmonium, 134
X^2 value, 51(fig.), 52, 62, 67, 74, 81(figs.)
Christenson, J. H., 169
Chu, S. Y., 175
Chwolson, O. D., 159
Cline, David, 150
Close, Frank, 215--216
^{57}Co, 55, 59, 78
^{60}Co, 241--243
Colella, R., 134, 135, 172

Collins, Harry, 4, 6, 187, 228
 experimenters' regress, 13--15, 34, 83, 184, 237--238
 and gravity wave detectors, 15--19, 34
 and gravity waves, constructivist view, 13--19, 20, 32, 230
 methodology, 14, 15
Compton scattered gamma rays, 60
Conjectural realism, 149
Constructive empiricism, 149, 155--159, 160
Constructivists. *See* Social constuctivists
Contradiction of parity, 1
Copper (Cu), 41
 beta decay, 43
 in scintillation counters, 99--11
Copper cryopanel, 55
Co2 trigger, 102--103, 104, 105(table)
Co3 trigger, 102, 104, 105(table)
Coulomb effects, 44, 45, 46(fig.), 207
Counters, 98, 100(fig.), 102, 111(fig.), 121
Coupal, D. P., 138
Cowsik, R., 195
CP (charge conjugation-parity) symmetry violation, 134, 163, 169
 as constant, 135
 and gravity, 135, 173
 tangent of the phase, 135--136, 137(figs.)
Cryopanels, 55, 58(fig.)

Dalitz pairs
 positrons, 104, 125
Data, and results, 20, 127, 172, 181(n30), 238
Datar, V. M., 44, 45(table), 50, 65(table)
Dawkins, Richard, 3
De Bouard, X., 135
De Haas, W.J., 273, 276
De Rujula, A., 50, 147(n27)
Deuterium, 212, 218
Dicke, R. H., 139, 142, 173, 175
Differential accelerometer, 264
Discovery, 163, 172--174, 179
Douglass, David, 23, 25, 29, 200, 202
Drever, R., 23--25
Drukarev, E. G., 46
Duhem-Quine thesis, 168, 169, 185, 278--279

Dunham, E., 171
Dydak, F., 163, 167--168, 178, 217

Eckhardt, D. H., 171, 176, 177, 178, 188, 189, 190, 191
Edge effects, 53
Edinburgh school, 164
Ehrenhaft, Felix, 159
Eigenstate, 59
Eigenvalue, 134
Eightfold way/particle classification scheme, 154, 156
Einstein, Albert, 139, 157, 273, 276, 277. *See also* General Relativity
Electromagnetic theory, 157, 162(n24). *See also* Bremsstrahlung
Electron, 149, 150
 antineutrino, 40
 aperture penetration, 75, 76
 back-diffusion, 75, 76
 capture, 49--50
 charge measurement, 158--159
 energy, 40, 44, 55, 59, 75
 gyromagnetic ratio, 273
 -muon universality, 119
 neutrino, 59, 110
 orbiting, 275--276
 scattering, 59, 64, 74, 75--78, 79(fig.), 80, 84, 209, 212, 218
 spin, 276
Electron beam helicity, 219
Electronic imaging, 129(n8)
Electronic tradition and selectivity, 98
Electron response function (ERF), 66, 74, 76, 254--255, 259--260
Electroweak interactions, 2, 87
Elliot, J., 171
Eman, B., 45--46, 48, 56
Emulsion experiments, 98
Energy dependence
 of the form factor, 118--125
 of K^0--K^0, 142--143, 174, 178
 of regeneration amplitude phase, 135, 136, 137--138
Energy pulses, 18
Entity realism, 149
Eötvös, R., 139, 142, 143, 144, 147(n27), 173, 175, 178, 192, 263

composition-dependent effect, 192--195, 196(figs.)
Eöt-Wash group. *See* Washington experiments, Fifth Force
Epistemological indicators, 83
Equivalence principle, 139
ERF. *See* Electron response function
Euclid, 3
Eurocentrism, and science, 1--2
Europhysics Conference on High Energy Physics (1991), 63
"Existence of a Subelectron, The?" (Millikan), 159
Experimental evidence, 4, 6--7, 14, 88, 163, 283
 and alternative explanation, 169--170, 183, 185, 186, 206, 248, 278
 and apparatus, 171--172, 179, 186, 237, 274--275
 certainty, 19
 and choice, 164
 epistemological strategies, 83--84, 91, 103, 169, 171, 172, 179, 184--185, 186, 187, 203, 212, 220, 227, 230--231, 243, 277--278, 283
 evaluation, 170--171, 172
 and evidential weight, 165--166, 170, 178, 182(n45), 213--214, 220, 231, 232
 and justification, 164, 165, 168, 176--178, 179, 183--184
 and theory of the phenomenon, 85, 163, 186
 validity, 19, 83, 168, 169, 186, 206, 217, 238, 279. *See also*
 Data, and results; *under* Calibration
 See also Discovery; Experimenters' regress; Pursuit
Experimenters' regress, 13--15, 83, 184
 and gravity wave detectors, 15, 16, 19, 20
Experiments, 97
 and theory, 273--276, 278--279, 280
 See also Experimental evidence; Instumental loyalty; Recycling of expertise

Fackler, Orrin, 178
Faller, J. E., 176
Falling steel balls, 7(figs.), 8

^{55}Fe, 52, 60, 61, 72, 73(fig.)
Fekete, E., 139, 144
Feminists, and science, 1
Fermi function, 45--46, 47
Fermion, 134
  masses, 58
Feyerabend, P., 276
Feynman, Richard, 183
Fifth Force, 6, 93--94(n30), 133, 134, 138, 139--140, 142--145, 146(n6), 163, 165, 171, 182(n43), 229, 230--231
  and calibration, 263--268
  discovery, 172--174
  justification, 176--178
  pursuit, 174--175
  and resolution of dscordant experimental results, 187--196, 230
Finite mass quantum, 138
Fischbach, Ephraim, 134, 135, 142--145, 163, 172, 173--174, 175, 177, 178
Fitch, V. L., 195
Float experiment, 176, 177, 193--196, 269
"Four of Five New Experiments Claim Evidence for 17-keV Neutrons" (Schwarzschild), 58
Franklin, Allan, 33, 155, 170, 283
Freedman, S., 78, 80
Frontov, V. N., 140
Fujii, Y., 133, 139, 140, 142, 143, 173

g (gravitational acceleration), 142, 143, 174
G (gravitational constant), 3, 8, 139--143, 173, 188
Galbraith, W., 135
Galileo, 7, 11(n20)
  replication of experiment, 7(figs.), 8, 176
Galison, Peter, 97, 273
Gamma (γ), rays, 59, 60, 90, 104, 112, 118(table), 121, 122--125
  anisotropy, 242--243
  conversion minimization, 119, 120
Gargamelle experiment, 88--89
Garwin, R. L., 26, 27--28, 30, 31, 32, 201, 202, 203
Gas molecules, hard sphere model, 156--157
^{71}Ge, 62, 65(table), 71

Geison, Gerald, 4--5
General Relativity, 35, 134, 139, 172, 191
  at quantum level, 134
  scalar-tensor alternative. See Brans-Dicke theory
Geodesy data, 140
Germanium crystal, 53, 54, 59, 63(table), 64(fig.), 80
Germany, Aryan science, 4
Gibbons, G. W., 140, 141--142
Glashow, Sheldon, 57, 58, 91, 175, 209
Gold, 139
Gravimeter, 188
Gravitation, 133--134, 191, 192. See also Fifth Force
Gravitational acceleration (g), 142, 143, 174
Gravitational constant (G), 3, 8, 139--143, 173, 188
Gravitational force, 143
  and resistive force, 8, 11(n20)
Gravitational radiation. See Gravity waves
Gravitational theory, 138--142, 172
  tests. See Fifth Force
Gravity waves
  calibration, 18, 19, 21, 32--34, 35--36, 237--238, 269
  calibration pulses, 21--22, 23--24, 31, 245, 246(fig.)
  computer simulation, 27--28, 30
  constructivist view, 6, 13--19
  detection, 19--32, 35, 231, 238, 244
  detection methodology, 14, 231
  detectors, 15, 16(fig.), 17--18, 196--197, 244
  energy fluctuations, 20
  experimental evidence, 6, 13--15, 31--34, 35
  literature, 14, 15
  resolution of discordant experimental results, 197--204
  time-delay data, 22--24(figs.), 246
  zero delay coincidence, 28--29
Greenland icecap boreholes, 189
Grenoble experiments, 65(table)
Gross, P. R., 3
Guard ring, 80--82, 86
Gullstrand, A., 159
Gyromagnetic ratio, 273

^{3}H. *See* Tritium
Hacking, Ian, 149, 153, 185
Hadrons, 213
Hansen, P. G., 46, 54, 56, 57
Harding, Sandra, 2, 6
Haxton, W. C., 44, 45, 48
Heavy neutrino. *See* 17-keV neutrino
Heisenberg, Werner, 157
Helmholtz coils, 267
Hetherington, D. W., 45(table), 50, 65(table), 66
High-energy physics community, 180(n11), 213
Hillside torsion balance, 176, 177(fig.), 194, 195, 266--267
Hime, A., 45(table), 53--54, 55, 56, 58--59, 63, 65(table), 66, 68, 69--72, 74, 75--78, 79, 86, 87, 90, 208, 209, 258, 259--260
Holding, S. C., 142--143
Howson, C., 155
Humanists, 1
Hydrogen, gravitational effects on, 134
Hyperphoton, 134, 135
Hypothesis, 133
  evidential basis, 163, 184
  plausibility, 169, 178, 278

^{125}I, 50, 65(table)
IB. *See* Internal bremsstrahlung
IBEC. *See* Internal bremsstrahlung in electron capture
IC. *See* Internal conversion
Infrared spectrum of organic molecules, 239--240
"In Old and New Experiments the 17-keV Neutrino Goes Away" (Schwarzschild), 79
Instrumental detection, 155
Instrumental loyalty, 91, 97, 126, 131(n48), 179, 183
Interferometer, 35
Internal bremsstrahlung (IB), 61, 62, 72, 73(fig.), 79
Internal bremsstrahlung in electron capture (IBEC), 49--50, 52--53, 56
  and kinks, 52, 53
  studies, 45(table), 60
Internal conversion (IC), 77, 79(fig.)

Interviews, 14
Inverse-square law, 140, 188
Ionization, 98
  and muon energy loss, 106
  and positron energy loss, 107
Iron foil, magnetized, 218
Isotopes, 45(table), 63(table), 64, 65(table)
ITEP experiments, 45(table), 65(table)

Jelley, N. A., 58--59, 63, 66, 69, 74, 80, 86, 209, 258
Justification, 163, 164, 176--178, 179, 183, 283

Kafka, Peter, 21, 23, 25, 28--29, 198
Kalbfleisch, G. R., 47, 64--65
Kaons (K^+ mesons), 99, 100, 150, 160
  decays, 101--102, 151, 152(fig.)
  and interactions, 137
  mass, 162(n18)
  momentum, 117
  properties, 151--152, 153
Kawakami, H., 80
Kepler, Johannes, 5
Kinetic energy, 121(fig.), 123
Kinetic theory, 156
Kinks, 40, 42, 44, 48, 52, 61, 62, 63, 65--66, 69(fig.), 73, 74, 79, 87, 204, 212, 252, 256
  simulation, 68
K_L decays, 134, 135
K meson energy, 134--135, 138, 139, 142, 147(n26)
  gravitational scale, 135
  short-lived, 135
K meson reality, 150, 152--154, 155--156
Knowledge viewpoints, 1--2, 3--5, 9(n1)
Koertge, Noretta, 278, 279
Konstantinowsky, D., 159
Koonin, S., 63
K^+ meson decay, 6, 33--34, 90, 98, 102--103, 170
  apparatus, 98, 99--102, 105(fig.), 106, 111--112, 119--120, 126--127, 172, 240--243, 274
  beamline, 99

detector, 98, 99, 151, 155
energy dependence of form factor, 118--125, 173
experiments, 97:
  $K^+_{\mu 2}$ branching ratios and muons, 98, 103, 104, 105--110, 120, 172, 240, 273
  $K^+_{\mu 3}$ branching ratios, 98, 105--110
  $K^+_{\pi 2}$ branching ratios and pions, 98, 103, 105--110, 120, 122
  K^+_{e2} branching ratios, 98, 110--118, 150--151, 172, 238, 240, 273--275
  K^+_{e3} branching ratios and momentum spectrum, 98, 102--105, 107, 108, 109, 115(table & fig.), 117--126
mean life, 101
modes, 106--110
Krotkov, R., 139, 142
K_s lifetime, 138
Kuhn, T. S., 172, 276, 277
Kurie plot, 39, 40(fig.), 42(fig.), 45, 48(figs.), 92(n5)
kink, 40, 46, 204, 205(fig.)
Kurlbaum, F., 277
K^0--K_0 parameters, 134, 135, 142--143, 144, 173, 174, 178
K^0_L meson decay, 33, 240
K^0_2 decay, 169--170

LaCoste-Romberg gravimeter, 188
Laser, 13
Laudan, L., 163
Law, J., 56, 57(fig.)
Law of Universal Gravitation, 3. *See also* Gravitational theory
Laws, 149, 154, 160, 161(n15), 188, 280--281(n1)
LBL experiments, 45(table)
Lee, T. D., 134
Left-right symmetry violation, 1, 173, 213
Leptons, 118, 139
Levine, George, 3
Levine, James L., 27, 28(fig.), 30, 201(fig.), 202
Levitt, N., 3
Lindhard, J., 46, 54, 56, 57, 253
Linear algorithm, 18, 21, 22, 31, 198, 199, 201(fig.), 203, 245, 246(fig.)
Lingua Franca (journal), 3

Liquid nitrogen cryopanel, 58(fig.)
Livermore group, 188, 189, 190
Ljubicic, A., 58, 61
Long, D. R., 139--140
Low Energy Tail (LET), 75
^{177}Lu, 64, 65(table)
Lummer, Otto, 277
Lunar laser-ranging experiments, 139
Lunar surface gravity, 140
Lynch, M., 171--172

Mackenzie, D., 168, 169
Madsen, J., 91
Magnetic field, 104, 120, 210
Magnetic spectrometer, 39, 41, 44, 65, 72, 73, 85
  for beta decay, 50, 52, 58, 204
  calibration, 41, 252, 255
  and kink detection, 68, 210
  shape-correction location, 64, 65, 70, 75
Magnetism, 276
Manipulability, 149, 160(n3)
Mann, Alfred, 97
Mann-O'Neill collaboration, 97, 125
Markey, H., 44, 45(table), 49, 50, 65(table)
Martensson, A. M., 225(table)
Maryland-Argonne collaboration, 22(fig.), 200(fig.)
Mass, as term, 276--277
Maxwell, James C., 157, 273
McHugh, M. P., 176
Mendeleev, D. I., 154
Mercury (element) plug, 119
Mercury (planet) flyby, 140
Microscope, scanning, tunneling, 153, 160(n3)
Mikkelsen, D. R., 140, 173
Miller, David, 216
Millikan, R. A., 155, 158--159, 160
Millis, R., 171
Milton, K. A., 47
Mines, gravity in, 140--141, 145, 173, 174, 177, 187, 189, 190
Molybdenum (Mo), 41
Momentum chambers, 100(fig.), 101--102, 106, 109, 115, 116, 119, 121
Monoenergetic electrons, 59, 76, 260
Monoenergetic neutrinos, 49

Monte Carlo simulations, 66, 73(fig.), 75, 78, 79, 86, 88--90, 104, 122, 123, 125, 209
Moriond Workshop (1987--1991), 58, 176, 177, 178, 188, 190, 233(n9)
Morrison, D., 63--64, 66(fig.), 90, 256--258
Morrison, M., 160(n3)
Moscow group, 221, 225(table)
Multichannel analyzer, 41(fig.), 251
Mumetal shield, 265
Munich experiments, 45(table)
Muons, 33, 99, 106, 150
 decay, 113, 119
 -electron universality, 119
 and ionization, 106
 momentum, 107(fig.), 108, 112, 120
 range spectrum, 114
 tri-, 216

Nature (journal), 215
Neglect of Experiment, The: Experiment, Right or Wrong (Franklin), 283
Nelson, Alan, 227--228
Neutrinos, 41, 42--43, 45, 47, 50, 137, 150
 anti-, 50, 215
 electron, 59, 110
 heavy, 43, 50, 61
 low-mass, 41
 mass, 43, 91(n1), 92(n10)
 monoenergetic, 49
Neutrons, 150
Newman, M. J., 140, 173, 178
Newman, R., 195
Newton, Isaac, 2, 3, 8, 133, 188
Newtonian mechanics, 2, 276, 277
^{63}Ni, 50--51, 56, 63(table), 65(table), 66, 73, 75, 77(fig.), 78, 254, 258
Nickles, T., 163
Niebauer, T. M., 176
Nobel Prize in Physics (1924), 159
Nonconservation of parity. *See* Contradiction of parity
Non-linear algorithm, 18, 21, 22--23, 31, 198, 199, 200, 203, 244, 246, 247--248(figs.)
Norman, E. B., 45(table), 57, 58, 80
Novikov, V. N., 225(table)

Novosibirsk experiments, 166, 167, 170, 216--217, 222, 225(table)

Observable phenomena, 156--157
O'Hanlon, J., 139, 140, 142
Ohi, T., 44, 45(table), 49(fig.), 50, 52, 64, 65(table), 66(fig.), 208
Ohshima, T., 80, 87, 212
Oklahoma experiments, 65(table), 71
Ω particle, 154, 156
O'Neill, Gerard, 97
Oscilloscope, 101
Overhauser, A. W., 134, 135, 172
Oxford experiments, 45(table), 65(table), 74, 84, 165--166, 167, 178, 214--215, 216, 217, 221, 260
 reanalysis, 75--78, 79(fig.), 208, 222, 223--224, 225(table)

Paik, H. J., 175
Panov, V. I., 140, 142
Paritino, 170
Parker, R. L., 189, 190
Particle-antiparticle symmetry, 134, 169, 173
Particle classification scheme/eightfold way, 154, 156
Particle detection efficiency, 98
Particle momentum spectrum, 116--117(figs.)
Particle physicists, 164, 183, 215
Particle properties, 155, 161(nn10&11)
Particles, strongly interacting, 162(n18)

Parton, 216
Pasteur, Louis, 4--5
Pekar, D., 139, 144
Periodic Table, 154
Perring, J., 134
Photon, 49, 54
Physics Today (journal), 58, 79
π (pi), 3
π^+ decay modes, 98
π^0, 119, 120(fig.), 121, 123, 124, 126

Pickering, Andrew, 6, 88--89, 90, 164, 178, 183, 184, 187
  and W-S theory, 165, 166, 167, 168--169, 170, 172, 213, 218, 228--229
Piezo-electric crystals, 15, 18, 197, 238
Piilonen, L., 68, 74, 86, 209, 258
Pinch, Trevor, 4
Pions, 33, 34, 99, 100, 106, 114, 120, 133, 150, 161(n8), 169, 170
  mass, 118
Planck, Max, 277
Pockels cell, 218
Popper, K., 91
Positronium, 134
Positrons, 33--34, 99
  decay, 113
  energy loss, 115, 119
  and ionization, 107
  momentum, 90, 102, 103--104, 107, 108, 109, 112, 114--115, 123, 125(fig.), 126
  range spectrum, 115, 125
Postmodernism
  and critics of science, 3, 4, 5
  and details of science, 9(n6)
  and knowledge, 1
PPA. *See* Princeton-Pennsylvania Accelerator
Princeton experiments, 45(table), 65(table)
Princeton group, 33, 240
Princeton-Pennsylvania Accelerator (PPA) experiments, 97, 99, 100
Principia (Newton), 2
Pringsheim, E., 277
Prism, 218, 219(fig.)
Protons, 99
Psychology, 163
Pulse-height defect, 53
Pursuit, 163, 164, 167, 168, 179, 179(n2), 183

Quantum chromodynamics (QCD), 162(n18)
Quantum gravity vacuum polarization effect, 140
Quantum level gravitational effects, 135, 138, 172
Quarks, 160(n3), 162(n18), 216, 217
QCD. *See* Quantum chromodynamics

Q value, 50

Radcliffe, T., 67, 256
Range chamber, 102, 106, 109, 111(fig.), 116
Rapp, R. H., 140
Rasmussen, Nick, 229
Realists and antirealists, 149, 157, 160
Recycling of expertise, 91, 97, 126
  and division of labor, 97
  and theorists, 128(n1)
Regeneration amplitude phase, 135--136
  energy dependence, 135, 136
Reichenbach, H., 163
Rise and Fall of the Fifth Force, The (Franklin), 283
Rochester-Bell collaboration, 21(figs.), 23(fig.), 198--199(figs.), 201(fig.)
Roll, P. G., 139, 142
Ross, Andrew, 2, 3, 4
Rubens, H., 277

Sandars, P.G.H., 216, 225(table)
Satellite data, 140, 142, 174
^{35}S beta-ray spectrum, 44, 46, 47, 49, 50, 55, 56, 62, 63(table), 64, 65(table), 66, 67, 73, 75, 76(table), 77(fig.), 78, 206, 210, 211(fig.), 252, 256
Scalar-tensor particles, 104, 139, 173
Scattering, 51, 60. *See also* Electron, scattering
Scherk, J., 139
Schreckenbach, K., 43, 65(table)
Schwarzschild, B., 58, 79--80, 84
Science
  Aryan, 4
  contexts, 163
  as "critical rationality," 91, 283
  fallibility of, 19, 232(n2)
  and hoaxes, 3, 184
  and humanists, 1, 3, 4
  and knowledge, 1--3, 4 5, 6, 183, 184, 187, 283
  logical structure of, 163
  methods of, 8--9, 185, 187, 283
  political agenda, 3, 4, 164, 183

INDEX 311

practice of, 6, 9, 157, 163, 164
social structure of, 2, 283
in Soviet Union (1930s), 4
and speculation, 179(n4)
and truth, 149, 160, 183
validity, 6. *See also under* Experimental evidence
"Science Wars," 2--3
Scientific Image, The (van Fraassen), 155
Scientific interest, 183
Scintillation counters, 99--100, 112
Sea floor and surface G measurements, 141
Selectivity, 98
Sellars, W., 149, 153, 156
Semileptonic decay, 135
17-keV neutrino, 2, 6, 48, 91
  and applied epistemology, 84
  criteria for good experiment, 67
  criticism of experiment, 46--47, 51--52, 252--253
  decay sources, 44, 50--51, 52, 155, 159, 206. *See also* Tritium
  detector energy calibration, 204, 250--252, 269
  detectors, 39, 44, 48, 52, 53, 55, 58(fig.), 59--60, 65(table), 68, 72, 73, 74, 78, 80, 81, 84, 91, 204, 209, 210, 248--249, 260--261. 263
  deviation spectrum, 57(fig.)
  discovered, 39--43, 48, 84--85, 204
  and energy ranges, 73--74, 75, 204, 207--208
  mixing probability, 77--78, 79, 208, 210(fig.)
  negative results, 39, 43--44, 45(table), 48--49, 50, 52--53, 56--57, 59--60, 61(fig.), 62--66, 74--79, 86--87, 204, 205, 206(fig.), 207, 209--212, 254--263
  positive results, 39, 43, 45(table), 53--56, 63(fig.), 66, 80, 86, 204, 208--209, 254
  reanalysis of experiments, 64--82, 86--87, 208, 209--211, 254--263
  recalculations, 48--49, 253, 255
  resolution of discordant experimental results, 204--212
  shape factor, 43(fig.), 51, 56(fig.), 60(fig.), 66, 68, 75, 77(fig.), 208, 209, 212
  threshold, 49, 51, 52

and theory, 85, 91
Sherlock Holmes strategy, 83, 169, 186, 212
Shower counter, 218, 219(fig.)
Shower spark chambers, 120, 121
Shrock, R. E., 42
Si(Li) x-ray detector, 40, 44, 53, 54, 58(fig.), 63(table), 78, 204, 210
Silicon, 59, 73
Silver (Ag), 41
Simpson, J. J., 39--43, 45(table), 47--49, 63, 65(table), 66(fig.), 78, 80, 82, 85--86, 90. *See also* 17-keV neutrino
SLAC. *See* Stanford Linear Accelerator Center
Snow, C. P., 1
Social constructivists, 4--5, 6, 165, 166, 167, 168--169, 170, 179
  and accepted theory, 227--228
  and atomic-parity violation, 6, 187, 227, 228
  and experimental evidence, 164, 184, 187, 227. *See also* Pickering, Andrew
  and gravity waves, 6, 187, 228
  and negotiation for resolution, 183, 227
  and science, 9(nn6&7)
Social Text (journal), 2--4
Sodium iodide anticoincidence shield, 60
Sokal, Alan, 3--4
Solenoidal magnetic field, 78
Solid-state detector, 39, 53, 58, 59, 60, 73, 78, 204
Soviet Union, 4
Space inversion symmetry, 169
Spark chambers, 98, 99--100, 111(fig.), 119
  shower, 120, 121
Speculation, 167, 179(n4)
Spero, R., 140
Spion, 170
Stacey, F. D., 140--141, 143, 171, 177, 189, 190
Stanford Linear Accelerator Center (SLAC)
  E122 experiment, 84, 165, 166, 171, 212, 213, 214, 218--220, 221(fig.), 229

theoretical papers on 17-keV neutrino, 96(n78)
Starlight intensity, 171
Statistical arguments, 186
Stokes' Law, 8, 264
Strange Weather: Culture, Science, and Technology in the Age of Limits (Ross), 2
Strikman, M. I., 46
Stubbs, C. W., 171
Submarine, gravity in, 140
Sur, B., 45(table), 63(table), 65(table), 86
Synchrotron, 100

Tadic, D., 45--46, 48, 56
Talmadge, Carrick, 134
TANDAR facility (Argentina), 71
Taylor, John, 8
TEA-laser, 13, 34
Telescope, 100(fig.), 101
Tensor field, 104, 138, 139
Tew, W. L., 177, 189, 190
Thallium, 217
Theories, 7, 133, 149, 155, 156--157, 162(nn18&24), 280--281(n1)
 and contradiction, 157, 185
 and discordant experimental results, 279--280, 230--231
 enabling, 273--276
 and experiments, 273--276
 justification, 163, 164, 167, 176--178, 179
 and pursuit, further investigation, 163, 164, 167, 168, 170, 174--175, 179, 179(n2)
 and speculation, 179(n4)
 undetermination of, by evidence, 185
 unified, 157--158
Thieberger, P., 171, 176, 177, 193--196, 230, 264, 266, 268
't Hooft, G., 133
Tokyo experiments, 45(table), 65(table), 73, 80, 231, 258--259
Torsion balance, 176, 177(fig.), 194, 195, 266--267
Tower experiment, 176, 177, 187--192, 193(fig.), 194, 264--266

"Transgressing the Boundaries: Toward a Transformative Hermeneutics of Quantum Gravity" (Sokal), 3
Trimuon events, 216
Tritium (^{3}H) beta (ß) spectrum, 40, 41, 42, 47(fig.), 44, 45--46, 47, 48--49, 50, 53, 55, 63(table), 65(table), 204
 BEFS, 63--64
 deviation spectrum, 57(fig.)
 energy calibration, 250--252
 neutrino. *See* 17-keV neutrino
 shape, 64
 use of, 271(n20)
Tuck, G. J., 142--143, 171
Two-fermion system, 134
Tyson, Tony, 21, 23, 25, 26, 198--199, 245, 246

University of California (Berkeley). *See* Berkeley group
University of Guelph (Canada), 45(table), 59, 65(table), 71, 78, 80

Van Fraassen, Bas C., 149, 150, 152, 155, 157, 159
V-A theory of weak interactions, 110--111, 118, 127, 215, 273
Vector, scalar, tensor interactions, 104
Venus (planet) flyby, 140
Viking time-delay, 139
Visser, M., 175
Visual tradition, 98, 155, 162(n19)
Vogel, P., 43

Wagoner, R. V., 139, 140
Wark, D., 45(table), 65(table)
Washington experiments, 84, 165--166, 167, 213, 214, 215, 216--217, 221--223, 225
 and Fifth Force, 176, 178, 193, 194, 195, 196(fog.), 230, 266--267, 268
Weak interactions theory. *See* V-A theory of weak interactions
Weber, Joseph, 13, 15, 196--204, 229, 230

computer program error, 25--26, 203
criticism of data, 27--30, 197, 198--199, 200--202, 203, 231, 247--248
experimental evidence, 31--34, 197, 202(fig.), 203, 244--245
gravity wave detector, 15, 16, 17, 20, 196--197, 198, 244
gravity wave detector calibration, 22, 23, 198--200, 203, 245--246
non-linear algorithm, 21, 22, 198, 199, 200, 203, 245--246, 247--248(figs.)
peak at zero time delay, 26(fig.), 29, 248(fig.)
Weinberg, Steven, 3, 135
Weinberg-Salam (W-S) unified theory of electroweak interactions, 2, 6, 87, 133, 156, 164, 178, 180(n15), 213
and discovery, pursuit, and justification, 164, 165--167, 169, 170, 171--172
resolution of discordant experimental results, 212--226, 231
Weisnagel, S., 56, 57(fig.)
Werner, S. A., 134, 135, 172
Whiting, B. F., 140, 141--142

Wide scan spectrum, 51(fig.), 52
Wieman, Carl, 226
Wolpert, L., 3
Woodward, J., 20, 155
W-S. *See* Weinberg-Salam unified theory of electroweak interactions

Xenon atoms, 153, 154(fig.), 160(n3)
X-ray detector, 40, 41
  energy calibrating, 41
X-ray energy, 49, 50

Yukawa, H., 133, 188(caption)

Zagreb experiments, 45(table), 65(table)
Zee, A., 139
Zerner, F., 159
Zlimen, I., 45(table), 52, 63, 65(table)
Zolotorev, M. S., 216
Zumberge, M. A., 190

Boston Studies in the Philosophy of Science

127. Z. Bechler: *Newton's Physics on the Conceptual Structure of the Scientific Revolution.* 1991
     ISBN 0-7923-1054-3
128. É. Meyerson: *Explanation in the Sciences.* Translated from French by M-A. Siple and D.A. Siple. 1991   ISBN 0-7923-1129-9
129. A.I. Tauber (ed.): *Organism and the Origins of Self.* 1991   ISBN 0-7923-1185-X
130. F.J. Varela and J-P. Dupuy (eds.): *Understanding Origins.* Contemporary Views on the Origin of Life, Mind and Society. 1992   ISBN 0-7923-1251-1
131. G.L. Pandit: *Methodological Variance.* Essays in Epistemological Ontology and the Methodology of Science. 1991   ISBN 0-7923-1263-5
132. G. Munévar (ed.): *Beyond Reason.* Essays on the Philosophy of Paul Feyerabend. 1991
     ISBN 0-7923-1272-4
133. T.E. Uebel (ed.): *Rediscovering the Forgotten Vienna Circle.* Austrian Studies on Otto Neurath and the Vienna Circle. Partly translated from German. 1991   ISBN 0-7923-1276-7
134. W.R. Woodward and R.S. Cohen (eds.): *World Views and Scientific Discipline Formation.* Science Studies in the [former] German Democratic Republic. Partly translated from German by W.R. Woodward. 1991   ISBN 0-7923-1286-4
135. P. Zambelli: *The Speculum Astronomiae and Its Enigma.* Astrology, Theology and Science in Albertus Magnus and His Contemporaries. 1992   ISBN 0-7923-1380-1
136. P. Petitjean, C. Jami and A.M. Moulin (eds.): *Science and Empires.* Historical Studies about Scientific Development and European Expansion.   ISBN 0-7923-1518-9
137. W.A. Wallace: *Galileo's Logic of Discovery and Proof.* The Background, Content, and Use of His Appropriated Treatises on Aristotle's *Posterior Analytics.* 1992   ISBN 0-7923-1577-4
138. W.A. Wallace: *Galileo's Logical Treatises.* A Translation, with Notes and Commentary, of His Appropriated Latin Questions on Aristotle's *Posterior Analytics.* 1992   ISBN 0-7923-1578-2
     Set (137 + 138) ISBN 0-7923-1579-0
139. M.J. Nye, J.L. Richards and R.H. Stuewer (eds.): *The Invention of Physical Science.* Intersections of Mathematics, Theology and Natural Philosophy since the Seventeenth Century. Essays in Honor of Erwin N. Hiebert. 1992   ISBN 0-7923-1753-X
140. G. Corsi, M.L. dalla Chiara and G.C. Ghirardi (eds.): *Bridging the Gap: Philosophy, Mathematics and Physics.* Lectures on the Foundations of Science. 1992   ISBN 0-7923-1761-0
141. C.-H. Lin and D. Fu (eds.): *Philosophy and Conceptual History of Science in Taiwan.* 1992
     ISBN 0-7923-1766-1
142. S. Sarkar (ed.): *The Founders of Evolutionary Genetics.* A Centenary Reappraisal. 1992
     ISBN 0-7923-1777-7
143. J. Blackmore (ed.): *Ernst Mach – A Deeper Look.* Documents and New Perspectives. 1992
     ISBN 0-7923-1853-6
144. P. Kroes and M. Bakker (eds.): *Technological Development and Science in the Industrial Age.* New Perspectives on the Science–Technology Relationship. 1992   ISBN 0-7923-1898-6
145. S. Amsterdamski: *Between History and Method.* Disputes about the Rationality of Science. 1992   ISBN 0-7923-1941-9
146. E. Ullmann-Margalit (ed.): *The Scientific Enterprise.* The Bar-Hillel Colloquium: Studies in History, Philosophy, and Sociology of Science, Volume 4. 1992   ISBN 0-7923-1992-3
147. L. Embree (ed.): *Metaarchaeology.* Reflections by Archaeologists and Philosophers. 1992
     ISBN 0-7923-2023-9
148. S. French and H. Kamminga (eds.): *Correspondence, Invariance and Heuristics.* Essays in Honour of Heinz Post. 1993   ISBN 0-7923-2085-9
149. M. Bunzl: *The Context of Explanation.* 1993   ISBN 0-7923-2153-7

Boston Studies in the Philosophy of Science

150. I.B. Cohen (ed.): *The Natural Sciences and the Social Sciences.* Some Critical and Historical Perspectives. 1994 ISBN 0-7923-2223-1
151. K. Gavroglu, Y. Christianidis and E. Nicolaidis (eds.): *Trends in the Historiography of Science.* 1994 ISBN 0-7923-2255-X
152. S. Poggi and M. Bossi (eds.): *Romanticism in Science.* Science in Europe, 1790–1840. 1994 ISBN 0-7923-2336-X
153. J. Faye and H.J. Folse (eds.): *Niels Bohr and Contemporary Philosophy.* 1994 ISBN 0-7923-2378-5
154. C.C. Gould and R.S. Cohen (eds.): *Artifacts, Representations, and Social Practice.* Essays for Marx W. Wartofsky. 1994 ISBN 0-7923-2481-1
155. R.E. Butts: *Historical Pragmatics.* Philosophical Essays. 1993 ISBN 0-7923-2498-6
156. R. Rashed: *The Development of Arabic Mathematics: Between Arithmetic and Algebra.* Translated from French by A.F.W. Armstrong. 1994 ISBN 0-7923-2565-6
157. I. Szumilewicz-Lachman (ed.): *Zygmunt Zawirski: His Life and Work.* With Selected Writings on Time, Logic and the Methodology of Science. Translations by Feliks Lachman. Ed. by R.S. Cohen, with the assistance of B. Bergo. 1994 ISBN 0-7923-2566-4
158. S.N. Haq: *Names, Natures and Things.* The Alchemist Jābir ibn Ḥayyān and His *Kitāb al-Aḥjār* (Book of Stones). 1994 ISBN 0-7923-2587-7
159. P. Plaass: *Kant's Theory of Natural Science.* Translation, Analytic Introduction and Commentary by Alfred E. and Maria G. Miller. 1994 ISBN 0-7923-2750-0
160. J. Misiek (ed.): *The Problem of Rationality in Science and its Philosophy.* On Popper vs. Polanyi. The Polish Conferences 1988–89. 1995 ISBN 0-7923-2925-2
161. I.C. Jarvie and N. Laor (eds.): *Critical Rationalism, Metaphysics and Science.* Essays for Joseph Agassi, Volume I. 1995 ISBN 0-7923-2960-0
162. I.C. Jarvie and N. Laor (eds.): *Critical Rationalism, the Social Sciences and the Humanities.* Essays for Joseph Agassi, Volume II. 1995 ISBN 0-7923-2961-9
    Set (161–162) ISBN 0-7923-2962-7
163. K. Gavroglu, J. Stachel and M.W. Wartofsky (eds.): *Physics, Philosophy, and the Scientific Community.* Essays in the Philosophy and History of the Natural Sciences and Mathematics. In Honor of Robert S. Cohen. 1995 ISBN 0-7923-2988-0
164. K. Gavroglu, J. Stachel and M.W. Wartofsky (eds.): *Science, Politics and Social Practice.* Essays on Marxism and Science, Philosophy of Culture and the Social Sciences. In Honor of Robert S. Cohen. 1995 ISBN 0-7923-2989-9
165. K. Gavroglu, J. Stachel and M.W. Wartofsky (eds.): *Science, Mind and Art.* Essays on Science and the Humanistic Understanding in Art, Epistemology, Religion and Ethics. Essays in Honor of Robert S. Cohen. 1995 ISBN 0-7923-2990-2
    Set (163–165) ISBN 0-7923-2991-0
166. K.H. Wolff: *Transformation in the Writing.* A Case of Surrender-and-Catch. 1995 ISBN 0-7923-3178-8
167. A.J. Kox and D.M. Siegel (eds.): *No Truth Except in the Details.* Essays in Honor of Martin J. Klein. 1995 ISBN 0-7923-3195-8
168. J. Blackmore: *Ludwig Boltzmann, His Later Life and Philosophy, 1900–1906.* Book One: A Documentary History. 1995 ISBN 0-7923-3231-8
169. R.S. Cohen, R. Hilpinen and R. Qiu (eds.): *Realism and Anti-Realism in the Philosophy of Science.* Beijing International Conference, 1992. 1996 ISBN 0-7923-3233-4
170. I. Kuçuradi and R.S. Cohen (eds.): *The Concept of Knowledge.* The Ankara Seminar. 1995 ISBN 0-7923-3241-5

Boston Studies in the Philosophy of Science

171. M.A. Grodin (ed.): *Meta Medical Ethics*: The Philosophical Foundations of Bioethics. 1995
    ISBN 0-7923-3344-6
172. S. Ramirez and R.S. Cohen (eds.): *Mexican Studies in the History and Philosophy of Science.* 1995
    ISBN 0-7923-3462-0
173. C. Dilworth: *The Metaphysics of Science.* An Account of Modern Science in Terms of Principles, Laws and Theories. 1995
    ISBN 0-7923-3693-3
174. J. Blackmore: *Ludwig Boltzmann, His Later Life and Philosophy, 1900–1906* Book Two: The Philosopher. 1995
    ISBN 0-7923-3464-7
175. P. Damerow: *Abstraction and Representation.* Essays on the Cultural Evolution of Thinking. 1996
    ISBN 0-7923-3816-2
176. M.S. Macrakis: *Scarcity's Ways: The Origins of Capital.* A Critical Essay on Thermodynamics, Statistical Mechanics and Economics. 1997
    ISBN 0-7923-4760-9
177. M. Marion and R.S. Cohen (eds.): *Québec Studies in the Philosophy of Science.* Part I: Logic, Mathematics, Physics and History of Science. Essays in Honor of Hugues Leblanc. 1995
    ISBN 0-7923-3559-7
178. M. Marion and R.S. Cohen (eds.): *Québec Studies in the Philosophy of Science.* Part II: Biology, Psychology, Cognitive Science and Economics. Essays in Honor of Hugues Leblanc. 1996
    ISBN 0-7923-3560-0
    Set (177–178) ISBN 0-7923-3561-9
179. Fan Dainian and R.S. Cohen (eds.): *Chinese Studies in the History and Philosophy of Science and Technology.* 1996
    ISBN 0-7923-3463-9
180. P. Forman and J.M. Sánchez-Ron (eds.): *National Military Establishments and the Advancement of Science and Technology.* Studies in 20th Century History. 1996
    ISBN 0-7923-3541-4
181. E.J. Post: *Quantum Reprogramming.* Ensembles and Single Systems: A Two-Tier Approach to Quantum Mechanics. 1995
    ISBN 0-7923-3565-1
182. A.I. Tauber (ed.): *The Elusive Synthesis: Aesthetics and Science.* 1996   ISBN 0-7923-3904-5
183. S. Sarkar (ed.): *The Philosophy and History of Molecular Biology: New Perspectives.* 1996
    ISBN 0-7923-3947-9
184. J.T. Cushing, A. Fine and S. Goldstein (eds.): *Bohmian Mechanics and Quantum Theory: An Appraisal.* 1996
    ISBN 0-7923-4028-0
185. K. Michalski: *Logic and Time.* An Essay on Husserl's Theory of Meaning. 1996
    ISBN 0-7923-4082-5
186. G. Munévar (ed.): *Spanish Studies in the Philosophy of Science.* 1996   ISBN 0-7923-4147-3
187. G. Schubring (ed.): *Hermann Günther Graßmann (1809–1877): Visionary Mathematician, Scientist and Neohumanist Scholar.* Papers from a Sesquicentennial Conference. 1996
    ISBN 0-7923-4261-5
188. M. Bitbol: *Schrödinger's Philosophy of Quantum Mechanics.* 1996   ISBN 0-7923-4266-6
189. J. Faye, U. Scheffler and M. Urchs (eds.): *Perspectives on Time.* 1997   ISBN 0-7923-4330-1
190. K. Lehrer and J.C. Marek (eds.): *Austrian Philosophy Past and Present.* Essays in Honor of Rudolf Haller. 1996
    ISBN 0-7923-4347-6
191. J.L. Lagrange: *Analytical Mechanics.* Translated and edited by Auguste Boissonade and Victor N. Vagliente. Translated from the *Mécanique Analytique, novelle édition* of 1811. 1997
    ISBN 0-7923-4349-2
192. D. Ginev and R.S. Cohen (eds.): *Issues and Images in the Philosophy of Science.* Scientific and Philosophical Essays in Honour of Azarya Polikarov. 1997
    ISBN 0-7923-4444-8

Boston Studies in the Philosophy of Science

193. R.S. Cohen, M. Horne and J. Stachel (eds.): *Experimental Metaphysics.* Quantum Mechanical Studies for Abner Shimony, Volume One. 1997   ISBN 0-7923-4452-9
194. R.S. Cohen, M. Horne and J. Stachel (eds.): *Potentiality, Entanglement and Passion-at-a-Distance.* Quantum Mechanical Studies for Abner Shimony, Volume Two. 1997
ISBN 0-7923-4453-7; Set 0-7923-4454-5
195. R.S. Cohen and A.I. Tauber (eds.): *Philosophies of Nature: The Human Dimension.* 1997
ISBN 0-7923-4579-7
196. M. Otte and M. Panza (eds.): *Analysis and Synthesis in Mathematics.* History and Philosophy. 1997   ISBN 0-7923-4570-3
197. A. Denkel: *The Natural Background of Meaning.* 1999   ISBN 0-7923-5331-5
198. D. Baird, R.I.G. Hughes and A. Nordmann (eds.): *Heinrich Hertz: Classical Physicist, Modern Philosopher.* 1999   ISBN 0-7923-4653-X
199. A. Franklin: *Can That be Right?* Essays on Experiment, Evidence, and Science. 1999
ISBN 0-7923-5464-8

Also of interest:
R.S. Cohen and M.W. Wartofsky (eds.): *A Portrait of Twenty-Five Years Boston Colloquia for the Philosophy of Science, 1960-1985.* 1985   ISBN Pb 90-277-1971-3

Previous volumes are still available.

KLUWER ACADEMIC PUBLISHERS – DORDRECHT / BOSTON / LONDON